D0392242

MARTIN HEIDEGGER

MARTIN HEIDEGGER

A Political Life

HUGO OTT

Translated by Allan Blunden

BasicBooks
A Division of HarperCollins*Publishers*

Published by BasicBooks, A Division of
HarperCollins Publishers, Inc.

Published simultaneously in the United Kingdom by
HarperCollins Publishers Ltd.
77–85 Fulham Palace Road
Hammersmith
London W6 8JB England

93 94 95 96 GL 9 8 7 6 5 4 3 2 1

Library of Congress Cataloging-in-Publication Data
Ott, Hugo, 1931–
[Martin Heidegger, English]
Martin Heidegger, a political life/Hugo Ott: translated by Allan Blunden.
p. em.
Translation of: Martin Heidegger.
Includes bibliographical references and index.
ISBN 0-465-02898-5
1. Heidegger, Martin, 1889–1976. 2. Philosophers--Germany--Biography, 3. Heidegger, Martin, 1889–1976--Political and social views. I. Title.
B3279.H4908613 1993
193--dc20
[B] 91-59018
CIP

CONTENTS

INTRODUCTION

For some time now I have been urged by many people to collect together my various articles on Martin Heidegger, hitherto published in rather obscure or 'provincial' journals, and bring them out as a book; or better still, to work them up into a more comprehensive study. The pressure on me intensified following the publication of Victor Farias' *Heidegger et le nazisme* in 1987, when my own essays attracted growing attention in the international debate that ensued. That attention has taken many forms, the most recent example being the very extensive and highly competent discussion in Thomas Sheehan's article 'Heidegger and the Nazis', which appeared in the *New York Review of Books*.[1]

It is indeed true that since I first began to investigate Heidegger's rectorship five years ago I have become increasingly absorbed by the whole subject.[2] It all began with the reissue in 1983 of Heidegger's rectorship address, 'The Self-affirmation of the German University', together with the first publication of *The Rectorship 1933–34: Facts and Thoughts* – the final and definitive version, so to speak, of the philosopher's apologia, edited by his son Hermann (Heidegger had instructed that it be published 'at the appropriate time'[3]). Its appearance was timed to coincide with the fiftieth anniversary of Hitler's seizure of power. Of course, to anyone who had read the interview in *Der Spiegel* – 'Only a god can save us now'[4] – this publication contained little that was new. All the same, the bold claim that here was a definitive and authoritative account of 'the facts' (a posthumous account at that) challenged the historian to check that claim against the available sources, defying him, as it were, to offer an alternative reading. A stone had been tossed into waters that had lain calm and undisturbed for a long time. People seemed to have forgotten all about the violent controversy stirred up by the young doctoral candidate Jürgen Habermas in 1953, in his extended review of Heidegger's recently published *Introduction to Metaphysics* in the *Frankfurter Allgemeine Zeitung*. The main point at issue there was Heidegger's philosophical assessment of National Socialism, and it provoked a debate that was fought out in the pages of the *FAZ* and *Die Zeit*, culminating in a letter to the editor of *Die Zeit* from Heidegger himself. The documentation by Guido Schneeberger, *Nachlese zu Heidegger. Dokumente zu seinem Leben und Denken* (Bern 1962) – an indispensable text – has been left to gather dust in the libraries, its author dismissed as an outsider.

The ripples from 1983 continued to spread. Things were further stirred up by the publication of the book by Heinrich Wiegand Petzet, *Auf einen Stern zugehen: Begegnungen und Gespräche mit Martin Heidegger 1929–1976*[5], which likewise came out during the 'anniversary year' of 1983. A biographical profile of Heidegger, it was the fruit of a very close association that went back many decades. The special significance of Petzet's account from the historian's point of view lies in the fact that it contains a number of statements authorized by Heidegger himself on his political involvement and the handling of his case after 1945. Furthermore, a good deal of the philosopher's apologia *Facts and Thoughts* has also been worked into the text, boosting the signal, so to speak. As I studied the 'political' period in Heidegger's life it soon became apparent that the 'facts' as asserted here would not stand up to scrutiny, and that the picture was increasingly subject to revision – to the point where what eventually emerged was the exact opposite.

But I very soon learned that anyone who approaches Heidegger with a critical mind, let alone who dares to question the received orthodoxy, is immediately branded as a hostile opponent. And to make matters worse, I had the temerity to announce my findings in two public lectures I gave at the University of Freiburg in 1984.[6] This was tantamount to sacrilege. Freiburg, the place where Heidegger had done his real work, had been 'desecrated' and made 'unholy', as I was told later by a distinguished American Heidegger scholar: not his own view, but the version that was being touted around in the USA. All my protestations to the contrary were of little avail. The damage had been done: I was a marked man, branded for life. I received a good deal of well-meaning advice. I was urged to let the matter drop, not least in order to avoid trouble within the University. And anyway, what right did I have, unschooled as I was in philosophy, to pass judgement in these matters? Just because I had the good fortune to be born too late ... even *that* old argument was trotted out. I need hardly add that I have encountered my fair share of open hostility as well – and I must expect more of the same.

The decision to pursue my investigations further, paying particular attention to the early period of Heidegger's life,[7] was based on the discovery that the factual evidence grew stronger and stronger the more I looked into it. And this despite the notoriously restrictive conditions: an indefinite ban on access to Heidegger's personal papers in the German Literary Archive in Marbach, no access to the personal papers of Bultmann (containing the correspondence between Rudolf Bultmann and Martin Heidegger) in Tübingen University Library, and other restrictions of a similar kind.[8] But there were enough other library collections to justify working towards a book on Heidegger, even allowing for the provisional nature of such an undertaking. A grant from the Volkswagenwerk Foundation in 1986 enabled me to seek out additional source material and study it at my leisure. One of my principal sources has been the correspondence between Jaspers and Heidegger, contained in Jaspers' personal papers (also in the German Literary Archive in Marbach), to which I was given access through the good offices of Hans Saner. I gather that this crucial correspondence is now about to be published.

In numerous archives (including the Federal Archive in Koblenz, the General Regional Archive in Karlsruhe, the Central State Archive in Stuttgart, and in Freiburg the State Archive, the Archiepiscopal Archive, the Municipal Archive and to a limited extent the University Archive – to name only the most important collections) and libraries (particularly the University Library in Marburg) I came across files and personal papers that revealed new information. In Marburg, for example, I had access to the papers of Dietrich Mahnke, to which my attention had been drawn by my colleague at Freiburg, Eduard Sangmeister. With his permission I was able to study the correspondence between Mahnke and Husserl. My research also produced some rather less edifying discoveries. In 1984 I stumbled upon the 'case' of Hermann Staudinger in the Freiburg State Archive, and was astonished to learn that in 1933 Rector Heidegger had denounced a distinguished scientist from his own university (Staudinger received the Nobel Prize for Chemistry in 1953) on the grounds of political unreliability, and had tried to have him removed from his post. Even back then I was troubled by the problem of Heidegger's mentality: finding the key to an understanding of such devious behaviour was by no means easy. The investigation of Heidegger's mentality is one of the primary concerns of this book. Much light is thrown on the subject by documents from private collections of papers.

Particularly in his earlier years, Heidegger was close to a number of men who played an important part in his life, but who receive no mention in his sketchy autobiographical writings: men like Engelbert Krebs, with whom he had been on friendly terms since 1912 – so much so that it was Krebs who conducted the marriage service in 1917 for Heidegger and his young bride, Elfride Petri. Certainly, this intimacy faded in time, to be replaced by an ever-growing distance. Also numbered among this early circle of friends was Ernst Laslowski, probably Heidegger's closest friend from 1911 onwards, who provided attentive and loyal support during the particularly difficult period from Heidegger's abandonment of his theological studies to the start of his academic career proper. The personal papers of Krebs – formerly preserved in the Department of Dogmatic Theology at Freiburg University, now housed in the University Library – became an important source for my study. No less important were the letters written by Laslowski to his friend (1911–1917), which Dr Wollasch kindly arranged for me to read in the library of the Deutscher Caritasverband in Freiburg. The diaries of Josef Sauer, Professor of Christian Archaeology and Patrology at the University of Freiburg, were another key source, not only for the crucial information they contain relating to Heidegger's rectorship period, but also for the light they cast on the early years: a number of Heidegger's early academic writings appeared in the *Literarische Rundschau*, published by the Herder Verlag and edited by Sauer. Heidegger and Sauer were in close contact as early as 1911; letters to Sauer from the young student, who was then girding himself for a major academic career, reveal the broad outlines of his early philosophical thinking. So when Heidegger set his sights on the rectorship in the spring of 1933 and

Sauer followed his progress with a critical eye, all this was taking place against the background of an acquaintance that went back more than two decades. Hence the depth of focus we find in Sauer's diaries. I am greatly indebted to Professor Sauer's nephew, Canon Sauer, for the opportunity to make use of the diary material here.

These latter figures were pre-eminently the soulmates of the 'Catholic' Heidegger. They belong within the more general context of the inferior status occupied by Catholics in the research and scholarship of Wilhelminian Germany. To rise above that disadvantaged status required enormous effort and determination. Shortly before Christmas 1923 the Marburg theologian Rudolf Bultmann wrote to his friend Hans von Soden, Professor of New Testament Exegetics in Breslau, and noted that his current seminar (on the ethics of Paul) was proving to be especially rewarding 'because our new philosopher Heidegger, a pupil of Husserl's, is one of the group. He comes from a Catholic background, but is a Protestant through and through.' He had an excellent knowledge not only of scholasticism (he went on), but also of Luther.[9] The path that Heidegger took in his Catholic period – only to be seen, at the end of it all, as a 'Protestant' – seems to me immensely significant. The book's first long chapter therefore sets out to explore all the ramifications of that path. The dominant presence in all this is Edmund Husserl, with whom Heidegger had a relationship of productive tension: admiration and friendship on the one hand, love-hate on the other. I was able to acquire many useful insights into this relationship thanks to the generosity of Professor S. IJsseling, the director of the Husserl Archive at the Catholic University of Louvain. Some early clues from the crucial years between 1917 and Heidegger's appointment at Marburg are to be found in studies by the American Heidegger scholars Thomas Sheehan[10] and Theodore Kisiel[11].

Relatively little discussion is devoted to Heidegger's Marburg period, important though it is. It seems to me that these years from 1923 to the summer of 1928 were something of an interlude in Heidegger's life; he put up with Marburg, which he did not like, in the knowledge that he could count on returning to Freiburg as Husserl's successor, and drew strength and comfort in the meantime from his sojourns up at the celebrated 'mountain hut' in Todtnauberg.

Being and Time was written in Todtnauberg – which also explains the famous dedication to Edmund Husserl as his mentor.

I was searching throughout for criteria that would enable me to understand Heidegger from the inside. One finds oneself walking a very fine line on these occasions. Where do the limits of discretion lie? At what point does one begin to invade personal privacy? I have, for example, omitted any discussion of Heidegger's relationship with Hannah Arendt when she was a student of philosophy at Marburg – which does *not* mean that this important relationship is of no consequence for Heidegger's biography. When and if we are given access to the key sources, this whole question will have to be looked into.[12] As things stand, however, I saw this as one of the natural limits of my inquiry, given the position with regard to the source material. On the other hand, I thought it essential to examine the development of the relationship between Heidegger and Elfride Petri, their marriage and the problems associated with a marriage between persons of different religious confessions, because this could give valuable insights into the biographical process. I was particularly interested in the things that Heidegger clung to: the fact that he clearly never broke free from the faith of his birth, that he lived all his life in the shadow of this conflict. Likewise the people that he clung to: the figure of Dr Conrad Gröber, for instance, the later Archbishop of Freiburg, who first led Heidegger into the paths of philosophical thought, and who reappears at every critical juncture in his life – and the figure of Edmund Husserl. Here I am indebted to Professor Karl Schuhmann of the University of Utrecht for many valuable suggestions in the course of our sustained academic intercourse, and not least for allowing me to read the extended version of his chronology of Husserl's life in manuscript form.[13]

My preparatory biographical study is structured according to a principle suggested by Heidegger himself, when he states that midway through his life (1935) he had to contend with two 'thorns in the flesh': the problem of religious allegiance, and the failure of his political career. However, it must be remembered that this was a continuing process, unfolding through time. Hence my attempt to bring out the broader continuities in Heidegger's life wherever their influence makes itself felt at the most intimately personal level.

Thus I have endeavoured, under the heading 'First indication', to explain the unbroken link between 1933 and 1945 by looking at the intense relationship between Heidegger and the historian Rudolf Stadelmann (hitherto largely overlooked): the revolutionary awakening of 1933, the 'historical occurrence' as defined by Heidegger's philosophy of Being, was not swept away in the general catastrophe, but remained as a task that confronted the Germans still, and was preserved, held fast, in the language of Hölderlin. Searching revelations of this kind throw a very specific light on the mental disposition of the philosopher.

My claim to be working 'towards a biography' of Heidegger may sound immodest, particularly as I have largely omitted any discussion of his place in the history of philosophy. But it ceases to be so when one considers the long and very troubled course of Heidegger's life. Commentators have repeatedly reminded us of Heidegger's own dictum, that his life as such is utterly uninteresting: the only thing that matters is his work. The sparse autobiographical memoirs have been published in various forms and are still viewed as an authoritative account, to which nothing needs to be added. The book by Heidegger's pupil Walter Biemel, for example – *Martin Heidegger in Selbstzeugnissen und Bilddokumenten* (Reinbek 1973) – remains very much at that level. Likewise Winfried Franzen, in his introduction to Heidegger's philosophy (particularly valuable also as a bibliographical tool), has been content to reiterate the authorized version.[14]

We have an abundance of memoirs from Heidegger's contemporaries, none more important than those written by Karl Jaspers[15] and Hans-Georg Gadamer.[16] Both men met Heidegger in the early 1920s and remained attached to him, albeit in very different ways. Also important are the pages of remembrances collected by Günter Neske, one of Martin Heidegger's publishers, in the volume *Erinnerung an Martin Heidegger*.[17] In a sense, Otto Pöggeler belongs with these contemporaries; he took an early interest in the ageing philosopher, publishing his monograph *Der Denkweg Martin Heideggers* in 1963[18] – an indispensable guide, particularly for the reader unschooled in philosophy. But all these accounts have been overtaken by the work of Karl Löwith, who in 1940 noted down from contemporary observation all that he experienced of and through Heidegger in the volume *Mein*

Leben in Deutschland vor und nach 1933, which was not published until 1986.

The reader of my book may well be surprised by the lack of references to Victor Farias. There are good reasons for this. It is not as if the book were unknown to me, in the published French version translated from the Spanish and German; on the contrary, my review in the *Neue Zürcher Zeitung*[19] has been translated into French and widely discussed in the French-speaking world.[20] But my own approach, based on the considerable number of articles I have laid before the academic world since 1983, and drawing on source material that has helped me to break substantially new ground is very different from that of Farias. For that reason I felt that I should continue to pursue my own line of research, especially since (so rumour has it) the forthcoming German edition of *Heidegger et le nazisme* will be notably longer than the French edition.

In France, where Heidegger's thought has taken root since 1945 in a very specific way and where the political persona of the philosopher has faded into the background, completely overshadowed by his intellectual significance, Victor Farias' book came as a bombshell that shook the academic world to its very foundations. It initiated a debate, primarily in the form of newspaper articles and essays, that can no longer be ignored. One of the most important communicators of Heidegger's thought to French audiences was the late Jean Beaufret of Lyons, who died in 1982, and to whom Heidegger addressed his 'Letter on Humanism' of 1946/47. Beaufret is now in bad odour – and consequently extremely suspect as a correspondent and associate of Heidegger's – following the publication of some letters written by him to the historian Robert Faurisson (University of Lyons II), the unspeakable champion of the 'Auschwitz lie' thesis. These letters, written in 1978, were published by Faurisson in the journal *Annales d'Histoire Révisionniste* (No.3, 1987). They express support for the work that Faurisson is doing, and encourage him to persevere with the same line of research. It was essentially the same line that he (Beaufret) had taken (he writes): but instead of putting his views in writing he had chosen to confine himself to the spoken word, to avoid being hounded by the mob. The authenticity of these letters from Beaufret cannot be doubted. So when, in the next issue of the same journal (No.4, Spring 1988), Faurisson dedicated his introductory essay 'A la mémoire de Martin Heidegger et de Jean Beaufret, qui m'ont précédé en révisionnisme', his reference to Beaufret is understandable enough; but his appropriation of Heidegger as a pioneer of the 'Auschwitz lie' thesis is of course perverse and without justification. That being said, the Beaufret letters do cast a very dubious light on the whole milieu in which the reception of Heidegger in France has thrived.

In November 1984, in an article for the *Neue Zürcher Zeitung* entitled 'The philosopher under a political cloud. Martin Heidegger and National Socialism', I summarized the position as I then saw it.[21] I concluded: 'Heidegger's efforts to play down the importance of his rectorship and make it appear as a thing of utter insignificance in the light of his resistance stance' must be regarded as a failure. Such efforts are unworthy of his great standing as a philosophical thinker.' Today, after a prolonged period of further study, I see no reason to revise my verdict.

I owe a considerable debt of thanks to the many people who took a critical interest in my work and assisted me in various ways. I can only mention a few here. At the beginning of 1986 I spent several weeks as a guest at the archabbey of Beuron, where I was able to complete the first draft of the manuscript. Many of my conversations with the resident monks proved helpful and illuminating. Herr Adalbert Hepp, my editor at Campus Verlag, who had been urging me for some time to write this book on Heidegger, has been a tolerant and co-operative working partner. My departmental colleagues at the University, who are normally engaged on serious and sober matters of economic history, gave generously of their time to assist with the proof-reading. My assistant, Herr Uwe Kühl, helped to co-ordinate the whole project. And finally I must thank my secretary, Frau Inge Wissner, who worked long hours to ensure that the manuscript would be delivered on time.

Merzhausen bei Freiburg,
in the late summer of 1988

HUGO OTT

PART ONE

Indications

First indication: 'The voice of the poet from his tower'

'Curious.' Only an hour before Stadelmann's letter arrived he had been thinking 'about historical self-awareness', and the writer of the letter had come vividly to mind: so writes Martin Heidegger on 20 July 1945 in his reply to Rudolf Stadelmann, Professor of Modern History and acting Dean of the Faculty of Philosophy at the University of Tübingen. The latter had written to the philosopher from the historic town of the poet Hölderlin, in those difficult days fraught with hopes and fears, to suggest that a place might be found for Martin Heidegger – then living in straitened circumstances – at Tübingen; two chairs of philosophy had become vacant, including the Chair of Systematic Philosophy, which Theodor Haering was now obliged to relinquish for political reasons. Heidegger's own position was at risk in Freiburg, which had been hard hit by the air raid of 27 November 1944. Word of this had quickly got around.

For Heidegger, however, this letter was more than just a solicitous inquiry from the man who had probably been his most loyal follower during the time of his rectorship in 1933–34 (Rudolf Stadelmann, a pupil of the historian Gerhard Ritter, was then a junior lecturer in Freiburg, with an interest in the philosophy of history). These lines from Tübingen 'spoke to me like the voice of the poet from his tower above his native river'. Heidegger now informed his faithful companion that he had spent the last six months 'in the land of my birth, some of it in thrilling proximity to the ancestral home of my fathers in the valley of the Upper Danube, beneath Castle Wildenstein', where his thinking had moved on a long way from mere interpretation of Hölderlin – he had analysed the poem '*Der Ister*' (i.e., the Danube) in a lecture given in 1942 – to 'become a colloquy with the poet', since for Heidegger his 'animated repose is the natural element of my thinking'. The letter almost turns into another lecture. It goes without saying that these few sentences are packed with quotations from Hölderlin and Rilke. And in his unspoken thoughts, no doubt, Heidegger repeated the observation that he had voiced explicitly in his 'Ister' lecture of 1942 (not included in the published edition of his works)[22]:

> It was perhaps inevitable that the poet Hölderlin should become the determining influence on the critical thought of one whose grandfather was born at the very time when the 'Ister Hymn' and the poem 'Remembrance' were written – born, according to the records, '*in ovili*' (that is to say, in a sheepfold on a farm), which lies near the bank of the river in the valley of the Upper Danube, beneath the lofty crags. Nothing is chance in the unseen history of poetic discourse. All is destiny.[23]

Now, in the summer of 1945, during the months of the political cataclysm, Heidegger was more than ever caught up in this self-mythologizing Hölderlin ideology, intoning darkly with solemn gestures. So Stadelmann might imagine what his inquiry meant to him, Heidegger: Tübingen! A youthful dream would be realized if he could 'give proper shape' to his 'real philosophical concerns' in the land of his birth, 'steeped in the element of Hegel, Schelling, and above all Hölderlin'. As he put it in a later letter to Stadelmann, the atmosphere of Tübingen held 'a powerful attraction' for him. He was also convinced 'that the spirit of the West will rise up once more from our Swabian land' – especially since Romano Guardini was due to come to Tübingen soon. What a constellation of talent! Although Heidegger had had no contact at all with the religious philosopher Guardini for many years–and particularly during the period of the Third Reich–now was an opportune moment to make himself known again. So Heidegger wrote to Guardini on 6 August 1945, hoping to interest him in coming to Freiburg, where the Chair of Christian Philosophy, appropriated by the National Socialists for their own ends, had now been reinstated.

The post was very slow in those days. Even so, Guardini took his time about replying, not writing until 14 January 1946, just as Heidegger's fate was being decided: 'How I should love to talk with you on all manner of things. It is so long since we last saw each other. I still have the clearest memories of my visit to Zähringen, and of your wonderful study'. And he sent his good wishes for the New Year already begun, 'which will doubtless decide our destiny one way or the other'.[24] Guardini was drawn to the big city again as a place of intellectual activity–which meant Munich, of course. But he had to bide his time until the gates of the Bavarian metropolis were opened wide again: he had to make do with Tübingen, a town that had been spared by the war, that was not a living reminder of the collapse like other German towns and cities, and where it was not *quite* so incongruous to be lecturing on aesthetics while children were starving in the street.

But let us follow the correspondence with Stadelmann a little further. True enough, Heidegger's future had not yet been decided in Freiburg, where there was little to hold him, even though they – 'not the Allies, but our own people' – had uncovered certain incriminating facts about his rectorship 'which was anything but an endorsement of the Party and Party doctrine'. Indeed, Heidegger went on, ever since he gave up the office in 1934 he had been subject to increasing harassment and vilification, 'narrowly escaping a worse fate'. Here he sounds the key notes that would come to dominate his future exercises in apologetics, and indeed this letter (followed by others in the coming months) is extraordinarily revealing in its intensity and intimacy – as a private document, between old friends.[25]

But what the philosopher wrote in his letter to Tübingen on that 20 July 1945 – a date of some significance in the *national* memory too – is part of a much wider picture. As he reflected 'on historical self-awareness, its nature and potential', Stadelmann had come vividly to mind. A word of explanation is called for. During the winter semester of 1933–34 Stadelmann, then a junior lecturer under Heidegger's rectorship, had given the first in a series of public lectures at the University of Freiburg entitled 'The role of intellectual life in the National Socialist state'. Delivered on 9 November, of all days the tenth anniversary of the march on the Feldherrnhalle, Stadelmann's lecture had as its theme 'the historical self-awareness of the nation'.[26] Here he sounds a paean to the National Socialist revolution, subscribing to an unbelievably naïve cult of the Germanic past, richly larded with quotations from Adolf Hitler's *Mein Kampf*. He proclaims a philosophy of history as myth ('The genius of the nation discerns itself not in its history, but in its myths'), every inch the assiduous disciple of Heidegger, even in the language he uses. So in referring to this lecture of 9 November 1933 Heidegger reflects now on historical self-awareness as though nothing at all had happened in the intervening twelve years: he ponders its 'nature and potential'. An extraordinary quotation to find in a letter dated 20 July 1945. It was Rector Martin Heidegger himself who had brought the winter semester lectures to a close on 22 February 1934, when he spoke on the subject of 'the necessity of science'.[27]

Now one could let such a letter pass, were it not for some concluding remarks that make one doubt the evidence of one's ears. 'Everyone now thinks of doom and downfall. But we Germans cannot go under because we have not yet arisen, and must persevere still through the night.' All around him were chaos and ruin, unspeakable suffering, homelessness and guilt: 'downfall' indeed, if the word has any meaning at all. And now *this* claim, born of his inmost thoughts, nurtured in ground upon which only the chosen few are permitted to tread, addressed to an intimate companion who had always understood Heidegger. Did Stadelmann, even now, share this unbelievable loss of reality with Heidegger, to whom he had once looked as a faithful follower in true allegiance, as to a leader or *Führer*? As this epistolary dialogue unfolded, memories may well have been stirred of the campfire ethos of that autumn of 1933, with its overtones of initiation and male bonding; memories, perhaps, of the notorious Todtnauberg academic summer camp [*Wissenschaftslager*] held within sight of the famous 'mountain hut', which Heidegger later recalls, selectively and sketchily, in the apologia that was published after his death in 1983. Here Stadelmann's name is not even mentioned, although this camp (to which we shall return in more detail later) loomed large both in the personal relationship between the two men and in their understanding of the National Socialist revolution. There is indeed something of a mystery here. The voice of the poet from his tower had become a protective, all-enveloping mantle. What Heidegger did not elaborate on to his would-be benefactor in Tübingen was exactly how he had ended up in the valley of the Upper Danube. In December 1944 Heidegger had left the ruins of Freiburg for his native town of Messkirch, and in March 1945 he had gone to join what was left of the Faculty of Philosophy from Freiburg University, most of which had fled before the advancing Western front into the valley of the Upper Danube to Castle Wildenstein and the surrounding villages, not far from the Benedictine abbey at Beuron. It was here, in this landscape, that Heidegger awaited the arrival of the French troops as the guest of a Prince and Princess of Sachsen-Meiningen. Within sight of the 'Ister', which has carved out its meandering bed through the limestone of the Swabian Jura, he entered into colloquy with Hölderlin. And when Heidegger returned to Freiburg it was this idyll of the Upper Danube valley, this safe haven for survivors, that he took away with him in his mind and dwelt upon, while nearly everywhere else in Germany the horsemen of the apocalypse were bringing destruction and ruin. He was living in the world that he had invoked in the forester's lodge at Hausen near Beuron 27 June 1945 before a motley audience that included the Prince and Princess of Sachsen-Meiningen, Faculty colleagues, students and domestic staff, taking as his text a quotation from Hölderlin: 'All our thoughts are concentrated on the things of the mind. We have become poor, that we might become rich.'[28] This was, incidentally, to be the very last lecture he gave as a full professor of philosophy.

In the subsequent course of the correspondence between Heidegger and Stadelmann, which has to be seen in the context of Heidegger's probable forthcoming suspension from office in Freiburg, the philosopher delivers himself of other fundamental statements. These have to do with the question of Heidegger's reception among French intellectuals and the relationship of his thought to the post-war German present.

In Heidegger's judgement a paradoxical situation had arisen: in France, and particularly in Paris, and among the ranks of the military government in Baden-Baden, he was esteemed as a 'fashionable philosopher', in favourable contrast to the narrow-minded and illiberal 'compatriots'. He was invited to contribute to leading French journals.

The French authorities, as he informed his Tübingen correspondent with barely concealed pride, recognized that in France his philosophical work 'guides and inspires people's thinking, and in particular the attitudes of the young in intellectual matters'. Only recently, he said, he had been urged by the *Revue Fontaine* to submit any unpublished writings or lectures or any material that had not yet been translated. Indeed, at the end of September 1945 the young lieutenant Edgar Morin, then on the fringes of the French Communist Party, had delivered a letter to Heidegger by hand from the editor of the *Revue Fontaine*, Max-Pol Fouchet, authorized by no less a person than the head of the press and information office in Baden-Baden, General Arnaud.[29] Heidegger was invited to give his views either on the current political situation or on the state of philosophy in France.

'I dare not promote our thinking in France unless I am simultaneously allowed to make my work available to the Germans' (letter of 30 November 1945). His own countrymen, meanwhile, wanted no more of him. The climate at the University of Freiburg became increasingly frosty. His own faculty was lying low, while in 'circles close to the Faculty of Theology and the Archbishop' there had been a change of heart as people realized 'that behind the alleged "nihilism" lies something quite different, since after all even the wise Meister Eckhart spoke of the nothingness of the godhead'. Included within these circles, however, were a number of companions from his early Catholic years, whose acquaintance we shall make more fully later. One such was Archbishop Conrad Gröber, his fellow countryman from Messkirch and a loyal supporter from the early days, who as a 'fatherly friend' had pointed him in the direction of philosophy, handing him 'rod and staff' for the journey. Now that many were abandoning him and he had been deprived of his chair, Heidegger beat a path once more, via many twists and turns, to the door of his Catholic origins, where at least he knocked – even though, in the event, he did not return to the bosom of the Church as a penitent.

In this complex isolation, all he could really do was react with defiance. He could afford to wait and bide his time, but the critical question was 'whether *the younger generation* and the *present intellectual situation of the Germans* can afford to wait'. (30 November 1945). As the prospect of material ruin stared Heidegger in the face in the weeks before Christmas 1945, he grabbed at his last thread of hope: recalling his earlier friendship with Karl Jaspers, he asked for a reference from his one-time colleague, with whom he had not had any contact for years and who had now returned in triumph to his chair at Heidelberg, where he was viewed as a latter-day *praeceptor Germaniae*. But the reference submitted by Jaspers just before Christmas (22 December 1945) pronounced a devastating verdict on his former friend and rival: he said Heidegger was unsuited as a teacher of young people since his mode of thinking was fundamentally unfree, dictatorial, uncommunicative, and would have a very damaging effect on the current generation of students.[30] Dismissed from his teaching post, Heidegger ended his correspondence with Stadelmann in Tübingen on 23 January 1946: 'I have the feeling that another hundred years of neglect are needed before people start to realize what Hölderlin's poetry holds in store.' And he closes with a few lines from Hölderlin's poem *Mnemosyne*:

> Long is
> The time, but the Truth
> Is fulfilled.

Second indication: The perpetual Advent

It was in May 1945, just a few days after Freiburg had surrendered without a fight to the French troops, that the superficial line of argument adopted by Heidegger in his own defence – and repeated many times since by writers on the subject – first took shape. As we have just seen, the professor was not in the university city during this difficult period of upheaval, but was whiling away the days in the valley of the Upper Danube, where he felt his real roots lay. But into the almost idyllic peace of the new-sprung Danube came the news that the house on the Rötebuck in Freiburg was to be requisitioned as a former Party residence: for Heidegger was viewed by the French military command as a '*nazi typique*'.

In her husband's absence, Frau Elfride Heidegger sought to avert the danger, insisting that the philosopher's involvement in the Third Reich could not possibly justify such measures. When he returned from his current posting, she said, Heidegger would give a fuller account of his activities.

This he did in July 1945, albeit now under the scrutiny of the *épuration*, when he was summoned to appear before the denazification commission of the University of Freiburg, which had been set up by the University at the instigation of the French. Heidegger was forced to defend himself on several fronts. In the months that followed he formulated one by one the exonerating arguments that were then recast, in November 1945, in a summary statement of his position. This in turn formed the basis for all his subsequent pronouncements, including the interview he gave for *Der Spiegel* in 1966, 'Only a god can save us now' – which was not published, in accordance with his wishes, until after his death in 1976 – and his memoir of the rectorship, published posthumously in 1983 under the revealing title *Facts and Thoughts*. In short, the philosopher Martin Heidegger remains with us after his death in a very special and particular way.

Nor is this an exclusively German drama; on the contrary, it was – and still is – being played out on a worldwide stage. There exists an extensive international network of Heidegger circles, organized latterly into Heidegger societies. These groups swing into action as soon as they get wind of an impending attack. So it came as no surprise to the initiated that *Facts and Thoughts* was rapidly disseminated in translation throughout the French-speaking and English-speaking worlds – and presented as the philosopher's final word to all those who see his political involvement in a different and more serious light.

It has been clear for some time that the tendency to play down the philosopher's political past could no longer be sustained in the light of certain research findings. Engaged as I was myself in a study of these matters, and motivated, as I said before, by that 1983 publication, I found that the reception of new knowledge based on sound research was a very slow, localized and selective process. In effect, whole areas remained completely sealed off and impervious to influence.[31]

Heidegger's line of argument in *Facts and Thoughts* is quite untenable both as regards the chronology and facts and the explanations he offers. For example, he did not, as he claims, end up as rector more or less by chance, sacrificing himself for the University of Freiburg in order to avert a worse catastrophe. On the contrary: a deliberate plan was devised by a small cadre of National Socialist professors at the University whereby Heidegger, who was now their man in dealings with the authorities in Karlsruhe and Berlin, was to be manoeuvred into the top job. Even the timing of his entry into the Nazi Party was dictated by tactical considerations.

Heidegger's links with the National Socialist movement went back some time, notably via the student groups of the National Socialist German Student League (NSDStB). He was acquainted, for example, with Gerhard Krüger, the leader of the German Student Union (which had come under *de facto* National Socialist control long before 1933), and with a Dr Stäbel in Karlsruhe, the area head of the NSDStB (South-West region).

The poet René Schickele (who describes himself as 'a French *citoyen* and a German poet'), who was living in Badenweiler at the time, registered the change in Heidegger and noted astutely in his diary (2 August 1932): 'In Freiburg University circles people are saying that Heidegger now consorts exclusively with National Socialists (can't believe it myself: must ask when I next get a chance).'[32] But what did these party political associations have to do with his philosophy? The most basic question of all, which has been asked over and over again.

For Heidegger the philosopher, the Advent weeks of 1932, the time of Christian hope in the coming of the Lord, brought with them the expectation of a secular turning point. On 8 December 1932 he wrote to Karl Jaspers in Heidelberg: 'Will it be possible, I wonder, to create a firm foundation and a place for philosophy in the coming decades? Are there men coming who bear a distant dispensation within them?'[33] Who might these men be, one wonders, 'who bear a distant dispensation within them'? And what might this mysterious 'distant dispensation' be? Jaspers could make no sense of all this. How could he, given that this rarefied, almost occult dimension to Heidegger's thinking remained a closed book to the more down-to-earth Jaspers? In the inaugural address given by the philosopher when he became rector on 27 May 1933 this 'distant dispensation' reappears three times as a key motif, interwoven with the idea of 'beginning' in the dawn of Greek philosophy.[34]

> The beginning continues to *be*. It does not lie *behind us*, as a thing that once was long ago, but stands *before and in front of us*. As the greatest part, the beginning has gone beyond all that is yet to come, and therefore beyond us too. The beginning has irrupted into our future, it stands there as the distant dispensation over us, waiting to claim its greatness again.

Now, in the revolutionary dawning of the new Reich, Heidegger was in no doubt as to who those men were who 'bear a distant dispensation within them'. He, the philosopher of the Beginning, of 'the Being to be revealed in its origins', was for that reason proof against error: and this is of central importance, since he, perhaps alone of all men, experienced the inner 'truth and greatness' of the movement at an intellectual level (1935: *An Introduction to Metaphysics*), when it was not apparent – or no longer apparent – to those who called themselves National Socialists. All the complaining and moaning, all the specious accusations of guilt that were thrown at him after 1945: these Heidegger dismissed in the statement published in 1983 as 'insignificant', 'fruitless muck-raking', 'less than trifling', thereby demolishing his critics once and for all.

For such superficial criticism misses the point, which is to see the centrality of the Advent motif in Heidegger's thought, a motif that runs throughout his work. This is the claim put forward by the philosopher in the handful of public statements that he made after 1945, such as the letter to the editor of *Die Zeit* that appeared in the summer of 1953. Those who had learned how to think, 'the hearers among the listeners', as he wrote, had understood him well enough in his lectures.[35] Another example is the draft of a letter (which was never sent) to the editor of the *Süddeutsche Zeitung*, dated 24 June 1950, in which he set out to correct serious allegations made against him in the Munich City Council and the Bavarian parliament.

Heidegger undoubtedly included under this heading his interpretation of Hölderlin's hymn, 'Wie wenn am Feiertage . . .', delivered as a lecture in 1940–41, in which he analyses the key lines, 'But now breaks the dawn! I tarried and saw it coming, / And what I saw, the sacred, let that be my utterance.' In stark contrast to the rectorship address of 1933 he deals here with the sacred, which in its coming establishes another beginning to another history. The sacred determines in the beginning, in advance, what the fate of men and the gods shall be, and it is this Advent that constitutes history, history proper, in Heidegger's eyes. But history is that rare and timely moment, that *kairos*, 'whenever the essence of Truth is determined in the beginning'. Truth, however, is the disclosure of Being. When Heidegger interpreted Hölderlin, as he did repeatedly after 1936, this always remained a history in the future, an Advent yet to come. He concludes the 1940/41 lecture thus: 'This utterance, as yet unheard, is stored up in the Occidental language of the Germans.'[36]

More mightily hewn yet are the sentences that Heidegger penned in a letter to Karl Jaspers of 8 April 1950 (a year previously the one-time friend, who had been living and teaching in Basel since

1948, had with some difficulty made contact with Freiburg in order to extend the hand of friendship to the dismissed philosopher):

> Despite everything, my dear Jaspers, despite death and tears, despite affliction and horror, despite privation and suffering, despite dislocation and banishment, what is happening in this state of *homelessness* is not nothing; *concealed therein is an Advent* whose distant beckoning we may just be able to divine in the faint stirring of the air, and which we must catch and hold fast for a future that no historical construction – least of all today's, which sees everything in terms of technology – will ever decipher.

1932, 1939–40, 1950: the dates form a continuum for anyone who can follow the line of thinking. For Jaspers such statements were the expression of an intuitive and poetic brand of philosophy. As 1932 drew to a close the Advent hope seemed to be coming true as the German universities made ready for the summer semester of 1933 and Heidegger, who now acted as the trusted agent of the small circle of National Socialist professors at Freiburg University in all dealings with government and Party agencies in Karlsruhe and Berlin, sought to find out what kind of changes in the university system were planned in the name of *Gleichschaltung*. On 18 March 1933, on his last extended visit to Jaspers in Heidelberg, Heidegger had left sooner than originally planned. Apropos of the headlong speed with which events were moving under National Socialism he had remarked, 'One must get involved' – as Jaspers records in his autobiography. Jaspers was taken aback, but did not press the point. It is unlikely that the subject as such was taboo in Heidelberg; otherwise the opening sentences of the letter Heidegger wrote to Jaspers on 3 April 1933, in which he thanks him for his recent hospitality, would make no sense: 'I was still hoping for some definite news about the plans to reorganize the universities. Baeumler is keeping quiet; from his brief letter I got the impression he was annoyed. Krieck in Frankfurt is not giving anything away either. And Karlsruhe is sitting tight.' All this suggests that the subject had been discussed at length – including the question of where philosophy would figure in the future scheme of things: 'Although a lot of things remain obscure and questionable, I feel more and more that we are emerging into a new reality, and that the old era has run its course. Everything depends on whether we can bring philosophy to bear in the right place and help it to do its work.'

And so the Advent of this year, 1932, ushered in the men who bore a 'distant dispensation' within them: the *Führer* and his fellow 'leaders'. It is now that the key names appear: Alfred Baeumler and Ernst Krieck, side by side with Martin Heidegger; Baeumler 'is keeping quiet' – apparently annoyed; and Krieck, too, is maintaining radio silence. These are names that will keep cropping up.

Third indication: Dialogue without communication

The image of the dreaming boy seems to have spoken strongly to Heidegger. It is an image used by Jaspers in a letter written to his former friend of many years in the spring of 1950 in which he sought to characterize the latter's response to National Socialism: 'Like a boy who is dreaming, who doesn't know what he is doing, who embarks as though blind and oblivious on an enterprise that looks so different to him than it is in reality – and then stands helpless before a heap of ruins, and lets himself be driven on' (20 March 1950)[37]. Jaspers felt such a characterization was in order after Heidegger had – at long last – come round to something like an acknowledgement of guilt. Writing to Jaspers in Basel – so near and yet so far – on 7 March 1950, Heidegger says that the reason he had not set foot in Jaspers' house since 1933 was not that there was a Jewish woman living there, but that he 'was simply too ashamed'. Not only had he not entered Jaspers' house since then, he had not been back to Heidelberg either, and he would not come until he and Jaspers could meet again on their 'old but still painful footing'.

However, the paths of the two philosophers were destined never to cross again, not in Freiburg, not in Heidelberg, not in Basel, nor anywhere else; not even on the station platform in Freiburg, where Heidegger had planned to be waiting when Jaspers passed through in the summer semester of 1950, on his way down the Rhine to lecture in Heidelberg. He asked Jaspers to let him know his time of arrival, so that they could 'at least shake hands again' on his way through. He surmised that Jaspers would not be intending to stop over in Freiburg. But Jaspers did not even tell him when the train was due: instead he retreated into a protracted silence, which he broke two years later only to announce brusquely, and unilaterally, that the philosophical dialogue between them – and hence dialogue *tout court* – was at an end. The fêted philosopher of Basel to whose door the whole world now beat a path, who seemed to have been spared calamity and crisis in his life and thought, who had not been discredited by political involvement – and therefore had no guilt to atone for: Karl Jaspers was not in the dock.

All that was now left between them was the exchange of formal greetings on birthdays, particularly the major ones. The 1950 exchange of letters had failed to restore communication between the two giants to the point where it could be sustained continuously and without significant interruption. Too much had come between these two men, who once thought of themselves as the twin standard-bearers of German philosophy, comrades-in-arms who stood together in 'a rare and independent alliance'; since 1922 Heidegger had been pressing Jaspers hard to 'throw yourself into philosophy and its possibilities as a study of first principles', and to undertake a 'historical critique of ontology that starts from its roots in Greek philosophy, and especially Aristotle' (Heidegger to Jaspers, 27 June 1922). Too much had now come between these two, who grew steadily closer after 1920: the younger, animated, but in the end withdrawn South German, and his older colleague from Oldenburg, stalwart but rather ponderous and aloof, who was finally caught up, overtaken, outrun and perhaps outclassed by the ambitious Heidegger, who became a world celebrity on the strength of *Being and Time*. In 1929 he debated with Ernst Cassirer in Davos. In 1930 he was offered the chair in Berlin that Jaspers himself coveted (congratulating him, Jaspers called it 'the greatest honour that can be bestowed on a university philosopher') – and could even allow himself the luxury of declining the offer. At the time Jaspers had hopes that Heidegger would come to the university at Heidelberg, to see if the two of them were capable of 'philosophizing and communicating, even in the most radical debate' – or if each would simply carry on in his old way (Jaspers to Heidegger, 24 May 1930). Yet they both took care to avoid such a pairing: it never went beyond vague declarations in letters.

Too much had happened since 1933 for Heidegger to be able to cleanse himself in Jaspers' eyes with a simple passing reference to 'shame'. More substantial explanations were needed to bring about clarification and redemption. But the liberating, redemptive words were never spoken. Instead the tissue of excuses manufactured by Heidegger in his own defence was paraded before his former friend as well – but here it missed its mark completely: Jaspers thought that by sending Heidegger his essay on the 'question of guilt' he would be able to enter into an instructive dialogue with him: 'What you said about "shame" often goes through my mind. It occurs to me that my old essay might be of interest to you, that you might really understand what I was getting at. So I'm sending it to you' (25 March 1950).

Heidegger responded with a petty-minded defence based on factual assertions that were partially inaccurate in themselves, and as a whole amounted to a pretty flimsy case. Responding to Jaspers' 'question of guilt' on a superficial level, Heidegger formulated the problem in relativistic terms: 'The guilt of the individual remains, and the more so the more individual he is. But the work of evil is not yet done. It is only now entering its universal phase.' The 'work of evil', for him, was Stalin and Bolshevist Russia. Each day Stalin was winning a new battle, without ever having declared war. Not everyone could see that, and it was not just the clear-sighted few – such as the philosopher Heidegger, who was 'for it now', just like 'the Jews and left-wing politicians' before him, when they were 'first on the list' – who now faced the common challenge: 'For us there is no avoiding the issue. Every word, every publication is a form of counter-attack, even though none of this is taking place in the "political" sphere as such, which has long since been superseded by other modes of being and now leads only a shadowy existence' (letter of 8 April 1950).

Heidegger hardly needed to relate all this specifically to Hitler – he does not mention the *Führer* once in the entire post-war correspondence with Jaspers – or to National Socialism either, since all that was only a prelude, at most; for it was only now that the work of evil was entering its universal phase. He had also been at risk in 1937 and 1938 – a low point in his life – and had become more clear-sighted in consequence, despairing as he sought to acquire his historical insight in the shadow of the approaching war that was already reaching out to devour the young men. 'And then came the persecutions of the Jews; everything was heading towards the abyss. We never believed in a "victory"; and if it had come to that, we would have been the first victims.' These are astonishing statements, whose credibility must remain open to question.

To have taken part directly or indirectly in these events at the beginning, as a dreaming boy, all unknowing – albeit caught up for a few short months in a kind of lust for power – engenders a sense of shame (wrote Heidegger) that grew from year to year, 'the more the evil truth came out'. But by resigning from the rectorship in a gesture of protest in February 1934 he had set an example, daring to do something that 'no one in our universities . . . dared to do', even if such an action by one individual could no longer make any difference in the face of the total regimentation of public opinion. (Once again, the meticulous demarcation of political involvement: thus far, and no further.) So he was all the more hurt (he goes on) by the campaign of hostility conducted against him since 1945/46, and conducted against him still – by a public that had no ears to hear what he was saying because the true hearers were too few in number.

Such an attitude, not to say affectation, had become second nature to Heidegger during those years. The draft of a letter to the editor of the *Süddeutsche Zeitung* in Munich, written in the summer of 1950 to answer attacks from the Munich City Council, is typical of many other statements. Heidegger had been invited to deliver a lecture at the Academy of Fine Arts[38], his first public engagement after 1945. Again he minimized his political involvement, dismissing it as a political error, to which the highest spiritual and worldly dignitaries had also fallen victim, the error of 'seeing in Hitler and his movement a regenerative force for our nation, and declaring my commitment to that'. He emphasized his inward opposition after he had relinquished the rectorship: 'Thereafter, that is, during the last ten years of my academic teaching career up until the autumn of 1944, I was engaged in an increasingly sharp intellectual conflict with and critique of the anti-intellectual foundations of the "National Socialist world-view".' Here too we find the anti-Communist ploy:

> Where crimes have been committed, atonement must be made. But how much longer will people continue to vilify publicly those who succumbed briefly, or not so briefly, to political error – and this in a country whose constitution permits anyone to be a member and militant supporter of the Communist Party? A strange blindness is thus conducing to the attrition and disintegration from within of the last remaining substantial powers of our nation.

Heidegger as the precursor, so to speak, of the *Radikalenerlass*.

This was the line of defence adopted by Heidegger in 1945 after the political and national catastrophe, and after his own long descent into the abyss – a defence that he worked ceaselessly to consolidate. Already, in the lengthy explanatory statement he addressed to the rector's office at the University of Freiburg on 4 November 1945, he casts himself in the role of resistance fighter:

> I claim no particular credit for the resistance I have kept up over the last eleven years. But when it is sweepingly asserted that many students were led into 'National Socialism' as a direct result of my rectorship year, then justice demands a recognition at least of the fact that in my lectures throughout the period from 1934 to 1944 I instilled an awareness of the metaphysical foundations of our age into thousands of students, and opened their eyes to the world of the intellect and its great tradition in the history of the Western world.

In 1950, however, the year with which we began, when the Western zones of occupation, now united under the aegis of the Federal Republic of Germany, were moving towards political and economic stability, the result of Heidegger's appeal against the harsh verdict imposed by the French military government finally came through. Contrary to his hopes, Heidegger, who wanted nothing less than his reinstatement as a full professor, which would have taken him out of his isolation, was dismissed from office and banned from teaching. He had to be content with the role of custodian of philosophical thought, whose task it was to hold out 'against dogmatism of every kind', albeit with little hope of exercising any influence. From his isolation, which he described as the only 'place in which thinkers and poets stand by Being to the best of human ability', Heidegger had already sent greetings to Jaspers (then teaching in Basel) in June 1949 when he learned that Jaspers had been trying to reach him by letter – without success: his first letter failed to get through directly to Freiburg. In fact Jaspers, for whom Heidegger 'had become an intellectual enemy on account of his public activities as a National Socialist,[39] remained inwardly ready, mindful of their association in the 1920s, to enter into dialogue with Heidegger – up until the early 1950s, at least. But intellectual enmity: there was the rub – that and Heidegger's recalcitrance.

Anticipating a more detailed discussion later as the larger picture unfolds, we will look briefly here at the document that Jaspers merely mentions in passing in his *Philosophical Autobiography*, keeping it for publication at a later date – namely the devastating reference he wrote on 22 December 1945 in the Heidegger affair.

> I had hoped to keep quiet and say nothing, except to close friends. I have held that view ever since 1933, when I resolved to remain silent after the terrible disappointment, in deference to my happy memories. This was made easy for me in so far as Heidegger himself, at our last meeting in 1933, either ignored awkward questions altogether or gave vague answers – particularly on the Jewish question – and thereafter ceased the regular visits he had kept up for the past decade, so that we have not met since. He carried on sending me copies of his publications to the end, but after 1937/38 he stopped acknowledging anything I sent him. I was particularly hoping that I wouldn't have to speak, that now, at last, I could keep quiet and say nothing.

Writing at the request of the denazification commission at the University of Freiburg and of Heidegger himself, Jaspers submitted his detailed report to the Freiburg biologist Oehlkers, a member of the commission who was married, like himself, to a Jew.[40] At the time Jaspers was not aware of the harshly dismissive treatment his own philosophical endeavours had received in Heidegger's 1936 Nietzsche lecture, in which the latter pronounced a verdict that went much further than a scholarly critique of Jaspers' Nietzsche interpretation: 'Because in his heart of hearts Jaspers no longer takes philosophical knowledge seriously, the real questions are no longer being asked. The practice of philosophy turns into a moralizing psychology of human existence.'[41]

Centrally preoccupied as he was with the question of guilt after the collapse, Jaspers pronounced a harsh verdict: Heidegger, whose philosophical standing he characterized in interesting fashion, must be called to account, since he was one of the few university professors who had actively helped to 'put National Socialism in the driving seat'. Heidegger should be barred from holding office and teaching for several years, but he should be 'provided with a personal pension' to enable him to pursue his philosophical studies. Jaspers felt that a review might be in order later, when and if Heidegger became a reformed character and the academic and political situation became more settled. But at a time when 'the education of the younger generation needs to be handled with the utmost responsibility and care', full teaching privileges should not be restored as yet.

> Heidegger's mode of thinking, which seems to me to be fundamentally unfree, dictatorial and uncommunicative, would have a very damaging effect on students at the present time. The mode of thinking itself seems to me more important than the actual content of political judgements, whose aggressiveness can easily be channelled in other directions. Until such time as a genuine rebirth takes place within him, and is *seen* to be at work within him, I think it would be quite wrong to turn such a teacher loose on the young people of today, who are psychologically extremely vulnerable. First of all the young must be taught how to think for themselves.

Heidegger was acquainted with significant portions, if not the entire contents, of Jaspers' report of 22 December 1945, which played a key role in determining the severity of his punishment. Jaspers did not find it easy to write such a report. It was only because Heidegger himself had suggested his name as a referee to the denazification commission that he felt he could not withhold his views from this body. Jaspers also had a hard time dealing with the yawning gulf of silence that had opened up between them over the years – he of all people, whose whole philosophy was based on communication. So before the move from Heidelberg to Basel he drafted a letter, dated 1 March 1948; like so many of Jaspers' letters it was never sent, but it formed the basis for his letter of 6 February 1949, which initially failed to reach Heidegger, as has already been noted. All in all it amounted to an honest effort on the part of the older man, at the peak of his career, to extend the hand of reconciliation to his defeated colleague, albeit not unconditionally: 'Unless something extraordinary happens between us, the darkness will always remain – which does not prevent us from exchanging the odd word on philosophical matters, and perhaps on private matters as well' (6 February 1949).

The extraordinary never happened, as we have already seen. The only concession made by Heidegger in the course of a lengthy correspondence was a reference to 'shame' – and even then only in relation to his own personal misfortunes.

Heidegger's formulations in the letter of 8 April 1950 show how hopelessly far he was from being able to write the handful of sentences that could have served as a basis for meaningful dialogue. Mindful of all this, Jaspers replied on 24 July 1952, more than two years after Heidegger's crucial letter of 8 April 1950 – crucial with reference to the question of guilt – disappointed that his correspondent had evaded the issue and refused to give an account of himself. Invoking the indelible distant past, when something of value had indeed passed between them in the philosophical dialogue of the 1920s, he subjected Heidegger's omissions to harsh criticism, which ultimately shows how irreconcilable their positions were, and how impossible it was to bridge the chasm that had opened up between them. Jaspers accused Heidegger of failing to respond on an important, and for him (Jaspers) indispensable point, namely the question of guilt, and of not pursuing his reference to 'shame'. 'What we both understand by "philosophy", what we aim to achieve with it, to whom we address ourselves through it, how it is connected with our own lives: these things, I suspect, spring from totally different origins in your case and mine.' The only way to clear the air was through intensive discussion, which in turn had to be based on a knowledge of each other's philosophical works. Here Jaspers puts his finger on one of the weak points in their relationship, namely their failure to keep up with each other's published work – a perennial failing since the 1920s.

Moving on from the personal level, Jaspers took up Heidegger's dictum 'the work of evil is not yet done' – the evil that 'is only now entering its universal phase', personified in the figure of Stalin. Reading this had alarmed and angered Jaspers:

> Several important questions present themselves. Does such a view of things contribute, by its very indeterminacy, to the spread of evil? Does not the seemingly grand sweep of such visions lead us to overlook the practical possibilities for action? How can you publish a positive verdict on Marxism somewhere, without at the same time stating clearly that you recognize the power of evil? Are we not all under obligation to tackle this power wherever we find it, and in the case of someone who speaks out, to tackle it by speaking in terms that are clear and concrete? Is not this power of evil in Germany the same thing that has been growing steadily, foreshadowed in the crimes of Stalin: the practice of throwing a veil over what is past and forgetting it?

Heidegger's prophetic and poetic brand of philosophy as voiced in his letter, invoking the spectre of a monster, had perhaps helped to 'prepare the victory of totalitarianism by cutting itself off from reality' – just as philosophy prior to 1933 had done much to prepare the ground for acceptance of Hitler. 'Can the political dimension, which you say has been superseded, ever disappear from the scene? And is this not precisely what we need to understand?' As for Heidegger's mystical dictum about a 'homelessness' in which an 'Advent' is concealed, Jaspers could not make head nor tail of this:

> My alarm grew when I read that. As far as I can see, this is nothing but an idle fantasy, of a piece with so many other idle fantasies that have mocked us by turns these last fifty years. Are you planning to set yourself up as a prophet, who has secret knowledge of transcendental truths, a philosopher who leads us away from reality, who offers us fictions that distract from what is possible and feasible? But the question is: where does someone like that get his authority and credentials from . . . ?

He portrays Heidegger as someone who leads men astray – if only 'away from reality'. Time and time again we encounter the charge that the allurements of his rhetoric have caused him to abandon the rigours of conceptual thought and become a kind of sorcerer.

However, Karl Jaspers had thoroughly misjudged his man if he expected a confession of guilt from Heidegger. This also explains why Jaspers' report of December 1945, so devastating in its impact on Heidegger's academic career, met with a blank response as far as Heidegger himself was concerned: it was a blow that failed to connect. 'Authority and credentials'? Such questions were of no concern to the philosopher of *Being and Time*, for whom guilt, responsibility and atonement belonged in a system of categories that had nothing to do with conventional morality. The only ethical category in which Heidegger was able to think or be understood – if indeed we can speak of ethics at all in his case – was that of 'obedience to Being'. The lines are already clearly drawn in the *Letter on Humanism* (1946):[42] anyone who sought, like Heidegger, to thrust forward in thought into the truth of Being, that is, into the 'forest clearing' where Being is illuminated, had found the place where the hidden god is present, where no care about 'directions for the life of action' ever penetrates; and by the same token such a thinker is untouched by whatever affects the actions of men in a technological age. So he is untainted, unburdened by guilt – in the conventional sense of the word 'guilt'. Karl Jaspers, it might be assumed, did not know his Heidegger well enough, otherwise he would have given up trying to press him before now. Or did he know him only *too* well? Did he realize that there was another Heidegger to be seen behind the prophet of the truth of Being, the Heidegger who had his being in the world, whose poetic style of philosophical discourse concealed more than it disclosed in its dark resonance?

At all events, Heidegger vouchsafed him no direct reply – what was there to say, thinking as he did? – and chose to remain silent. But in the letter he wrote on 19 February 1953 to congratulate Jaspers on his seventieth birthday he set the seal on his legacy, modestly taking his place alongside the academic public at large:

> And there will be one who tries to follow the path you have taken, while simultaneously examining his own. He will recall the years of fellowship and the painful incidents, accepting as the will of destiny the differences in their philosophical enterprises, as both men seek, in a troubled and vacillating world, to draw attention to essentials by asking questions. Such questioning may be so persistent that it ends up by questioning itself, in an attempt to discover whether, for all the differences in each other's paths of thought, there remains a neighbourly affinity characteristic of that proximity in which all men stand to one another, ultimately unknowable and impenetrable to each other's gaze, sharing only a loneliness born of the same cause and calling. Please accept this greeting from a wanderer. It carries with it the wish that you may continue to find the strength and confidence to help your fellow men, by your deeds and works, to attain to the clarity of what is real.

It is a well-known fact that Heidegger amassed an extensive, not to say vast, correspondence, conducting a dialogue with many different people – though with varying degrees of persistence and intensity, of course. This may serve to qualify the importance of his correspondence with Jaspers to some extent, but it remains of special interest in that it began very early, in 1920, after the two men had met in the circle gathered around Edmund Husserl in Freiburg. Its particular significance derives from the fact that once Heidegger had accepted the chair at Marburg, in 1923, he used to stop off in Heidelberg whenever possible on the journey between Freiburg/Todtnauberg (the mountain hut) and Marburg, staying as a guest of the Jaspers family. Often these visits lasted several days, as described by Karl

Jaspers in his *Philosophical Autobiography*, and, as the correspondence between the two giants of German philosophy delightfully documents, Heidegger found his philosophical discussions with Jaspers enormously enriching. For sheer intensity the letters are probably without equal, and they are informed by a warm intimacy that extended into the private sphere as well. All this may explain why the issues as such have been defined with reference to these epistolary sources, and why direct quotation from the letters has such a power to illuminate and clarify: we are dealing here with material of exemplary importance. It is open to question whether Heidegger's letters to other correspondents contain disclosures of equal moment. Take for example the letter he wrote to Jaspers in 1935, at a time when he felt very much alone and isolated, seeking to re-establish contact after a long silence of more than two years. Here we find him groping around in uncertainties and generalities:

> On my desk is a file labelled 'Jaspers'. Every now and then I slip another piece of paper inside; there are unfinished letters in there too, notes on my first attempt to get to grips with Volume III of the *Philosophy*. But I haven't worked them into any sort of shape as yet. And now here are your lectures, which I would guess are the prolegomena to your *Logic*. I thank you warmly for this salutation, which cheered me *enormously*; for my isolation is virtually complete. Somebody told me you were working on a Nietzsche book; I note with delight how strongly the tide of creativity continues to flow in you, even after the *magnum opus*. With me, on the other hand, it is more a kind of laborious groping in the dark; it's only in the last few months that I have been able to get back to the work I left unfinished in the winter of 32/33, during my sabbatical; but it is only a sketchy outline. For the rest, I am finding the two great thorns in my flesh – the struggle with the faith of my birth, and the failure of the rectorship – quite enough to have to contend with (1 July 1935).

Apart from the isolation, the Pauline allusion to thorns in the flesh (2 Corinthians 12, v.7) – 'the struggle with the faith of my birth, and the failure of the rectorship' (and in that order) – may well surprise some readers. One can understand Heidegger mentioning the 'failure of the rectorship' – however *that* phrase is to be understood – at this point in time and in this context, writing to his former intimate Jaspers. But what are we to make of 'the struggle with the faith of my birth'? Both were 'thorns' that he had to 'contend with'. But did Heidegger *ever* get the better of them?

The 'faith of my birth' to which Heidegger alludes was Roman Catholicism.

PART TWO

'The struggle with the faith of my birth'

From Messkirch to Freiburg

The little town of Messkirch in Baden where Heidegger was born, the former seat of the local assizes, lies in staunchly Catholic country, nestling in a harsh, forbidding landscape with a rich and colourful historical past, where a number of small principalities once rubbed shoulders and converged.

Chief among them were the former lords of Zimmern, whose legacy is preserved in the art and architecture of Messkirch – notably the castle and the renowned late Gothic parish church of St Martin (completed in 1526). The 'Master of Messkirch' had created eleven altarpiece paintings for this church, of which only the painting on the high altar, *The Adoration of the Magi*, remains in place today. In 1627, in the middle of the Thirty Years' War, the Princes of Fürstenberg annexed the territory and town of Messkirch. Under their patronage the church of St Martin was later remodelled after the Baroque style (1770–1776); the painted altar panels executed by the 'Master of Messkirch' found their way to the palace of the Fürstenbergs at Donaueschingen, where they can be admired to this day.

Adjoining Messkirch is the territory of the Hohenzollerns, likewise Catholic, owing its ecclesiastical allegiance to the archdiocese of Freiburg – as does the barren and forbidding landscape of the high Messkirch plateau, and the fertile lakelands that stretch out to the west, blessed with a milder climate. And in the centre of this area around Lake Constance lies the former episcopal city of Constance itself. When the Grand Duchy of Baden was founded by the grace of Napoleon Bonaparte, Constance lost its pre-eminent status, not least because the bishopric was transferred to Freiburg; but it remained the cultural centre of the region. Above all it offered opportunities for higher education, available also to gifted Catholic boys from the country, through the archiepiscopal grammar school for boarders, which had been successfully re-established in the early 1880s when the *Kulturkampf*, which had been conducted with particular virulence in Baden, began to abate.

A development with important consequences for the Messkirch church district was the revival of the monastery at Beuron in 1863. Founded by the Augustinian canons, the monastery had lain empty since the secularization and was then in an advanced state of disrepair. A group of Benedictines moved in, and with the aid of an endowment from the Dowager Princess Katharina von Hohenzollern-Sigmaringen they brought a new lease of monastic life to the languishing buildings. Here they established a new focus of spiritual, intellectual, and above all artistic life, which had a profound impact on the area and on the many individuals who now looked to Beuron for inspiration. Among these was Martin Heidegger, who was well acquainted with the monastery and with the sons of St Benedict from the days of his youth, and he returned there repeatedly. He established many links with Beuron, which was consecrated to St Martin, and was a welcome guest there, singing for his supper by giving lectures to the small circle of resident monks. It was here that he gave the first reading of 'The Essence of Truth' in 1930, and 'Augustinus: Quid est tempus? Confessiones lib. XI' in the same year or 1949. In Beuron, at least, he was not viewed as an outcast.

Messkirch, Constance, Beuron, Freiburg: these are the places where Martin Heidegger felt at home all his life, and from which he could not take his leave. A provincial milieu, certainly: but this was no ordinary province.

This is staunchly Catholic country, but receptive to new intellectual currents and influences. Here, to the east of Lake Constance, the enlightened brand of Catholicism, with its own specific spirituality and liturgy, that was propagated during the early decades of the nineteenth century by Ignaz Freiherr von Wessenberg, intercessor for the see of Constance, fell on fruitful ground, even if its influence was confined to the well-educated and the affluent. As the nineteenth century unfolded, the religious sphere gradually came to reflect a whole range of social distinctions, which only became fully apparent with the emergence of Old Catholicism in the last three decades of the century. The origins of Old Catholicism go back to the firm rejection of the dogmatic declarations of the First Vatican Council of 1870 on the subject of papal primacy and papal infallibility. In the southern territories of the Grand Duchy of Baden the town of Messkirch, along with Constance, became an important local centre of the Old Catholic movement, and here the confessional divide led to a complete split into two *political* camps. The smaller but affluent liberal party and the large but poorer Catholic party produced a kind of two-class society in Messkirch in the years following 1870 – a division that was fully reflected in the life of the local Church.[43]

Encouraged by the liberal policies of the Baden government, which had looked with special favour on the Old Catholic movement during the years of the *Kulturkampf* and promoted it at every opportunity, the Old Catholic community in Messkirch was granted rights of joint use in the Catholic church of St Martin's. The Catholic Church authorities in Freiburg could not accept this solution on principle, with the result that the town's Catholics eventually withdrew from their traditional house of worship. In 1875 they acquired a temporary place of worship through the purchase of a fruit store or fruit barn from the Fürstenberg estates, which stood close to the castle and not far from the town church, and with the aid of the talented monks from Beuron the interior was transformed into a welcoming, makeshift church. It was here that Heidegger's father performed his duties as sexton, although he was no longer able to live in the old sexton's house, which had been allocated to the Old Catholics. And it was here too, in this makeshift house of God painted and decorated by the monks from Beuron, that Martin Heidegger was baptized in 1889. One wing of this temporary church housed the ground-floor workshop of master cooper Friedrich Heidegger, where little Martin often used to go, 'well provided with words of warning from my mother, and pulling my little blue trolley behind me', as he was later to recall. This was the tiny world of Heidegger's early childhood, succinctly evoked by him in his 1949 essay *The Country Path*.[44]

When the numerical discrepancy between Catholics and Old Catholics became too glaringly obvious (the ratio was around 3:1) and the fierce passions aroused by the *Kulturkampf* had subsided, the representations made by the Catholic parish of St Martin's in

Messkirch to the government of Baden were duly rewarded with success. In 1895 the late Gothic parish church, together with all its patrimony and property (including the sexton's house on the Church Square, in which the Heidegger family now took up residence) were restored to the town's Catholic congregation. They moved back into their church with solemn ceremony on 1 December 1895 – the first Sunday in Advent – and celebrated their first mass there after the long years of exile. For the Catholics of Messkirch it was a deeply moving occasion; likewise for the six-year-old Martin, whose family circumstances were undergoing such a dramatic change. His brother Fritz reports that the outgoing Old Catholic sexton gave the keys to little Martin, because he could not face surrendering them to his successor in person. The exposure to Old Catholicism undoubtedly remained an enduring childhood memory of church life, given that it had dominated the religious life of Messkirch for several decades and led to the creation of far-reaching social distinctions, not to say discrimination. However, we have somewhat anticipated the later course of events, so will confine ourselves here to biographical information about the Heidegger family.

Martin Heidegger was born on Thursday, 26 September 1889 in Messkirch, in a small house on the 'Graben' that was pulled down in the 1890s, the first child of Friedrich and Johanna Heidegger. Friedrich Heidegger, born on 7 August 1851 in Messkirch, a master cooper and sexton of the Catholic parish since 1887, was married on 9 April 1887, when he was nearly 37, to Johanna Kempf, who came from the village of Göggingen that lies a few kilometres to the east of Messkirch. In those days the post of sexton carried an annual salary of 500 marks – quite a tidy sum when one considers the opportunities for supplementing it with casual earnings. To be sure, the sexton was required to perform various special duties, and the position was essentially full-time.[45]

Johanna Kempf, born on 21 March 1858, was brought up on a farm that had been in her family for centuries. Her father, Anton Kempf, was born in Göggingen on 7 July 1811, and died there on 3 July 1863; her mother, Justina Jäger, was born in the same village on 25 September 1818 and baptized the next day on the festival of St Justina. She died in Göggingen on 17 April 1885.

Martin Heidegger never knew his grandparents on either side of the family, for his father's parents also died before he was born. His grandfather and namesake Martin Heidegger, a master cobbler in Messkirch, was born on St Martin's day, 11 November 1803, in Leibertingen (which lies midway between Beuron and Messkirch), and died in Messkirch on 8 November 1881. Walburga Rieger was born in 1815 in Gutenstein (in the district of Messkirch), and died on 5 April 1855 in Messkirch, still a young woman; her son Friedrich was barely four years old. The philosopher's grandparents on his father's side moved to Messkirch later on in life, though they came from the immediate vicinity. All branches of the family remained loyal to the Roman Catholic faith – the faith, incidentally, to which Martin Heidegger claimed to belong throughout his life.[46]

The family on his mother's side was principally formed by several centuries of continuous residence on the farm in Göggingen. An ancestor of Heidegger's mother, one Jakob Kempf, had been granted feudal tenure of the farm back in 1662 by the Cistercian convent of Wald, near Pfullendorf. For generations it remained in the possession of the family as a hereditary fief. It was a sizeable estate, comprising upwards of fifty acres of arable land and several acres of pasture and woodland. In the wake of the liberalizing agrarian reforms carried out in his day, Heidegger's grandfather, Anton Kempf, was able to purchase the farm in 1838/39 for the sum of 3800 *gulden*. In 1839, now an independent farmer with his own freehold, he married Justina Jäger, who came from a family of innkeepers in Göggingen. This marriage produced nine children, including Heidegger's mother Johanna.

The young Martin Heidegger spent a good deal of time with his relations over in Göggingen, especially since he had a cousin there of almost exactly the same age: Gustav Kempf, who later became a Catholic cleric and schoolmaster, and spent parts of his school and university career in the company of his cousin Martin – as we shall see later.[47] In Heidegger's early childhood the farm at Göggingen, with its yards and open spaces, and the nearby schoolhouse with its own school garden, were among his favourite playgrounds. The two boys led 'a carefree life', as Martin Heidegger wrote in 1972 following the death of his cousin Gustav, 'with no thought of the two world wars that were to come (the second of which was to claim the lives of the heirs to the Kempf family farm). In 1972 Heidegger took comfort amidst all the pain and grief in a maxim that he attributed (wrongly) to the old Church Father Hegesipp: 'A joy it is to linger among the familiar domestic things of our forefathers and to study their words and deeds in fond remembrance.' The letter shows how closely bound up were the memories of the ageing philosopher with an important part of his early life and his birthplace.

In Messkirch Heidegger grew up with his sister Marie and his brother Fritz. His parents were neither rich nor altogether poor, typical lower-middle-class people, writes the philosopher's brother Fritz Heidegger, his 'one and only brother', in a delightful letter that describes the congenial atmosphere in the parental home. 'In material terms our parents were neither poor nor rich; they were comfortably off in a lower-middle-class way; we knew neither want nor plenty; 'saving' came high on the list of priorities in our household: bright new coins, as rare as genuine pearls, were for many people the "heart of all things".' And Fritz Heidegger does not hesitate to spell out the social contrasts that existed in his day.[48]

As luck would have it, there is a contemporary study on the economic situation of artisans and tradesmen in Messkirch 'with special reference to blacksmiths, cartwrights and saddlers'. This was undertaken in 1896 by the 'Association for Social Policy', which commissioned a series of 'studies on the situation of the skilled trades in Germany with particular reference to their ability to compete with big industry'. Of the 130 enterprises surveyed in Messkirch, 83 skilled tradesmen were tax-rated at between 500 and 2000 marks, which placed them in the lower ranks of the social order. In the context of the confessional and political divides already described, the pattern of economic development was also reflected in banking and financial institutions. The little town had a 'red' (liberal) savings bank and a 'black' (Catholic) bank. 'The former', notes the author of the report, 'had the larger turnover, the latter the larger membership.'

This was the social and economic category in which the Heidegger family found itself. Its financial circumstances were assessed in 1903 at 2000 marks in real estate and 960 marks in taxable income, which was enough for a family of five to live on, but it did not leave a penny to spare to send gifted children to a school that led on to higher education – not unless it happened to be located in the town or within easy daily reach. The educational facilities that Messkirch had to offer consisted of a higher-grade elementary school or *Bürgerschule*. Such schools were sometimes comparable to a junior grammar school or *Progymnasium*, but only if the special subjects taught at German grammar schools, notably Latin, were acquired elsewhere. Given the regional distribution of classical grammar schools in the Grand Duchy of Baden – and this was the only type of school that could be considered for a gifted Catholic boy of humble origins, destined as he was for a career in the Church – it was common practice to prepare boys from the country for the 'Quarta' or 'Untertertia' (the seventh or eighth year classes at the grammar school), and then to throw them in at the deep end, straight into the grammar school proper, where they would soon learn to swim. This was the route taken by Heidegger, too – which explains how a highly gifted boy from humble circumstances was led into the paths of philosophical thought. Part of the explanation also lies in the importance of the Catholic Church in those days as a powerful organizing force in society.

In the autumn of 1903 the talented, bright and also athletically active Heidegger, then just fourteen, finished his eighth year at the elementary school in Messkirch and entered the next year's class – the 'Untertertia' – at the grammar school in Constance, enrolling as a pupil of the 'Konradihaus', as the archiepiscopal seminary for grammar school boarders was popularly known. The easy transition to the next class in the grammar school had only been accomplished with the active support of Messkirch's then parish priest, Camillo Brandhuber. He had discovered Heidegger's talent and done everything in his power to see that it was properly nurtured. By giving the boy Latin lessons the parish priest opened the door to Constance.

Born in 1860 in Sigmaringen – on Hohenzollern soil, therefore – Camillo Brandhuber had been appointed to his post in Messkirch in 1898. He was a colourful personality, a highly talented man of the people and popular speaker in the tradition of the Centre-party clerics who were so numerous in his day. Later he was to make a political career for himself in his Hohenzollern homeland. Having served as parish priest in Hechingen from 1906, Camillo Brandhuber was elected a member of the Prussian Assembly for Hohenzollern in 1908, a position he held until 1918; and in the wake of the November revolution of 1918 he became a member of the Hohenzollern Communal Assembly, serving as its president until 1922. In the few biographical *aperçus* he has left us, Heidegger fails to mention the figure of this priest once; but the truth of the matter is that he owed him a very great deal.

In Constance the young grammar school student, newly enrolled at the Konradihaus, discovered that the headmaster of this archiepiscopal establishment was a fellow countryman. Dr Conrad Gröber, born in Messkirch in 1872, was a graduate of the elite Collegium Germanicum in Rome, and destined for the very highest office in the Church. Small wonder, then, that the Rome-educated Dr Gröber enjoyed enormous prestige within his native parish.[49] As a local man, Gröber was well acquainted with the circumstances of the Heidegger family – he came from a similar social background himself – and had aided the efforts of the parish priest, Father Brandhuber, to obtain the young Heidegger's transfer from the local elementary school to the grammar school in Constance. The career of a future priest was at stake, after all. It should be emphasized once again that there was nothing unusual about this kind of advancement in those days. How else were gifted young men from the country to gain access to higher education, especially as most of them came from humble, not to say very humble social backgrounds?

In 1903, as it happened, Gröber was busily engaged in writing a history of the grammar school in Constance which was due to celebrate its 300th anniversary in 1904. Admittedly he had confined his attentions to a selected portion of the school's illustrious past, namely the period from 1604 to 1773 when it had been a Jesuit institution. The abolition of the Society of Jesus in 1773 marked the end of an era in Constance, just as it did in Freiburg and Mannheim and other cities in southern Germany. After their re-establishment the Jesuits never took up this academic tradition again: the conditions that had made it possible, and above all its material basis, had been too completely swept aside. Yet in biographical essays on Heidegger one constantly reads that the philosopher attended 'the Jesuit grammar school' in Constance, and then transferred to the 'Jesuit grammar school' in Freiburg. Since 1773 the grammar school in Constance had been an independent foundation that subsequently came under the aegis of the city of Constance – as was the case in Freiburg too, where Heidegger was later to complete the last three years of his grammar school education.

From 1903 onwards, as Heidegger himself attested on many occasions, Conrad Gröber became an important figure in his life, possibly the most important – and on a number of counts. First of all the material difficulties had to be resolved: as a boarder at the archiepiscopal seminary, the Konradihaus, Heidegger was expected to pay for his board and lodging. Mindful of the family circumstances in this instance, Dr Gröber applied to the archiepiscopal church authorities in Freiburg, under whose overall control the Konradihaus lay, for full remission of all boarding fees; but his submission was rejected, with the result that Heidegger's annual fees were assessed at 100 marks. But thanks to the efforts of his parish priest back home in Messkirch, Father Brandhuber, he was awarded an annual grant of 100 marks from the local Messkirch endowment fund during his first year at Constance, which was partially offset against the contribution his father was required to find, leaving an actual charge of 50 marks. A little later the so-called 'Weiss Grant' that Heidegger had been awarded was increased to 300 marks. This sum was used to pay the boarding fees in full, thus removing the financial burden from his parents.

For a young man who is fully acquainted with his family's financial straits, but who is also aware of their willingness to make sacrifices in the hope that the priest-to-be, the Lord's anointed, will requite all their efforts, it is irksome to be dependent on their financial support – and a release when there is no longer any need to accept their sacrifice. It seems altogether likely that Heidegger thought along these lines, even if we have no direct evidence to prove it. This interpretation is confirmed by a document from 1945, the critical year when the bottom seemed to have dropped out of Heidegger's life. As he faced the prospect of having his home and belongings, and more especially his library, confiscated in the summer of 1945 because of the political role he had played under National Socialism, Heidegger wrote a very moving letter to the then Acting Mayor of Freiburg: 'I come from a poor and simple home, and had to struggle through my years as a student and a junior lecturer with many sacrifices and privations. At home we always lived a simple life, so I do not need a lecture in what it means to think and act in a socially responsible manner.'[50]

Committed to simplicity all his life, Heidegger always acknowledged his humble social origins and the debt he owed to them. His modest way of life was influenced both by his family upbringing and by his time at the Konradihaus in Constance. The headmaster from Messkirch, Conrad Gröber, guided his pupils with a firm hand, always taking a benevolent interest in his young countryman Heidegger and encouraging him wherever possible, particularly in matters intellectual. This close personal relationship endured even after Gröber had left the Konradihaus in 1905 to take charge of a city parish in Constance. He continued to follow his former pupil's progress even when Heidegger moved to Freiburg in 1906, after completing his third year at the Konradihaus.

Messkirch remained the lasting point of contact. It was here that Gröber and Heidegger met repeatedly in later years. In the *curriculum vitae* that he submitted to the Faculty of Philosophy at the University of Freiburg in 1915, in connection with his forthcoming habilitation [the process of qualifying as a university lecturer: translator's note], Heidegger wrote: 'I owe an important debt of gratitude to the intellectual influence of the then headmaster of the boarding-school for boys in Constance, Dr Konrad

Gröber, now a parish priest of that city.' Heidegger has also told us of Gröber's key role in directing his attention towards philosophy:

> In 1907 a paternal friend from my native town, the future archbishop of Freiburg, Dr Conrad Gröber, gave me a copy of Franz Brentano's dissertation 'On the Manifold Meaning of Being according to Aristotle' (1862). The many extended quotations from the Greek served as a substitute for the edition of Aristotle that I did not yet possess, though a year later it would be sitting on my desk on loan from the school library. The question that stirred but dimly, hesitantly and helplessly in my mind then – the problem of the singleness of multiplicity in Being – remained throughout many vicissitudes, labyrinthine wanderings and perplexities the enduring inspiration for my treatise *Being and Time* that appeared two decades later.

So wrote the philosopher in the brief autobiographical sketch he wrote for his admission to the Heidelberg Academy of Sciences.[51] Since 1907 (as he tells us in another context) the dissertation by Franz Brentano had been a rod and staff to him in his first fumbling attempts to enter the domain of philosophy.[52]

The three years in Constance evidently had a very fruitful effect on the young Heidegger's intellectual development. Brilliant teachers at the grammar school, the rich landscape around Lake Constance: an early poem, 'An evening walk on Reichenau Island' (first published in 1916) shows how vivid were the memories of his Lake Constance homeland, so packed with interest and stimulation. At school, after all, he rubbed shoulders with gifted pupils drawn from a large catchment area – boys like Max Josef Metzger, two years his senior, whom Heidegger never actually mentions by name, but who was already trying his hand at writing in the senior classes of the school during Heidegger's time in Constance, as well as organizing lecture evenings in the Konradihaus. After his ordination as a Catholic priest Metzger later pursued a highly individual career, becoming a pioneer of the ecumenical movement, deeply committed to working for peace, and a staunch patriot whose concern for the fate of Germany brought him before Freisler's 'People's Court'. He was executed on 17 April 1944 in Brandenburg (Havel)-Görden.[53]

Assisting the headmaster, Dr Gröber, was a clerical prefect with special responsibility for the junior classes, appointed in 1905. This was the future headmaster of the Konradihaus, Matthäus Lang, to whom Heidegger wrote a very warm letter on 30 May 1928 from Marburg, thanking Lang for his congratulations on his appointment to succeed Edmund Husserl in Freiburg. As early as 1928 the philosopher was formulating the synthesis of his thought and his native land, underlining the key role played by the boarding-school in Constance:

> I think back with pleasure and gratitude to the beginnings of my student career at the Konradihaus, and I become ever more aware of how closely all my efforts are bound up with my native soil. I can still remember clearly the trust I came to feel for you as the new prefect, a trust that has endured, and that made my time in the seminary one of joy. Since then I have inquired frequently about you, your work and life at the school. Now I should have more opportunities of visiting you and looking back at the world in which I started out as a young 'Untertertianer'. From there to *Being and Time* seems a long and tortuous road. And yet everything shrinks into insignificance, when I compare what has been achieved with what was purposed. Perhaps philosophy demonstrates more clearly and forcibly than other disciplines what *beginners* we all are! That's what the practice of philosophy means in the end: being a beginner. But then, insignificant as we are, as long as we keep faith with our inner selves, and try to do what we can from within, then even the little that we do accomplish must be to the general good.

He signed the letter: 'Your former pupil, Martin Heidegger.'[54] Though now a world-famous philosopher, Heidegger used the German word 'Zögling', corresponding to the Latin term 'alumnus', that was customarily applied to boarders at the archiepiscopal grammar school. The word encapsulates a whole tradition of Catholic pedagogy, as Heidegger was very well aware, and his use of the term here, in 1928, is slightly playful. If this letter were the only documentary evidence we had from Heidegger's hand, we would have to assume that the author of *Being and Time* regarded his Catholic roots as the essential condition of his work, never to be forsworn – and that there was an intimate connection between the Konradihaus in Constance and the *magnum opus* of the philosopher, which had already made him world-famous. His teacher Matthäus Lang, for whom Heidegger felt a warm personal attachment, was widely regarded as the strong proponent of a strictly orthodox line. 'One of Lang's particular aims with the young persons committed to his charge was to inculcate a true devotion to the Church, a love of the Church, indeed . . . The authority of the Church was paramount in Lang's eyes. Once she had spoken, there was nothing more to discuss. Her decree was absolute.' Thus was Matthäus Lang characterized in an obituary published after his death in 1948.[55]

In strange contrast – though it comes as no real surprise to anyone who knows the inside story – is the remark that appears in a letter written by Heidegger to Karl Jaspers in Heidelberg on his return to Freiburg in 1928. Since his departure in 1923, he wrote, the University had 'become more dyed-in-the-wool Catholic than you would ever believe' – whatever that might have meant in concrete terms. In Heidegger's first account of his experiences as a lecturer in Freiburg, written on 10 November 1928, he talks of 'spies' planted in the auditorium. He had occupied this 'outpost', which his 'innermost conviction' told him was 'lost'; 'the Catholics have made "lamentable" advances, and young *Catholic* lecturers are now in place everywhere.' The Faculty of Philosophy at Freiburg had deteriorated 'significantly', he claimed: the only difference was 'that I *no longer "hide"* in my philosophical work.'

Catholicism offered such an easy target – or at least, it did when writing to such a sympathetic listener. Of course, we know that the old wound was festering still, the wound that had been inflicted on Heidegger as a Catholic junior lecturer. But again, we are anticipating events that came much later.

In the autumn of 1906, at the end of the 'Untersekunda' or tenth school year, Heidegger transferred from Constance to

Freiburg, where he attended the highly respected Bertholdgymnasium and became a boarder at the archiepiscopal seminary of St Georg in Freiburg. The move seems slightly surprising; normally one would have expected him to finish school and graduate in Constance, if only because this was closer to home for the grammar-school boy from Messkirch. Speculations that the young Heidegger was undergoing some sort of spiritual crisis, and needed a change of scene on that account, are quite unfounded. Heidegger remained close to his classmates from Constance throughout his life, even though they had not been through the final years of school together. Even in his old age he was a regular guest at the annual old boys' reunions in Constance – as we read in a memoir by the city's official archivist, Helmut Maurer.[56] No: what prompted the move to Freiburg was the material and financial situation of Heidegger's parents.

Back in the sixteenth century there had been a distinguished citizen of Messkirch, one Christoph Eliner, who matriculated at the University of Freiburg in 1538 and finally obtained his doctorate from the Faculty of Theology. He served as dean of the Faculty of Theology for many years, and from 1567 held the office of rector for several semesters. On 5 January 1575, in the shadow of death (Eliner died on 15 January and was interred in the university chapel of Freiburg Cathedral) he executed a detailed will, whose provisions included the establishment of an endowment fund for two students in the amount of 2700 *gulden*, carefully invested in copyhold deeds at a rate of five per cent, yielding an annual sum of 130 *gulden*. The purpose of these studentships was to enable the recipients to study for a doctorate in theology. The level of grant was scaled according to the student's progress. This endowment survived through all the troubled times, being adjusted at intervals, of course, to reflect changing economic and social circumstances, and at the time that interests us here the rules governing the award of a studentship were as follows: the qualifying criteria, in order of priority, were still the same (family connections, Messkirch, the former county of Zimmern), the subject of study remained Catholic theology, and grants could be made to boys in their senior years at grammar school, provided they had reached the 'Untersekunda'. Applications had to be submitted to the Messkirch parish council. The award of the studentship was therefore independent of the University, although the Freiburg Faculty of Theology did have the right to involve itself in the selection process. By the beginning of the twentieth century the studentship was worth 400 marks. But: the awards were conditional on attendance at the grammar school in Freiburg and enrolment thereafter at Freiburg University.

Hence the upheaval that Heidegger went through in 1906, moving to a new school in another city in order to qualify for the studentship. Freed thus from material worries, he continued his studies in a fresh academic environment, adapting himself to the different atmosphere of the boarders' seminary in Freiburg. We know from the reminiscences cited above how fruitful Heidegger's school-days in Freiburg proved to be: his first reading, in 1907, of the dissertation by Franz Brentano (1838–1917), the Viennese philosopher and founder of a 'descriptive psychology or phenomenology' (1888), as well as the teacher of Edmund Husserl; his intensive study of Aristotle in the seminary; and the broadening of his horizons that the university city of Freiburg must have brought. The teaching at the Bertoldgymnasium – originally a Jesuit foundation, but run by the city of Freiburg as a non-Jesuit institution since 1773 – was stimulating and demanding, as Heidegger states in his *curriculum vitae* of 1915. As he approached final graduation, the grammar-school boy who had long since acquired the habits of independent thought pursued his own academic interests, not content simply to follow the school's prescribed syllabus. The ultimate goal of his studies, after his brilliant performance in the school-leaving examination in the summer of 1909, was and remained Catholic theology – or in other words, the road to the priesthood.

Initially Heidegger chose not to take the most obvious and direct route, which would have been to seek ordination via enrolment as a student of the Archiepiscopal Seminary for Roman Catholic priests in Freiburg. This is already apparent from the testimonial that the headmaster of the Freiburg seminary, Professor Leonhard Schanzenbach (who had also been Heidegger's teacher for religion and Hebrew) gave him on his departure from the St Georg residence. In his report, which he produced for every candidate in theology, he writes:

> Martin Heidegger was born in Messkirch on 26 September 1889, the son of the local sexton. He left the grammar school and seminary in Constance and entered the 'Obersekunda' at this school, being obliged to change schools in order to qualify for receipt of an Eliner studentship. He is gifted, diligent and of good moral character. He had already attained a certain maturity when he came to us, and he was used to studying on his own initiative; indeed, his studies in German literature, an area in which he proved to be extremely well read, were sometimes pursued at the expense of his other subjects. Since he is quite sure he wishes to pursue a theological career and favours the life of a religious order, he will probably apply for entry to the Society of Jesus (10 September 1909).

The expectations of Heidegger's headmaster proved to be correct. There were no Jesuit organizations as yet within the territory of the German Reich, but on 30 September 1909 Heidegger entered the novitiate of the Society of Jesus at Tisis near Feldkirch in the Austrian Vorarlberg, admitted by the Provincial of the day, P. Thill. On 13 October 1909, however, the young candidate was allowed to leave again, for unspecified reasons, as we can see from the register of the novitiate at Tisis, now kept in the novitiate of the Upper German Province of Jesuits at Nuremberg. The novitiate period proper, which involves the taking of vows, is preceded by a candidature of fourteen days, during which time candidates are not yet robed and only permitted to share in the life of the community to a limited extent. Heidegger left the novitiate exactly at the end of his two-week candidature. According to a well-documented rumour current among Jesuits, Heidegger complained of pains in his chest while hiking up on the mountain pastures near Feldkirch; so he was discharged from the order on account of his weak physical constitution – a highly plausible explanation, as will appear below.[57] In other words, the decision to leave was not Heidegger's but the Jesuits'. A sound physical constitution and the stamina to match were the basic prerequisites for life in a religious order – and equally so for anyone aspiring to the secular clergy. The postulant Heidegger had been given his first warning: his health was not up to it.

Heidegger then applied immediately for admission as a candidate at the theological seminary in Freiburg, founded under the patronage of Carlo Borromeo and known as the Collegium Borromaeum. He was duly accepted, and in the winter semester of 1909 he began his studies in Catholic theology at the University of Freiburg – which still left him time enough, as he has testified, for the philosophical studies that formed part of the syllabus anyway. So from his very first semester at the Freiburg Borromaeum, he tells us, the two volumes of Husserl's *Logical Investigations* were sitting on his desk, on loan from the university library, where they were requested so rarely that he was able to renew them more or less at will. Among his theological teachers he was particularly impressed by the dogmatist Carl Braig, a proponent of systematic theology, who had a decisive influence on the development of Heidegger's thought (all this on the testimony of Heidegger himself). In his final year at grammar school Heidegger had already come across Braig's treatise *On Being: An Outline of Ontology*, first published in 1896, whose reference section contained 'lengthy extracts from Aristotle, Thomas Aquinas and Suarez, as well as the etymologies of terms for basic ontological concepts' – tools of thought that Heidegger himself would later use.

Scattered throughout his sparse autobiographical writings Heidegger has left some grateful reminiscences of this teacher. Carl Braig, 'the last in the tradition of the Tübingen speculative school, which by engaging in a debate with Hegel and Schelling has conferred status and breadth on Catholic theology', had impressed him, he states, by the incisiveness of his thought. 'It was from his lips, when he let me walk with him sometimes, that I first heard of the importance of Schelling and Hegel for speculative theology, as opposed to the doctrinal system of scholasticism. In this way the tension between ontology and speculative theology as the basic framework of metaphysics first engaged my attention and interest,' notes Heidegger, in an essay written in honour of the publisher Hermann Niemeyer, in which he charts the road that led him to phenomenology.

These few indications will give some idea of the intellectual horizons that opened up before the young student of theology. But in the *curriculum vitae* he wrote in 1915 in connection with his habilitation procedure, the text of which will be quoted in full later, Heidegger presents a very different picture. Braig is not mentioned, but he does acknowledge the influence of Hermann Schell, whose works on apologetics had answered some of his questions. He had made a private study of the scholastic textbooks, and by this means acquired a certain formal training; but he failed to find there what he was looking for. Thomas Aquinas, Bonaventura and finally the *Logical Investigations* of Husserl all had a decisive influence on his academic development.

Later on the philosopher summed up the relationship between his thought and Catholic theology – and the scholarly studies on this subject are legion – in this succinct formula: 'Without these theological origins I should never have found my way to philosophy. But our origins are always before us.'[58]

Little is known of the daily life of the theology student. The seminarists lived and worked within a highly structured discipline, their days ordered by the regular round of mass, prayers, lectures, private study and tutorials, with specific hours set aside for rest and recreation – all of which was familiar enough to a boy who had lived the life of a grammar-school boarder. Whether Martin Heidegger was on close terms with any of his fellow students I have been unable to establish, except in one instance. His only personal friend on the course was Friedrich Helm, a very cultivated but shy young man, who later became court chaplain and private secretary to two archbishops of Freiburg, Thomas Nörber (deceased 1921) and Carl Fritz (deceased 1931).

The student of theology: early works

It should be possible to gauge and assess the intellectual development of the young theology student from his early publications. However, there is something rather strange and surprising about our knowledge of these texts. When we examine the numerous bibliographies of Heidegger's works, we find that the post-1912 writings – from the period *after* his full-time theological studies – are catalogued in minute detail. The famous Volume 13 of the complete works, *Adventures in Thought* [*Aus der Erfahrung des Denkens*] (1983) contains only a very slender *oeuvre* from the years 1910 and 1911: a brief article on the unveiling of the monument to Abraham-a-Sancta-Clara at Kreenheinstetten in August 1910, and three poems, 'Sterbende Pracht' ['Dying glory'], 'Ölbergstunden' ['Gethsemane Hours'] and 'Wir wollen warten' ['Let us tarry'] – all published in the *Allgemeine Rundschau*, the ultra-conservative Catholic weekly put out by Armin Kausen in Munich.[59] These pieces constitute a kind of authorized version, approved by Heidegger himself towards the end of his life. But why the omission of the other works from his time as a theology student? Have they somehow escaped the attention of those zealous investigators of Heidegger's early life, appearing as they do in obscure publications and sometimes under a pseudonym or cipher, though never apocryphally? Or did they fall victim to the censorship of the ageing Heidegger? Whatever the explanation, these texts should and must be taken into account, since they furnish us with a good, and above all accurate, picture of Heidegger the student of theology.

These pieces are to be found almost exclusively in the pages of *Der Akademiker*, the journal of the German Association of Catholic Graduates. First appearing in November 1908, this publication served as a link between the Catholic student fraternities – 'the voice that proclaims our high Christian ideals in all areas of student life', as the initial editorial announcement put it.[60] The initiative for founding this Catholic journal for students came from the circle associated with the *Allgemeine Rundschau*. In addition to the religious and academic component, *Der Akademiker* also turned its attention to social and charitable concerns. Consequently the youthful author Heidegger found himself in the company of Romano A. Guardini and Oswald von Nell (-Breuning), a student of mathematics and the pioneer of Catholic social teaching. Certainly at that time, when the issue of German modernism was being hotly debated within the Catholic Church, the journal stood firmly on the side of Pope Pius X, defending the authority of the Church in the field of academic theology and indeed in the humanities in general.[61]

A text that figured prominently in the Catholic press of the day was the book by Friedrich Wilhelm Foerster entitled *Authority and Liberty* (1910), in which a philosophy founded on the authority of the Church is defended against a philosophy of ethical-religious individualism. Seen against this background, the review published by Heidegger as a theology student in the May 1910 number of *Der Akademiker* is of special interest (and readers would have already been primed by a preview of Foerster's introduction that appeared in the journal). Here Heidegger takes the official line all the way, coming out firmly against an 'unfettered autonomism'. The whole philosophical approach was 'eloquent testimony to the inestimable value of this book'. The twenty-one-year-old student Heidegger is totally at one with his Church: 'And in order to keep faith with its eternal store of truth the Church is right to strive against the destructive forces of modernism, which remains blind to the utter contradiction between its modern view of life and the ancient wisdom of the Christian tradition.' Indeed, for the zealous young student of theology Foerster does not go far *enough* in his critique of modernism. Nevertheless: 'He who has never strayed into the paths of error and never let himself be deceived by the false appearances of the modern spirit, he who in true, profound and unwavering abandonment of the self seeks to journey through life in the radiant light of truth, to him this book will be a source of great joy, and an astonishingly vivid reminder of the felicity that comes with the possession of the truth.'

Reading Foerster's book, Heidegger 'recalls with pleasure a saying by the great Görres: dig deeper, and you will find Catholic soil'.

The underlying Catholic orthodoxy that informs this review also appears in other important pieces by Heidegger from this period, such as the essay 'Per mortem ad vitam: thoughts on Jørgensen's "The lie of life and the truth of life"' which appeared in the March 1910 number. This is a study of the Danish writer, poet and essayist Johannes Jørgensen (1866–1956), who abandoned his early Darwinism and naturalism to convert to Catholicism in 1896 and became Professor of Aesthetics at the Catholic University of Louvain in 1913. The conversion experiences undergone by the Dane, who was for Heidegger 'a modern St Augustine', were a source of powerful fascination and led him to reflect on the themes of personality, self-realization, ethical standards, commitment and unfettered freedom. What was it that impressed the young Heidegger, and what were the maxims by which we should live our lives?

> There is a great deal of talk these days about 'personality'. And philosophers are always coming up with new concepts of value. As well as critical, moral and aesthetic categories, they now work with 'personality assessment', particularly in the field of literature. The *person* of the artist moves into the forefront of our attention. So now we hear a great deal about 'interesting persons': Oscar Wilde the dandy, Paul Verlaine the 'drunken genius', Gorky the great vagabond, Nietzsche the superman – 'interesting persons' every one. And if one of them suddenly realizes, in the hour of grace, that he has been living a lie in his bohemian existence, and smashes the altars he has raised to false gods and becomes a Christian, this is dismissed as 'boring' and 'sickening'. Johannes Jørgensen was one who took this step. It was not a craving for excitement that drove him to conversion, but a profound and utterly serious conviction.

With this resounding indictment he begins his essay, an attack on 'a distorted, mendacious philosophy', which he contrasts with 'the restless seeking and patient synthesis, the final step towards the truth'. The decadence of individualism as a false standard to live by is here unmasked, the 'psychology of the freethinker' branded as a sickness unto death. The theologian Heidegger strikes a note of high pathos: 'And if you would live a spiritual life and attain your salvation, then die, extinguish all that is base within you, be about your business with supernatural grace, and you shall be resurrected. And so our poet-philosopher rests now in the shadow of the Cross, resolute and bright with hope: a modern St Augustine.' Here is a Jørgensen of one mind with the young Heidegger, since he has 'uncovered the great, indestructible links with the past' and dwells among the mystics of the Middle Ages, while 'his poet's heart, drunk with peace' burns for St Francis, *il poverello*.

The following spring, in the March 1911 issue of the magazine, Heidegger offers some advice 'for the philosophical guidance of students'. Philosophy for him is *philosophia perennis*, 'in truth a mirror of the eternal' – in contrast to the unwholesome contemporary obsession with subjective world-views. Here Heidegger vigorously champions the objectivity of strict logic, the 'immovable, eternal limits of logical principles'.

He concedes that strict logical thought needs a certain foundation of 'ethical strength', the 'art of self-composure and self-expression'. For the Catholic student, a thorough grounding in apologetics and the adoption of a true and genuine philosophy were absolutely vital. The present fashion, however, was for people to tailor their philosophy to suit their lives, instead of the other way round.

> And as they flit to and fro, sampling different philosophies like so many delicacies, turning it into a kind of game, there wells up within them, for all their conscious awareness and self-satisfaction, an unconscious desire for clear-cut, definitive answers to the ultimate questions of Being, which flash upon them from time to time with such suddenness, only to lie unresolved for days on their tormented souls, which are without purpose or direction.

As an aid to students interested in philosophy, Heidegger reviews a number of relevant titles, all originating within the Catholic camp. He gives special prominence to the little book by the Jesuit Friedrich Klimke entitled *Major problems in philosophy* (1910), with the help of which it is possible 'to find one's way in some sort through the maze of "modern philosophies" of various hues and persuasions, separating the wheat from the chaff and making use of the former.' In his role as philosophical critic, Heidegger the student of theology feels at home in the tabernacle of Catholic philosophy, and particularly in the area of epistemology, the most 'fundamental science' of all, for which he recommends the book by Josef Geyser, *The Foundations of Logic and Epistemology: A study of the forms and principles of true knowledge* (1909). Heidegger adopts a similar tone in his brief review of *The need for God* (1910), a monograph by Otto Zimmermann, SJ, which appeared in the *Korrespondenz* on 15 May 1911. This is an essay in apologetics that looks at the problem of proving God's existence from a new point of view, namely by adapting it to modern ways of thinking. The reviewer Heidegger welcomes this approach but with some reservations: 'It may even be wise, in the face of the modern assault on conventional proofs of God's existence, to develop our system in the direction advocated by Zimmermann. For many an educated man of our day who is an earnest seeker after truth, this little book could well serve as a powerful bulwark to protect his faith in God.' The *Korrespondenz* was the official organ of the Akademischer Bonifatiusverein, an association in which Catholic students, both fraternity members and others, laymen and theologians, had come together to further specifically Catholic interests in philosophy, the natural sciences and social policy.

Reflected in this literary mirror we can discern the figure, however shadowy, of Martin Heidegger the student of theology.

The enforced termination of Heidegger's theological studies and the start of his academic career as a philosopher

In the winter semester of 1910/11, as Heidegger entered the second year of his course in theology (his third and fourth semesters), Gustav Kempf, his cousin on his mother's side and his playmate from early youth, also enrolled as a student of theology. Already they had spent some years together as fellow pupils and boarders at the grammar school in Constance and at the Konradihaus. But the winter semester of 1910/11 brought with it certain complications regarding Heidegger's health. The young student had simply been working too hard. In addition to his not insignificant workload in theology he was also immersing himself in the study of philosophical systems and exploring the great legacy of Greek and medieval texts, learning the language of philosophical thought. All of this taxed the physical strength of the slightly built theologian. In mid-February, before the winter semester was over, Heidegger had to break off his studies when a medical examination once again revealed nervous heart trouble 'of an asthmatic nature' – a complaint from which the young student had probably been suffering at least since the autumn of 1909, when he stayed as a postulant at the Jesuit novitiate of Tisis near Feldkirch. Heidegger had to struggle with this condition throughout his life. The report by the head of the seminary, Dr Bilz, is brief and to the point: 'Martin Heidegger, a second-year student, was obliged to terminate his studies in mid-February following a recurrence of his nervous heart condition. With our permission he returned to his home town. He was strongly advised to discontinue his studies entirely until he is fully recovered.'[62] The resident physician at the theological seminary, Dr Heinrich Gassert, had recommended on 16 February 1911 that Heidegger should be sent home for several weeks of 'complete rest'. In the event, however, Heidegger was given leave of absence for the entire summer semester of 1911, which he spent at home in Messkirch. Finally, on the advice of his superiors, he abandoned the study of theology altogether: 'My earlier heart complaint, brought on by too much sport, now recurred with such violence that I was told there was very little prospect of my being able to serve later in the Church', as he writes in his *curriculum vitae* of 1915.

This enforced period of convalescence, aggravated by the uncertainty about his future career and financial cares that came with a change of academic course, was among the most difficult episodes in Heidegger's life. As a student of theology he had continued to receive the Eliner endowment grant, which ensured a smooth transition from school to university. But Heidegger was no longer eligible for this grant if he ceased to study theology (and at that time the study of Catholic theology was inseparably linked to the career aim of becoming a Catholic priest).

What began as a temporary break soon led to the final and complete abandonment, much against Heidegger's will, of his theological studies, and hence of any plans to enter the priesthood. The effects of this on the future course of his life can scarcely be overestimated.

The decision-making process with which Heidegger had to wrestle in the coming months can be traced in the letters from his closest friend at the time, whom Heidegger got to know in February 1911. This was the young history student Ernst Laslowski, a native of Kreuzburg in Upper Silesia, who was working in Freiburg with Heinrich Finke, Professor of History (Catholic Chair) at the University.[63] This correspondence remains a valuable source of information right through until 1917, not least because Heidegger himself soon came under the protection of Heinrich Finke.[64]

But first let us review the situation as it presented itself in the spring of 1911 – a situation that Heidegger described as absolutely hopeless in a letter to his friend Laslowski, in which he sought the latter's advice. In reply, Laslowski conceded that the position was serious – but denied that it was hopeless (letter of 20 April 1911).

Three proposals put forward by Heidegger were discussed between them. The first was for Heidegger to begin a course of study in mathematics, on which he did actually embark in the winter semester of 1911/12, with a view to sitting the state examination. The second was for him to pursue an academic career in philosophy, which immediately raised the question of financial support. Once he had been awarded his doctorate, he could count on a grant from the Albertus Magnus Association or the Görres Society for the Advancement of Science. Laslowski's advice, if Heidegger was minded to pursue this path, was to get in touch with the Catholic philosopher Clemens Baeumker, then teaching in Strasbourg, and to enrol as his pupil. Since Baeumker was chairman of the Görres Society's philosophy committee, he would be well placed to support a grant application.[65]

The third possibility considered by Heidegger was to continue his theological studies. Money would no longer be a problem, and he would soon obtain a post that would leave him plenty of time to prepare for his habilitation. Under the circumstances, Laslowski strongly advised his friend to pursue his theological studies. He should concentrate on apologetics, study at Freiburg under the dogmatist Carl Braig and the specialist in apologetics Heinrich Straubinger, obtain his doctorate in due course as curate of a municipal parish in Freiburg, and then move out to a rural parish to broaden his experience. The Würzburg writer on apologetics Hermann Schell[66], whom Heidegger acknowledged in the *curriculum vitae* he submitted in connection with his habilitation procedure in 1915, was put forward as a role model. His friend urged him to re-read Schell's autobiography.

All these arguments relate to established patterns of behaviour: this was the normal course of an academic career in Catholic theological circles. The vast majority of postulants came from humble backgrounds, and were invariably imbued with strong feelings of inferiority *vis-à-vis* their social superiors. In the letters exchanged between the two friends the problems of those who came from the poorer classes – 'sons of artisans, peasants, schoolteachers at the very most' – are constantly being aired. What later came to be recognized as a genuine social issue – the shortcomings of the Catholic education system – is here seen in immediate personal terms: 'If only your father could support you for the three to five semesters you will need to get your doctorate and prepare for your habilitation, I'm sure the money could be found from somewhere.' But Friedrich Heidegger, cooper and sexton of Messkirch, was in no position to do so, and the crisis in his son's career brought about by the problems with his health presented Martin with a bleak choice: if he did not continue with his theological studies he would not receive any grants. The 'lowly' circumstances from which Heidegger emerged, thence to rise to the 'secret of greatness': this antithesis or dialectic can be seen at work in his later writings, notably in his interpretation of Hölderlin ('Germania', 'The Rhine'). 'The lowly has its own constancy and endurance, the dull obstinacy of the never-changing everyday, which is constant only because it resists all change, and must resist it.' Only one who has 'surrendered to the power of history' with understanding and awareness knows that there is something greater and higher above him.

> The ability to accept something greater above oneself is the true secret of greatness. The lowly cannot do this, even though the lowly is most easily and obviously the furthest removed from greatness. But then the lowly desires only itself, i.e., to be lowly, and its secret is no secret at all, but only a ruse and the peevish cunning that belittles and denigrates everything that is not on its own level, in order to bring it down to that level.[67]

No other solution was in sight. Heidegger was a troubled man.

During the summer of 1911, a time of convalescence which he spent in Messkirch apart from an interlude for hydrotherapy in Bad Wörishofen (that was the plan, at least), the solution took shape in his mind. A continuation of his theological studies was by now out of the question; so Heidegger decided to embark on a course of study in mathematics at Freiburg University starting in the winter semester of 1911/12, with the intention of working towards the state examination.

It was during the spiritual crisis of early 1911 that the little poem 'Gethsemane Hours' was written. It was published in the

Allgemeine Rundschau on 8 April 1911, the Saturday preceding Passion Week. As far as I know, nobody has made the connection between this poem and Heidegger's personal difficulties at this time – if indeed it has been the object of critical attention at all:

Gethsemane Hours

Gethsemane hours of my life,
in the dim light
of doubt and despair
how oft have you seen me!

My tearful cries were never in vain.
My youthful being,
Weary of lamentation,
Trusted only in the angel of mercy.

To these lines should also be added a hitherto unknown but nonetheless published poem, 'On Still Paths', which appeared in the July number of *Der Akademiker* under the signature '-gg-', and in which Heidegger's state of mind in the interlude between his studies emerges very clearly:

On Still Paths

When the lights of summer nights wash
Over the white birches on the heath,
When the wan moon's glow
Hangs over all like a crown of jewels –
Then opens wide the heart,
All lamentation done,
And memories seek me out
From distant days
Of blissful joy:
But the heady scent of fiery roses
Has long since wrapt me about
Love's sepulchre . . .

Freiburg i. Br. -gg-

We have a further reflection of Heidegger's mood in a third poem from that summer [private bequest]:

July night
Songs of eternity
You sing to me once more.
You draw my soul hence
Into the silent forest fastness
Plunge me into endless spaces
Closer to the divine.
July night
Sorceress
Mistress of the
Homesick heart.
Forasmuch as the sun today
Expired early in the fields,
Forasmuch as the fruits of the day
Were consumed in twilight gloom:
Forasmuch as the song-weary
Finch falls silent
And coldly frets
The sullen night wind,
Forasmuch as the lindens
Hearken to the dying song
Forasmuch as the leaves shiver
As though I were parting from you –
The bitter question rises up within
My fearful heart: O happiness, is
Thy bride called 'lamentation'?

The dark side of a depressive sensibility is immediately apparent in all these lyrical attempts – and small wonder, given the hopelessness of his situation. But meanwhile Heidegger was taking steps to establish contact with the Catholic philosopher Clemens Baeumker by drafting his first small article for the *Philosophisches Jahrbuch* (1912), repeatedly interpreted by critics and scholars of the early Heidegger as his first work.

Contacts were also established with the Freiburg theologian Josef

Sauer, then Extraordinary Professor for the History of Art and Christian Archaeology at Freiburg University, but more importantly the editor of the Catholic literary journal *Literarische Rundschau für das Katholische Deutschland*, published by the Herder Verlag. Sauer was to become one of the young Heidegger's most important patrons, recognizing his outstanding gifts and gently nurturing his talent. It is worth noting the personal nexus that was formed here: this early encounter between the two men, and then the key role that Josef Sauer (himself rector of Freiburg University until 15 April 1933) played in Heidegger's own rectorship.

In the autumn of 1911, when Heidegger was back in Freiburg again, he offered the editor of the *Literarische Rundschau* a more substantial piece of work entitled 'Recent investigations in logic'. Sauer published it a year later in three instalments, and these studies have likewise attracted (and continue to attract) a great deal of attention from historians of philosophy everywhere.[68] It may not be inappropriate, therefore, to document in more detail how Heidegger – now at the end of his first semester as a student of mathematics and the natural sciences – chose to explain his research interests to the theologian Sauer, outlining his own basic philosophical position in the process. In a letter of 17 March 1912 – which was also written to mark the name day of Josef Sauer: the feast of St Joseph, whose cult had strong popular roots, falls on 19 March – the young Heidegger expressed his sincere affection for his spiritual mentor, and at the same time set out his own academic agenda.[69]

> Dear Professor,
>
> Please permit me, Sir, to offer you my warmest congratulations on your forthcoming name day.
>
> May God grant you the strength and grace to pursue your scholarly studies for many more years to come, that you may continue to work wholeheartedly towards the one true goal of promoting the religious and cultural development of our Church.
>
> Thus will your tireless labours be accomplished in the spirit

> of your patron saint, whom we honour as the special protector of our Catholic Church.
>
> At the same time, Sir, I feel the need to thank you sincerely for your gracious readiness to encourage me at all times in word and deed. If I may venture to speak here of my own humble efforts, I can tell you that my present work is almost complete. In essence it is only a preliminary study, intended as a point of departure for embarking on an extensive investigation into mathematical logic. If the whole undertaking is not to become a sterile exercise in faultfinding, a scholastic exposure of contradictions, then the problem of time and space must at least be brought close to a preliminary solution by applying to it the principles of mathematical physics. That task is made the more difficult by the fact that the theory of relativity has thrown everything in physics into a state of flux at present. At the same time, the study of logic has recently sought to merge with the general theory of objects, which in turn serves to simplify the investigation again very considerably. In short, my chosen field of study is itself in such a state of turmoil that it would be premature for me to adopt a firm position – quite apart from the fact that I do not yet feel competent to do so . . .

Outlined here is a research programme of extraordinary substance and startling modernity, not larded with fashionable catchwords but resting on solid foundations; a programme that the young Heidegger was unable to pursue, in the event, because circumstances obliged him to turn his attention to a topic from scholastic literature – as we shall see later. In the end this ambitious speculative project yielded only the trial lecture on 'The concept of time in historical studies,' which Heidegger gave on 27 July 1915.[70]

If we pause here and attempt to summarize the course of events so far, then it appears certain that what Heidegger's headmaster, Professor Schanzenbach, had said in his school-leaving report two years earlier was still valid in 1911: Heidegger was determined to pursue a theological career. Whether the enforced change of direction had a traumatic effect on the student must remain an unanswered question. Perhaps a good few of Heidegger's later statements, even before 1933, which are suggestive of a profoundly anti-clerical attitude, together with many subsequent anti-Church pronouncements, are to be explained in the light of what happened in 1911. We must not lose sight of this possibility, bearing in mind that all his life Heidegger probably felt deeply equivocal, not to say caught in a dilemma, with regard to the faith of his birth: a fact that is centrally relevant to the subsequent course of his philosophical career.

For the winter semester of 1911/12, therefore, Heidegger decided to resume his studies in the Faculty of Mathematics and Natural Sciences. In the semesters that followed he attended lectures and classes in mathematics, physics and chemistry, but did not sit the final examinations in any of these subjects (which had probably never been his serious intention anyway). In addition he went to lectures and seminars given by Arthur Schneider, who held the chair of Christian Philosophy (II), and Heinrich Rickert, who held the chair of Philosophy (I). Heidegger's chief interest was philosophy, and he turned his attention at this time to modern philosophical trends without abandoning his basic convictions, which were those of Aristotelian-scholastic philosophy. He sought to combine Greek and medieval philosophy with modern logic, and his dissertation on 'The theory of propositions in psychologism', supervised by Arthur Schneider, was written from this perspective. A couple of years later he would adopt a similar approach for his habilitation thesis.

From the summer semester of 1912 onwards Heidegger received a grant of 400 marks administered by the University of Freiburg, which relieved him of his most pressing financial cares. Certainly Heidegger was no stranger to financial cares at this time, and for the rest of the year his friend Ernst Laslowski was working particularly hard to drum up loans from private backers so that Heidegger could study for his doctorate. The grant that Heidegger received from the summer semester of 1912 onwards was evidently not enough to get by on, and the extra income from private tuition failed to bridge the gap. Laslowski knocked at the door of many a well-respected figure, most of them clerics – without success, until in the winter of 1912/13 he managed to persuade an old boy from his Catholic student fraternity 'Unitas' in Breslau to advance a loan.

He wrote repeatedly to Heidegger, dispensing words of comfort and encouragement, portraying him as the great white hope of German Catholic philosophy.

The articles that Heidegger had just published in the *Literarische Rundschau* made a great impression, not least in Rome, where Laslowski's studies had taken him in 1912. When Laslowski wrote to Freiburg on 20 January 1913 to tell Heidegger that the loan was on its way, he gave expression once again, in his customarily fulsome style, to his vision of the brilliant career that awaited his friend: 'My dear fellow, I have the feeling that you are destined to become one of the truly great, and the universities will be falling over each other to get you. Anything less would be inadmissible.' Of course, 'Catholicism doesn't fit in at all with the whole modern philosophical system'. In twenty years' time Heidegger would have to make a statement on this subject – 'preferably from a lecture platform in Berlin, whose effect must be "epoch-making" (but in the *right* sense of that term!)'. He warned Heidegger against publishing anything else in the *Philosophisches Jahrbuch* of the Görres Society, since too many eyes were now upon him. He should not allow himself to be pigeon-holed by the bigwigs and shoved straight into the 'Catholic' category. 'I think it would be good if you were to surround yourself with an air of mystery for a time, to arouse people's curiosity. Things will be easier for you after that.' Laslowski then runs through the list as he has done so many times before: the Catholic chairs of philosophy, the career openings, the future opportunities. The right place for Heidegger to begin was probably Munich. 'Because I suppose you'll have to start out as a Catholic. But this really is a confoundedly vexed question, you know.' Much would depend on the first book, the dissertation.

During his stay in Rome in October 1912 Laslowski had met the Freiburg lecturer Dr Engelbert Krebs, a priest of the archdiocese of Freiburg and a member of the Faculty of Theology. Krebs had earlier spent an extended period studying at the Campo Santo Teutonico, where he had held a post as chaplain. The German lecturer offered to act as guide to the young history student, showing him the sights of Rome. For his part, Laslowski spoke effusively of his friend Heidegger, and predicted that it would not be long before Krebs received a visit from Heidegger in Freiburg. He also learned in passing from Krebs that 'the way to get through the habilitation procedure is to canvass the support of one particular professor; you can't do it on your own. So you probably ought to keep in with Geyser or one of the others' (Letter from Rome, dated 25 October 1912)[71].

Heidegger followed his friend's advice. Although working on his dissertation under Arthur Schneider, who taught at Freiburg, he made contact at the beginning of 1913 with Josef Geyser, then teaching at Münster, sounding him out about his future. But Geyser was not particularly forthcoming. He simply advised Heidegger to sit the 'state examination in the philosophical subjects', after which he might think seriously about taking up editorial work in scholastic philosophy. This piece of advice was thrown to the winds by Heidegger, but the fact remains that within the space of a few months he had corresponded with two of the leading Catholic philosophers of his day, namely Clemens Baeumker and Josef Geyser, with whom he would be competing a few years later for the chair of Catholic philosophy at the University of Freiburg.

On 26 July 1913 Heidegger took his doctoral examination before the Faculty of Philosophy and passed 'summa cum laude'.[72] His dissertation supervisor, Arthur Schneider, was unable to do anything more for the hopeful young graduate, having meanwhile accepted the offer of a post at the Reich University in Strasbourg. The young doctor of philosophy was faced with an entirely new academic situation. On the one hand he was no longer obliged to take into account the views and preferences of his doctoral supervisor Schneider, with whose philosophical approach he had little enough in common – though for formal reasons the work had to be supervised by the Catholic professor; on the other hand, there was no way of knowing whether a successor would soon emerge at Freiburg, and if so, who. Might it even be Heidegger himself?

So Heidegger could have ended up in something of a vacuum, had it not been for Professor Heinrich Finke, Privy Councillor and holder of the Catholic-endowed chair of History, who was also the dominant figure within the Faculty of Philosophy – and Engelbert Krebs, the aforementioned lecturer in dogmatics in the Faculty of Theology at Freiburg, who was eight years Heidegger's senior. A few weeks before he was awarded his doctorate, Heidegger had paid Krebs the visit for which Laslowski had paved the way in Rome. From July 1913 a relationship of genuine, if at times strained, friendship developed between the two men. The meeting with Krebs proved immensely important and fruitful for Heidegger, even though there is no mention of it in Heidegger's autobiographical notes – or perhaps precisely for that reason. Their close rapport also spilled over into the private domain.

Krebs kept a full and meticulous diary, enabling us to reconstruct the years from 1913 to 1917 with a considerable degree of accuracy. The diary entries document Heidegger's academic development, but also the mood of rivalry between the two friends, both of whom had their eye from the beginning on the chair of Christian philosophy that had fallen vacant through Schneider's departure. For Krebs also had good reason to hope that he would be offered the position, since he had been appointed in the winter semester of 1913/1914 to deputize for the vacant chair: an unusual arrangement in itself, since the normal procedure was to try and appoint deputies from within the same field of study. Indeed, it would not be too much to say that the two aspiring candidates were already crouched on their blocks, listening intently for the sound of the starting-gun that must come sooner or later. In the approaching race it rather looked as if the theologian Krebs, already a fully-qualified university lecturer, was starting with an advantage. On the other hand, his clerical status might be seen as a handicap, which was all the greater in that Krebs had sworn the oath against Modernism exacted by the Vatican, and was no longer regarded by most members of the Faculty of Philosophy as an independent scholar. The account that follows is largely based on Krebs's diary, as the quotations indicate, although repeated reference will also be made to Heidegger's correspondence with his friend Laslowski.

Let us turn first to Heinrich Finke. He was an internationally respected scholar, author of studies on the medieval Councils, an acknowledged expert in Spanish history, an influential member of Freiburg's Faculty of Philosophy and an energetic patron of young Catholic scholars. That there was a great deal to be done in this area, given the shortcomings of the Catholic education system, was generally acknowledged in his day. Finke now took a close personal interest in Heidegger as an outstanding philosophical talent, and intervened decisively in the development of his career.

Before moving to Strasbourg his dissertation supervisor Schneider had strongly urged the newly fledged doctor to proceed to habilitation, emphasizing the need for Heidegger, who was a systematist, to turn his attention to the medieval philosophers. He had also set the machinery in motion to secure an adequate habilitation grant. But thereafter the responsibility for encouraging this up-and-coming academic talent rested with the historian Finke, who took Heidegger under his wing, reckoning on the fact that the chair would remain vacant for some time to come: for he had big plans for the brilliantly gifted student from Messkirch. The remaining full professor of philosophy, Heinrich Rickert, played only a marginal role in the development of Heidegger's academic career – not least in terms of his philosophical influence – and was content to leave the field to Finke. In November 1913 the latter urged Heidegger to write on some aspect of the history of philosophy, whereas Heidegger himself wanted to do research 'into the logical nature of the number concept', based on his studies in mathematics and the natural sciences, as we have already seen.

Earlier that year, during the semester in which he took his doctorate, Heidegger had attended a four-hour lecture course given by Finke under the title, 'The age of the renaissance'. However, some sort of relationship must have existed prior to this, if only through Heidegger's friend from Breslau, Laslowski, who had been studying under Finke. But he seems not to have been on close terms with the Privy Councillor, given that in the autumn of 1912 hc had asked Laslowski to make representations to Finke about securing financial support. 'I'd rather not write to Finke. I know him too well. There wouldn't be any point. I'd rather wait until the article has appeared in the *Literarische Rundschau.*' In the meantime, however, Finke had noted Heidegger's talent and added him to his list of protégés.

Finke's first thought was to secure a basic source of income. He advised Heidegger to sit the state examination to qualify as a grammar-school teacher, and only then, from this established position, to turn his attention to the pursuit of an academic career. But then he supported the grant option after all. As was noted earlier, Heidegger's dissertation supervisor, Arthur Schneider, had set the grant application machinery in motion, with an approach to Freiburg's suffragan bishop and dean, Justus Knecht. The grant in question was funded through an endowment administered by the metropolitan chapter of the archdiocese of Freiburg. As dean of the cathedral Knecht had the most influence in these matters; he was also very well disposed towards the young Heidegger and willing to further his career in a decisive way.

It is worth pausing to consider the matter of the grant a little further, because the circumstances surrounding it give us a useful insight into Heidegger's mentality. The grant, for which he applied to the Freiburg cathedral chapter following the initial approaches referred to above, was funded through an endowment established in 1901/02 by Constantin von Schaezler and his sister Olga, under the title 'The Constantin and Olga von Schaezler Foundation in honour of St Thomas Aquinas'. The foundation was endowed in the amount of 200,000 marks, which was no mean capital fund, given that it would be worth twelve times that figure at today's values. The award of grants from this foundation was contingent upon the strict observance of the doctrine of St Thomas Aquinas in philosophy and theology. The explanation for this rigidly neo-scholastic tendency lies in the religious convictions of the von Schaezlers, who were scions of the celebrated Augsburg family of bankers and industrialists. They had converted to the Catholic faith in the 1850s, causing a great stir in Augsburg and Bavarian Swabia; and as is often the case with converts, they observed and lived out their new faith with a special passion and intensity.[73] In his first application of 20 August 1913 Heidegger wrote in the following terms:

> The obedient undersigned makes bold to submit a humble request to the Reverend Cathedral Chapter of Freiburg im Breisgau for the award of a grant from the Schäzler Foundation. The obedient undersigned intends to devote himself to the study of Christian philosophy, and to embark on an academic career. Since the writer lives in very modest circumstances, he would be deeply obliged to the Reverend Cathedral Chapter if it would please the same to award him a grant from the aforementioned Foundation for the period during which he is preparing his habilitation thesis.

In his letter approving the grant the suffragan bishop Justus Knecht reminded Heidegger of the wishes of the founders and the purpose for which the endowment had been established: 'Trusting that you will remain true to the spirit of Thomist philosophy, we are pleased to award you a grant of 1000 marks from the Schaezler Foundation for the academic year 1913/14.' Heidegger, who received the grant for three consecutive years, always asserted that he would justify the trust placed in him 'in the service of Christian-scholastic philosophy and Catholic ideology' – or as he put it in his letter of 13 December 1915, in which he applied for the grant to be extended for a third year: 'The obedient undersigned ventures to think that he can show something at least of his lasting gratitude for the valued trust placed in him by the Reverend Cathedral Chapter by dedicating his scholarly lifework to the task of harnessing the intellectual and spiritual potential of scholasticism to the future struggle for the Christian–Catholic ideal.' When Heidegger wrote these words he was already a junior lecturer in philosophy, but he had also been conscripted for military service on the home front, attached to the Postal Control Office in Freiburg. The First World War had been raging for a full year.

For the moment, let us accompany Heidegger along the path towards habilitation, recalling the words that Krebs used to sum up the initial impression made on him by Heidegger in the summer of 1913; 'An acute mind, modest but assured in his demeanour'. Heidegger explained to his host that the greatest impetus to his studies in logic had come through the works of Husserl. Krebs, to whom the phenomenological method was still very alien, was so impressed by this conversation that he saw in Heidegger the future incumbent of the chair of Christian philosophy at the University of Freiburg. 'A pity he was not this far on two years ago. We could do with him now.' This acute mind was also extremely ambitious, especially as Heinrich Finke made no secret of the fact that he favoured him for the vacant chair even before the topic of his habilitation thesis had been decided on. This is evident from an entry in Krebs's diary for 14 November 1913:

> This evening between five and six he [meaning Heidegger] came to see me and told me how Finke had urged him to do his thesis on some aspect of the history of philosophy, and that

> Finke had clearly given him to understand that as long as the chair remained vacant Heidegger should seek to qualify as a lecturer as soon as possible, thereby making himself available as a candidate. So it may be that in my present caretaker role I am simply keeping the chair warm for Heidegger – coincidentally a schoolfellow of my own brother Hans.

We noted earlier that Krebs had been appointed by the Faculty of Philosophy to deputize for the vacant chair, with effect from the winter semester of 1913/14 that had just begun. The preparation of his lecture course (Krebs gave four hours of lectures on logic and noetics) caused him considerable difficulties, which he overcame with Heidegger's help. The young candidate for habilitation discussed the lectures with Krebs, and indeed went on to initiate his dogmatist friend into the philosophical doctrines of Husserl's phenomenology, while for his part Krebs, an expert in scholastic philosophy, shared the riches of his knowledge with Heidegger. Krebs frequently talked things over with Heidegger, 'in order to arrive at a clearer understanding of problems. I read out to him what I am planning to say in the lecture, then discuss it with him. He helps me more than perhaps he himself realizes.' It was a mutual process of give and take, the more so as the ties of friendship grew stronger between them. In fact Krebs sought to interest his friend Heidegger in doing some work on a treatise on logic by Meister Dietrich of Freiburg, looking at its place in the history of philosophy – an area in which Krebs himself had done a great deal of research.

Heidegger meanwhile, under strong pressure from Finke, had turned his attention to the subject of Duns Scotus, albeit with some reluctance, since he would rather have worked on the logic of the number concept. The latter was an area in which Heidegger 'felt thoroughly at home, because he has a thorough grasp of higher mathematics (infinitesimal and integral calculus, group orders (??)) and that sort of thing', noted Krebs in November 1913.

In his habilitation thesis, 'The categories and doctrine of meaning in Duns Scotus', on which he was working in earnest from the spring of 1914 onwards, he had laid the foundations of a particular interpretative approach, whereby scholastic patterns of thought were subjected to a phenomenological interpretation inspired by the work of Husserl. At the same time this piece of work also contains the requisite allusions to the philosophical thought of Rickert, who (after all) was in charge of the habilitation procedure as the specialist subject representative on the committee. Admittedly Rickert asked Krebs to draft the principal report: 'When he [Heidegger] submitted it [the thesis], I read it through for Rickert at the latter's request and wrote a report on it, on the basis of which it was accepted. As I read it, however, I had Heidegger sitting right there beside me, and we discussed all the difficult or problematic passages as we went along.' While the cosiness of such a scene may in itself appear surprising, the close ties of friendship between the two men make it understandable enough. On their frequent walks together they talked long and hard on every subject under the sun, including the almost intolerable strain imposed on Catholic theologians as a result of the great 'Modernist debate' that came to a head during the pontificate of Pius X. This created a dilemma that Krebs, for one, felt very keenly. At the end of 1912 Krebs had sworn the so-called oath against Modernism required of Catholic theologians – and admitted as much whenever anyone asked. He had his answer already prepared: 'I regard the oath against Modernism as an unmerited vote of no-confidence by Pius X, which simply represents a formal tightening-up of the existing constraints imposed by dogma.' For the rest he regarded the issue as a matter of conscience, which had nothing to do with the appointment of university professors.

We also know this from a letter Heidegger had written to his friend Krebs in the summer of 1914, at a time when a further tightening-up of papal-ministerial doctrine on the normative foundations of Catholic theology and philosophy was in prospect: henceforth the *Summa Theologiae* of St Thomas Aquinas was to be adopted as the sole guide in these matters. In formal terms this ruling applied only to mainland Italy and the surrounding Mediterranean islands, but its effect was much wider than that. It also applied only to Catholic theologians; but Heidegger, while not a theologian as such, saw himself bracketed within the Catholic system of academic scholarship that felt threatened by this new hard line from the Vatican. He struggled against the extra-philosophical loyalty that was expected of him as a good Catholic. In the aforementioned letter to Krebs of 19 July 1914, in which he reports on the progress of his habilitation thesis, Heidegger pokes fun at the recently promulgated *motu proprio* of Pius X, in which the tightening of the reins was clearly apparent. Densely packed with allusions of various kinds, the letter serves to illustrate the closeness of the relationship between the two scholars:

> Honoured and esteemed Herr Doktor,
>
> Many thanks for your card. Since I get too many interruptions in the department, I have withdrawn. Last week I got stuck in my work once again. On Wednesday I am going to see Rickert, to try and get hold of his report. I've had to sacrifice my holidays, since Husserl's *Phenomenology* is really giving me a hard time in the later sections, and I don't want to be accused of misunderstanding, as Messer and Cohn recently were. I hope to be able to send off my essay on the subject by the end of the month. In my leisure hours I take out your lecture notebook – though I really need to know about the lectures you are giving now, so that the two things are not running along side by side in a totally unconnected way. Do the surrounding islands also include us? The *motu proprio* on philosophy was all we needed. Perhaps you, as an 'academic', could propose a better procedure, whereby anyone who feels like having an independent thought would have his brain taken out and replaced with Italian salad.
>
> And for philosophical requisites they could set up vending machines in the railway stations (free of charge for those who don't have any money). I have received dispensation for the duration of my studies. Would you be so kind as to put my name on the list as well?
>
> Before we know where we are you will have evolved into *homo phaenomopius*, and will be demonstrating the metaphysics of motion *ad oculos*. Perhaps we can find time soon to go for a walk together, when we can talk about the lectures on logic. Please accept the warmest wishes of your grateful admirer,
>
> M. Heidegger

These formulations point to larger and deeper issues. For we are dealing here with one more stage in Heidegger's search for a personal identity. In the papal *motu proprio* of 1914 St Thomas Aquinas, *doctor angelicus*, was proclaimed as the sole and absolute source of doctrinal authority within the Catholic Church. But Heidegger himself had very close material ties, as it were, with St Thomas and Thomism through the von Schaezler studentship established in honour of St Thomas Aquinas, which required him to 'remain true to the spirit of Thomist philosophy'. We can readily imagine the inner dilemma in which the young Heidegger must have found himself: the child of humble parents, dependent once again on the Catholic Church for financial support, as he had been in his schoolboy and student days, well aware that he was expected to toe the line.

Without wishing to get into the realms of depth psychology, it is clear enough from later pronouncements that the events of 1913 and the years that followed left their traumatic mark on the young Heidegger. We shall have ample occasion to note that fact in the course of the narrative. At this point we might usefully interpolate a highly characteristic passage from the Parmenides lecture he gave in 1942/43, which serves to illustrate and illuminate how the 'Roman element' remained a source of vexation and irritation to Heidegger. The essential realm of the *aletheia* – for Heidegger the central and quintessential concept and *locus* of his thought – has been 'blocked' (he says) 'by the mighty rampart of the Roman – and Romish – character of the truth'. For Heidegger the power of the Church in the shape of the *sacerdotium* (the supreme papal power and might) is the successor to the worldly power of the Roman *imperium*:

> The 'imperial' power appears in the guise of the curial power wielded by the papal curia in Rome. The pope's dominion is likewise founded on the power to command. Command is of the essence of Church dogma. Consequently the latter reckons equally with what is 'true' in the 'true believer' and with what is 'false' in the 'heretic' or 'unbeliever'. The Spanish Inquisition is one embodiment of the Roman-curial *imperium*.

It seems to me that Heidegger's early learning experiences, particularly in relation to conflicts such as this, have to be seen as the background and foundation of his central intellectual preoccupations.

Work on Heidegger's habilitation thesis was not disrupted by the outbreak of the First World War. Although he was called up on 10 October 1914 he was discharged a few days later on account of his heart condition, and for the time being, as a reservist, he was left in peace by the military. He could enjoy the luxury of uninterrupted work, safely removed from the mortal perils of front-line fighting. It was not until 18 August 1915 that he was called up again, having qualified as a lecturer in the meantime; and after a good four weeks in a military hospital (13 September to 16 October 1915) in Mülheim/Baden, where he was treated for neurasthenia and heart disease, he was assigned to censorship duties at the Postal Control Office in Freiburg with effect from 2 November 1915, with the status of a *Landsturmmann* or member of the territorial reserve. The posting took place on batallion orders. This period of mail censorship – those assigned to these duties were a very mixed bunch of Freiburg tradesmen, women conscripted for labour, men pronounced unfit for garrison service – remains totally shrouded in mystery as far as the available sources of information are concerned. We know that *all* suspicious mail was opened: not just the letters home from the front, but also (and most notably) correspondence addressed to neutral foreign countries. In Freiburg there has been repeated speculation that Martin Heidegger the army censor obtained access to important information by this means, and in particular was able to read the correspondence of his colleagues.

Before we go into the rest of Heidegger's military career and the problems arising out of his particular mentality, let us revert briefly to the habilitation procedure, quoting the full text of the personal résumé that Heidegger wrote in his own hand. Much of what we have already touched upon is summarized and brought into sharp focus in this document, which contains a particularly clear statement of the aspiring candidate's philosophical position – and that in the form of a direct and more or less contemporary testimonial, which has not been distorted or idealized in retrospect.[74] At the same time a knowledge of the résumé's contents contributes a great deal to our understanding of Heidegger's school and university career.

Résumé: I, Martin Heidegger, born 26 September 1889 in Messkirch (Baden), son of the sexton and master cooper Friedrich Heidegger and his wife Johanna, née Kempf, attended the higher-grade elementary school in Messkirch until 1903. From 1900 onwards I received private tuition in Latin, which enabled me to enter the 'Untertertia' class at the grammar school in Constance in 1903. A decisive intellectual influence on me at that time was the then headmaster of the boys' boarding-school, the present parish priest Dr Conrad Gröber in Constance. After completing my 'Untersekunda' year in the summer of 1906 I attended the Bertholdgymnasium in Freiburg im Breisgau until I obtained my school-leaving certificate in the summer of 1909. In my first year in Freiburg the emphasis in mathematics shifted from simple problem-solving towards a more theoretical approach, and my natural liking for this subject now became a really serious interest, which soon extended to physics as well. I also derived a lot of stimulation from my classes in religion, which prompted me to read widely on the biological theory of evolution. In my final year at school it was primarily through the lectures on Plato given by Professor Widder, who died a few years ago, that I was introduced in a more conscious way to philosophical problems, albeit not yet with any theoretical rigour. On graduating from the grammar school I entered the University of Freiburg im Breisgau in the winter semester of 1909, where I remained without interruption until 1913. Initially I studied theology. The lectures in philosophy prescribed at the time failed to satisfy my needs, so I set out to study the scholastic textbooks on my own account. They gave me a certain formal training in logic, but in philosophical terms they failed to give me what I was looking for – and what I had already found, in apologetics, in the works of Hermann Schell. Alongside the little *Summa* of St Thomas Aquinas and individual texts by Bonaventura, the decisive influence on my academic development was the *Logical Investigations* of Edmund Husserl. The same writer's earlier work, the philosophy of arithmetic, also served to place mathematics in an entirely new light for me. My exhaustive study of philosophical problems, on top of the regular course-work I was required to do, meant that after three months I was suffering from the effects of severe overwork. My earlier heart complaint, brought on by too much sport, returned in such force that I was told there was very little prospect of my being offered a position in the Church. Consequently I enrolled in the Faculty of Natural Sciences and Mathematics for the winter semester of 1911/1912. My philosophical interests were not diminished by the study of mathematics – on the contrary: since I was no longer required to attend the prescribed lecture courses in philosophy, I was able to range more widely in my choice of philosophy lectures, and in particular I was able to attend the seminars given by Professor Rickert. In the new faculty I learned first of all to recognize philosophical problems as problems, and gained an insight into the nature of logic, the philosophical discipline that still continues to interest me the most. At the same time I acquired a proper understanding of modern philosophy since Kant, which I found was covered only very sketchily in the scholastic literature. My basic philosophical convictions remained those of Aristotelian-scholastic philosophy. In time I came to see that the ideas contained in it must permit of – and indeed demanded – a much more fruitful interpretation and application. In my dissertation on 'The theory of propositions in psychologism', which addresses itself to a central problem in logic and epistemology with simultaneous reference to modern logic and fundamental Aristotelian-scholastic propositions, I attempted to find a basis for further investigations. On the strength of this work I was admitted to the oral examination for my doctorate by the Faculty of Philosophy at the University of Freiburg, which I duly passed on 26 July 1913. My reading of Hegel and Fichte, my close study of Rickert's 'Limits of conceptualization in the natural sciences' and the investigations of Dilthey, and not least the lectures and seminars given by Professor Finke, all combined to destroy completely the antipathy for history that my taste

> for mathematics had previously fostered in me. I realized that philosophy must not allow itself to be one-sidedly guided either by mathematics and the natural sciences on the one hand, or by history on the other – although the latter, in the form of the history of ideas, has incomparably more to offer philosophers. The current upsurge of interest in historical aspects thus made it easier for me to undertake a more detailed study of medieval philosophy, which is acknowledged as essential for a thorough synthesis of scholasticism. For me, initially, this was less a matter of establishing historical links between individual thinkers than of using modern philosophical methods to arrive at an interpretative understanding of the theoretical content of their philosophy. This is how I came to write my study, 'The categories and doctrine of meaning in Duns Scotus'. This work in turn gave me the idea for a comprehensive account of medieval logic and psychology in the light of modern phenomenology, which would also take into account the historical significance of individual medieval thinkers. If I am fortunate enough to be admitted to the service of scholarship and teaching, I shall devote my life's work to the realization of these projects.

The habilitation procedure culminated in the trial lecture given on 27 July 1915, for which Heidegger had chosen as his topic 'The concept of time in history'. For his motto he had chosen a quotation from Meister Eckhart: 'Time is subject to change and diversity: eternity is single and undivided.' The full text of the quotation can be found in the thirty-second sermon of Meister Eckhart, entitled 'Consideravit semitas domus suae et panem otiosa non comedit'.[75]

Heidegger, born into humble circumstances, had achieved his great aim: he had attained the status of a university lecturer, qualified for an academic career that seemed on the point of opening up before him. To his friend Laslowski, meanwhile, he offered as his 'motto for university lecturers and aspiring university lecturers' a quotation from a letter written by Erwin Rohde to Friedrich Nietzsche: 'There is no morass more calculated to turn even the boldest of pike into a bloated, full-blown, healthy frog than the conceit of the university academic' (3 January 1869).

Having been assigned to military censorship duties at the Postal Control Office in Freiburg, Heidegger began a two-hour lecture course in the winter semester of 1915/16 on 'Principles of ancient and scholastic philosophy'. The lectures were well attended, he wrote to his friend Laslowski. He was chiefly preoccupied with organizing the printing, and more particularly the financing, of his habilitation thesis, and was still awaiting the decision, expected any time now, on the appointment of a successor to fill the still-vacant chair of Christian philosophy. In fact there was nothing to prevent a successor from being appointed forthwith, since Rickert would be moving to Heidelberg to succeed Windelband for the start of the summer semester of 1916; but it was an open secret in Freiburg that Rickert had sought to keep the Catholic-endowed companion chair vacant for as long as possible, in order to swell the numbers at his own lectures. Rickert's strategy also suited Heinrich Finke, who, so it was supposed (and correctly) by those in the know, was keeping the position open for Heidegger.

But there had been a lot of misunderstandings already, and relations between Krebs and Heidegger, despite their underlying friendship, had been growing more strained of late. The theologian Krebs, who had been deputizing for the vacant chair of Christian philosophy since the winter semester of 1913/14 and had been charged by Rickert with the task of reporting on Heidegger's habilitation thesis, did all he could to force a decision on the vacant chair. He wanted to know if he still stood any chance himself – not least in the light of Heidegger's imminent habilitation. So Krebs got in touch with the government official who was dealing with the matter at the ministry in Karlsruhe (12 March 1915); it was an unofficial, private approach, but his intentions were clear enough. He stated that as from the forthcoming winter semester (1915/16) he would no longer be available to deputize for the chair. He was no longer prepared to be a party to the ongoing cover-up of an impossible situation. He let it be known that within the Faculty of Philosophy there was a majority in favour of his appointment, even though he was a theologian himself. 'So there is no longer any obstacle in the way of making a permanent appointment to the chair, quite apart from the fact that there are plenty of candidates to form a short-list, what with Geyser in Münster, Ettlinger in Munich and Dyroff in Bonn.' In the end it became clear that the real opposition to the appointment of the theologian Krebs came from Privy Councillor Finke. Krebs's approach to the authorities in Karlsruhe rapidly became known in Freiburg; Heidegger also found out about it, Krebs having sent copies of his letter to several persons in Freiburg. As a possible candidate, about to qualify as a university lecturer, Heidegger's name had not been mentioned.

Heidegger took Krebs's intervention as a personal attack upon himself. He immediately told Laslowski what had been going on: time teaches one, as he put it, to see all kinds of people for what they really are. Rickert had been playing a kind of double game, from what he thought was a cleverly contrived position. Finke was clearly in an awkward predicament; adherence to his principle of 'no theologians in the Faculty of Philosophy' might well be his own undoing. Heidegger's friend in Breslau sought to reassure him: 'The best way to deal with Finke and Rickert is just to stay cool, calm and collected. Even if they did appoint Krebs, there would still be a place for you there as a lecturer, and if not there, then somewhere else. Of course, Freiburg would have been best, because that's where the prospects are most favourable.' If Krebs was selected, it wouldn't be such a bad thing after all. 'It would actually be an advantage as far as your own scholarship is concerned' (Letter of 15 May 1915).

What a comfort they must have been, these reassuring letters from his friend Laslowski in Breslau. He stuck with Heidegger through thick and thin and took a deep interest in his work, even in small matters. Only recently, for example, he had arranged for 'a little poem' of Heidegger's to be published in the *Heliand* – 'under a pseudonym'. The March 1915 number of this 'monthly journal of religious life for educated Catholics' duly contained the following three stanzas under the simple pseudonym 'Martin Heide':

Consolation

The sun shines
But one small hour.
Early must it die.

Love weeps –
The lea of life
A field of broken shards.

As God wills!
Along eternal paths
The angels go a-wooing.

Whatever mental or emotional state Heidegger was expressing here, the fact is that since December 1913 he had been 'secretly engaged' to a young woman from Strasbourg who was related to a junior customs official in the government service of the 'Reichsland' of Alsace-Lorraine. This engagement to 'Margaret' was clearly beset by constant strains and difficulties, and was broken off in November 1915. Prior to this Heidegger's fiancée had been suffering from a serious lung complaint, which required treatment in Davos. At all events, in his consolatory letter of 21 November 1915 Laslowski alluded to Heidegger's poem of the previous spring, arguing that his friend Martin had had to make this sacrifice for the sake of his work. 'I watched you growing day by day, until you had far outgrown the sphere in which "love" and "happiness" are able to flourish. I have known for a long time that you will have to tread paths – *have* to tread them, if you ever want to reach your goals – where "love" must freeze to death.'

It was at this time, in the days and weeks leading up to Christmas 1915, that the shape of things to come in philosophy at Freiburg University began to emerge. Heidegger followed the course of events with keen interest – not surprisingly, since his own future would in part be determined by what happened. The appointment of a successor to Rickert, the last representative of idealistic philosophy at Freiburg, was hastily pushed through; it was time to put an end to the long-standing vacancy. The Philosophy I position was to be filled in time for the summer semester of 1916. Under the decisive leadership of its dean, Professor Finke, the Faculty appointed the distinguished phenomenologist Edmund Husserl (then teaching in Göttingen) to the senior position, acknowledging him to be 'the most outstanding scholar and teacher'.[76]

On reading of Husserl's prospects for the Freiburg post, Laslowski had anxiously asked Heidegger, 'What is he like personally? Generally speaking the Austrians are extremely amiable and approachable . . .' – to which Heidegger was able to give the reassuring reply: 'He lacks the necessary breadth of vision', thereby pin-pointing the fundamental difference between his own academic personality and that of Husserl (January 1916).

But the correspondence between the two friends is shot through with concern about Heidegger's future career (6 December 1915). His rival Krebs was out of the running in so far as he was certain to get a professorship in dogmatics in his own Faculty of Theology. The main thing now was to prevent 'some philistine' from pushing his way in. 'Once he's been appointed with tenure he'll just sit tight like an erratic block and refuse to be moved.' Laslowski warned Heidegger, as he had so many times before, not to make any public pronouncements on the subject of scholasticism:

> *Please* be careful, especially *now*, in what you say about scholasticism. I wouldn't be giving you such an avuncular piece of advice if you yourself hadn't already hinted, in your last-but-one letter, that certain gentlemen were pricking up their ears. And you know yourself how pathologically hypersensitive theologians are, and how highly developed is their 'sense of responsibility' when it comes to intriguing against someone they consider 'unsound'. Your critique will come quite early enough for the people concerned.

By way of warning Laslowski cites the case of Franz Xaver Kraus, an example to which he keeps returning.[77] If we compare this discussion of the 'scholasticism issue' with the undertaking that Heidegger gave in his letter to the cathedral chapter of 13 December 1915 (i.e., at the very same time) – namely that he would dedicate his life's work to 'the task of harnessing the intellectual and spiritual potential of scholasticism to the future struggle for the Christian-Catholic ideal' – the charge of opportunism is impossible to refute.

This opportunistic behaviour on Heidegger's part, which became known to his patron Finke, probably explains why the Privy Councillor withdrew his support from the young Heidegger when it came to the crucial 'encounter' – the hour of truth. Since Finke's private papers have not survived, we can only speculate about the subsequent course of events and the factors that influenced the final decision. Matters may have been complicated by the fact that Husserl took up his appointment as Rickert's successor in Freiburg at the beginning of the summer semester of 1916; the young Heidegger had already made some study of his work, but so far had had no personal contact with him. Be that as it may, the appointments board met at frequent intervals during the summer semester. Heidegger's name was mentioned, but his was no longer the only name – and it was certainly not at the top of the list. The Faculty of Philosophy finally reached a decision at the end of June 1916, appointing Josef Geyser, currently a full professor at Münster.

Born in 1869, Geyser was twenty years older than Heidegger, who was not yet twenty-seven, and whose youth and lack of teaching experience – and of course the absence of a substantial body of published work – barred his immediate access to a full professorship. At the meetings of the board the terms of appointment to the vacant chair were very precisely defined: they wanted a professor 'for philosophy *with special emphasis on the history of medieval philosophy*'. This decidedly restrictive job description, which excluded a speculative philosopher of the modern school from the outset, might almost have been formulated with the specific object of keeping Heidegger out. This definition was patently devised in the interests not only of Husserl, who would hardly have welcomed a strong rival, but also of Finke, who dropped his former protégé on the grounds that he was 'unsound on scholasticism'.

Heidegger's tentative approaches to Edmund Husserl after his appointment to the chair got him nowhere. In May 1916 he had offered Husserl the chance to look at his habilitation thesis, which had yet to be printed, but the great phenomenologist's lack of interest in the as yet unknown young lecturer was plain to see. It is doubtful whether Husserl took much note of Heidegger's work at this time. It was Finke, rather than Husserl, who directed the deliberations of the appointments board.

At all events, once the board's work was done and the die had been cast – against Heidegger – Husserl was quite willing to meet him. In a letter of 21 July 1916 Husserl confesses that he has not had a chance to have another look at Heidegger's work, and since his recollections were a little faded he was doubtful whether he could say anything very useful about it. Apart from that, he was extremely busy. Clearly, Heidegger's work had made no lasting impression on him, and he was content merely to observe the polite formalities.

The decision of the appointments board on 23 June 1916 was nothing short of a disaster for Heidegger. The board could only agree on a so-called 'short-list of one', explaining its decision in terms that make its aversion to Heidegger abundantly plain:

> The appointment of a candidate to the second chair of philosophy presents special difficulties at the present time, given the essential stipulation that only candidates whose academic abilities in research and teaching are beyond question may be considered for the post. The shortage of suitable candidates of lay status (and only lay candidates may be considered) is so acute that after mature consideration the Faculty finds itself able to recommend only one name.

So much for all the encouraging appeals from Heinrich Finke.

There was no place at Freiburg for a thinker as individual as Heidegger. How could there be, when the official view of the future held that 'philosophical studies at our university would be significantly enriched if by this means they could become a cradle of philosophical–historical scholarship specializing in the relatively unexplored period of scholasticism.' This was precisely *not* Heidegger's field: in fact it was the kind of philosophical scholarship he abhorred.

The board dealt the young lecturer a further blow by declining to recommend him even for an associate professorship. If Professor Geyser in Münster were to turn down the offer of the chair at Freiburg 'the Faculty would not be in a position, given the aforementioned lack of suitably qualified candidates, to recommend an alternative appointment even to the rank of associate professor.' Heidegger was good at most for a stopgap appointment, 'whereby teaching duties in philosophy may be assigned to a junior lecturer on a temporary basis'. Under such circumstances, the board agreed, there would be no need to favour someone from outside 'over our own Martin

Heidegger, who shows great promise'. In that event he would indeed be given a temporary teaching appointment. The Senate promptly qualified Heidegger's prospects by declaring that 'in the opinion of the Senate Dr Heidegger should only be considered if Professor Geyser declines the appointment despite the most strenuous efforts to secure his acceptance'. But the inner circle of the Görres Society had taken good care to ensure that the professor from Münster *would* make the journey south to Freiburg. For Geyser Freiburg was only a stepping-stone on the way to his ultimate goal of Munich, which he duly reached in 1924. For its part the Faculty wanted to go for a well-established figure who occupied a recognized position in the tradition of Christian philosophy as a 'critical precursor of a realistically oriented philosophy based on Aristotelian principles' – although Husserl did not think much of Geyser, having once described him to Finke as an 'insignificant compiler'. But then, the spectacle of senior professors who can brook no rivals is familiar enough in university circles; they guard their territory as jealously as any stag.

Certainly Heidegger must have been completely shattered by this latest development. After all, by placing his trust in Finke he had good reason to hope that he would become a full professor while still a young man, thereby escaping from the financial cares with which he had had to live for so many years – and which were now to be his lot still. But then he did not know Privy Councillor Finke or his tactical thinking when it came to recommending a Faculty appointment on the time-honoured principle of back-scratching . . .

It must have been a bitter letter indeed that Heidegger wrote to Finke when the board was nearing the end of its deliberations and the balance had tipped against him. The letter itself has not survived, but its contents may be inferred from Finke's reply of 23 June 1916, the day of the board's final meeting. In the 'frank exchange of views' which evidently took place Heidegger felt that Husserl was prejudiced against him and failed to appreciate his true merits, but Finke brushed these concerns aside. Husserl was well aware of Heidegger's merits. The Privy Councillor went on to console the frustrated junior lecturer by pointing out that he was still a young man, with his whole future before him. He must not lose heart just because things didn't work out immediately. Age had its privileges – and Geyser *was* twenty years his senior. But Heidegger's name had been discussed, so he was certainly not forgotten. Finke's final words of encouragement were, 'Work, work and more work!'

At a later date (8 April 1917) Finke assured his protégé that he had high hopes of him: 'one major theistic speculative philosopher' was more important than any number of Christian-Catholic philosophers working on historical lines, he averred, thereby directing Heidegger towards the philosophy of religion, where he would be able to do pioneering work for Catholicism. This was scant consolation for the offended young lecturer. From where he was standing the affair looked like a conspiracy instigated by Krebs, who had turned Finke against him. He sought consolation once more in Laslowski, who was aware of the situation in Freiburg and had the following words of comfort for his friend: 'Finke and his crowd feel under an obligation to Krebs', he wrote in a letter from Silesia of 17 September 1916, 'and they don't want to offend him. They're afraid of you. It's all based on purely personal motives. They're simply incapable of making an objective judgement. I see it all so clearly. The question is whether you are going to take the hint and give in to these people. I should think myself too proud to oblige them so readily.' At this time Heidegger was hoping for the offer of a chair at Tübingen. He should respond to his Freiburg opponents with 'a mixture of contempt and pity': 'A bunch of petty-minded schemers and intriguers, bourgeois family men, timid souls. They're professorial has-beens, for heaven's sake!' A rag-bag of mixed emotions comes into play here: Frau Finke, the Privy Councillor's wife, figures in it somewhere, along with all the animosities generated by disappointed hopes.

It was in this same summer of 1916, while Heidegger was still on 'military' service at the Freiburg Postal Control Office, that he met Elfride Petri, a student of economics at Freiburg University, whom he married in March 1917. As we know from the poem 'An Evening Walk on Reichenau Island' (published in 1916 and reprinted in Volume 13 of the *Complete Works*), Heidegger spent some days on the island in Lake Constance with Elfride Petri and her friend Gertrud Mondorf. Ernst Laslowski, who was in Freiburg during the summer semester of 1916, had also met the two girls. A new human dimension was opening up before Heidegger, invested with a special significance of its own: for his fiancée came from the family of a high-ranking Prussian officer and belonged to the Lutheran faith. This confessional mix undoubtedly had an important bearing on Heidegger's progressive alienation from Catholic circles. That this alienation, rooted in personal experiences of this kind and fed by disputes with academic representatives of the Catholic world, began in 1916 appears clear enough. At all events, later statements of an anticlerical nature made by Heidegger – and they are legion – can readily be traced back to these early experiences. One such statement comes in a letter that Heidegger wrote as rector in February 1934 to Dr Stäbel, the *Reichsführer* of the German Student Union, in which he deplored the lifting of a ban placed on a Catholic student fraternity by the local student leader:

> Such a public victory for Catholicism in this of all places [Freiburg] must on no account be allowed to stand. It represents a damaging blow to our whole enterprise, *the worst that could possibly be imagined* at the present time. I have an intimate knowledge of the circumstances and personalities here that goes back many years . . . People *still* haven't realized how Catholics operate – and one day that will cost us dear.[78]

Such uncontrolled invectives, veiled in the stilted language of philosophical discourse, can only be explained by reference to the events described above.

The summer of 1916 inflicted serious psychological damage on the young lecturer Heidegger, the traumatic effects of which were to last a lifetime. It was the decisive blow. We need only recall his rejection by the Jesuits on the grounds of poor physical health, and later by the archiepiscopal authorities in Freiburg on the same grounds: and now he had to suffer this kind of treatment at the hands of Catholic academics. These events presaged the first 'turn' or 'change of direction' [*Kehre*]: not a change of intellectual direction, but a turning-away from Catholicism, the Catholic system, or however else one chooses to describe it.

Heidegger found himself forced to look for a new point of

reference. Formally he remained the protégé of Finke, more especially since Geyser, who would technically have been his superior, did not move to Freiburg until the summer semester of 1917. Nor is there any evidence that Geyser and Heidegger had anything to do with each other – not during Geyser's time at Freiburg. For the present Heidegger deputized for the vacant chair during the winter semester of 1916/17; his friend Krebs had meanwhile obtained an associate professorship in the Faculty of Theology, and was no longer interested in an extension of the caretaker post he had held for the past three years. According to Krebs's diary, Heidegger lectured on 'Basic problems in logic'. He attracted a wide audience from the secular faculties, but met with little understanding among theologians, 'as he uses a difficult terminology and expresses himself in a way that is too complicated for beginners!'

Heidegger found his new point of reference in Husserl, who, as Finke had intimated to him in the summer semester of 1916, now had a full and just appreciation of his abilities. The relationship between the two philosophers was not particularly close to begin with, although they exchanged a fair number of letters in 1916; in 1917 the correspondence then dropped off, only to resume more vigorously in 1918, when Heidegger found himself turning more and more to Husserl.

A useful yardstick of their developing intimacy is furnished by the letters exchanged between Husserl and his Marburg colleague Paul Natorp in October 1917 against the background of efforts to fill the associate professorship in philosophy at Marburg, which had fallen vacant on the appointment of Georg Misch to a post at Göttingen. Heidegger was a possible candidate for this post, which in future was to be more strongly biased towards medieval philosophy. 'He was the favourite by a long way', and for this reason Natorp wanted to check with his friend Husserl, who after a year and a half at Freiburg had presumably had time to get the measure of Heidegger's philosophical talent.[79] Natorp recognized that Heidegger's writings contained 'notable things', promised 'greater things to come', and were distinguished by 'a remarkable breadth and freedom of intellectual approach'. But what about his teaching record? And could one really be quite certain that he was not blinkered by confessional allegiances? Naturally enough, Natorp also wanted to know how Husserl himself rated Heidegger as a person and as a scholar.

Husserl replied in rather vague terms but at length. For the present Heidegger's time was much taken up with his military duties (at the Postal Control Office); consequently Husserl had not yet had sufficient opportunity – after a year and a half in Freiburg! – to get to know him better, in order to form a 'reliable opinion' of his 'personality and character'. But neither did he have anything negative to report on him. Natorp had been particularly keen to know whether Heidegger's Catholic sympathies were a factor to be reckoned with. Indeed, the prospect of a Catholic professor of philosophy at Marburg, the setting for the 'Marburg Colloquy' of 1529 between Luther, Zwingli and Melanchthon, and the site of the first-ever Protestant university (founded in 1527) – it was a little hard to imagine.

Husserl's conclusions were as follows. It was clear that Heidegger had certain 'confessional allegiances', since he was 'under the protection, so to speak, of my colleague Finke, our "resident Catholic historian"'. Consequently Heidegger's name had been mentioned along with others the year before, when the appointments board was considering candidates for the chair of Catholic philosophy – another post where the emphasis was to be shifted towards medieval philosophy. Finke had treated him 'as an eminently suitable candidate on confessional grounds'. On the other hand Heidegger had married a Protestant a few months before, 'who to the best of my knowledge has not yet converted to Catholicism'. For the rest, Husserl felt that Finke's protégé was really too young 'for our position', and 'still lacking in maturity', even for the post of associate professor. His book on Duns Scotus was very much 'a first work', which testified to 'a lively mind and a considerable talent', and was 'undoubtedly a very promising beginning for a historian of medieval philosophy.' Of course, he had been unable to acquire sufficient teaching experience as yet, because he had been conscripted into the 'military postal service'. He had heard conflicting reports about his teaching, some 'very good', others 'unfavourable' – which had to do with the fact that Heidegger was trying 'to make his name in systematic philosophy' and had therefore given 'lectures that were systematic rather than historical': he was still 'struggling to find his feet on the basic problems and methodology'. He had worked hard on his own initiative to develop a closer understanding of phenomenology, having long since abandoned the line taken by Rickert. 'He seems to be going about it in a serious and thorough way.' More than that he could not say.

It is abundantly clear from Husserl's comments that the older phenomenologist had taken only fleeting note of the young Catholic philosopher, and relied more on rumour and second-hand reports than on his own reading of Heidegger's writings – including those early, forceful studies of Husserl's own *Logical Investigations*. Their personal relations also remained on a strictly formal footing. For Husserl, Heidegger fitted into the pigeonhole labelled 'Catholic philosophy', and into the compartment marked 'Under the protection of Heinrich Finke' (with all that that implied): and what *that* meant, first and foremost, was a faint odour of 'dubious scholarship'.

The observation that Finke had treated him 'as an eminently suitable candidate on confessional grounds' at the meetings of the appointments board was – put in those terms, in that context – thoroughly damning. The fact is that in the autumn of 1917 Husserl was not yet prepared to endorse his colleague Heidegger. Consequently Heidegger ranked only third on the Marburg list of candidates, as he did again in 1920 (it was not until 1922 – under an entirely different set of circumstances, of course – that he finally made the breakthrough). Before the end of October 1917 Natorp had written back to confirm that Heidegger had been placed only third in view of 'his youth and his limited field of study'. But he remained a hopeful prospect for the future.

Husserl's detailed characterization contains a number of clues that bear directly on our present inquiry. For the issue here is Heidegger's confessional allegiance: Heidegger the Catholic, the one-time postulant who sought entry to the Society of Jesus, the student of Catholic theology and lecturer in Christian (which is to say, Catholic) philosophy (even though his teaching qualifications only specified 'philosophy'), whose staunchly Catholic origins – his father having been sexton in Messkirch – were clear beyond a shadow of doubt. First of all it should be emphasized again that Heidegger was definitely moving towards Husserl and his phenomenology, even though the latter had put him down as Finke's man. Husserl for his part had at least registered the fact that Heidegger was married to a Protestant, who had not – as yet – embraced the Catholic faith.

At the beginning of 1917 Heidegger's fiancée had hinted to his Silesian friend Laslowski at the possibility of marriage, thereby plunging this hapless neurasthenic into a state of deep disquiet:

> My dear Martin: if only I could be with you at this time! I don't know what it is, but I cannot feel entirely happy about what Fräulein Petri told me in her letter. It would be wonderful if I were proved wrong. But I beg you to be careful! Wait until we are together again. I'm really very worried for you, particularly in a matter of such enormous importance as this. You understand my meaning when I ask you not to make a hasty decision (Letter of 28 January 1917).

These confused ramblings were the product of Laslowski's complicated mental state, penned by one who did not want to lose his friend Martin to a woman, especially when the situation was aggravated by confessional differences.

But the decision to enter into marriage had already been taken. What difficulties these confessional differences created for the members of Heidegger's family need hardly be emphasized. His humble parents, to whom the very idea of a mixed marriage must have seemed alien in the extreme, were hard put to comprehend the step their son was taking. But their prospective daughter-in-law, who did after all come from such a distinguished family, had intimated her intent to embrace Catholicism. On 21 March 1917, at ten o'clock in the morning, the bridal couple, 'Dr Martin Heidegger, university lecturer and a member of the territorial reserve', and 'Thea Elfride Petri, student of political science at the University of Freiburg', were married in the university chapel by 'Professor Dr Engelbert Krebs, deputizing for the military chaplain Monsignore Wächter'. Krebs had learned of the engagement from Heidegger only that same month, when he was told that Fräulein Petri intended to call on him. It was a plain wartime wedding, which took place very quietly, without pomp or display. Heidegger's bride came from a Lutheran family, and for a while she toyed with the idea of converting to Catholicism; but Krebs advised against it. He told her not to make any hasty decisions, but to wait until after the wedding. Such a serious step should be carefully weighed.

At that time Heidegger's father-in-law was a colonel on half-pay. One of the witnesses to the marriage was a student of philosophy named Heinrich Ochsner, who came from the small town of Kenzingen in Baden. A former theology student, two years Heidegger's junior and a friend of his – as we know from fragmentary memoirs, since published[80] – he was clearly a man well able to hold his own in scholarly debate; well-attested rumour has it that he acted as a kind of guiding spirit for Heidegger. The administration of the sacrament of matrimony on 21 March 1917 took place under the specific dispensation of a wartime wedding, since the bridegroom was classed first and foremost as a territorial reservist, and the Catholic cleric responsible for these matters, the divisional chaplain Monsignore Johannes Wächter, a priest of the Berlin provostry and military chaplain for the Freiburg garrison since 1907, had delegated his authority to Heidegger's friend in the priesthood. For this reason the customary public announcement was not necessary. The church wedding remained a *de facto* secret.[81]

The ceremony took place in the university chapel of the Cathedral of Our Lady in Freiburg, a setting rich with historical associations and once the resting place of university professors, though its former function was probably largely forgotten. The heyday of Freiburg University as an unequivocally and exclusively Catholic institution had ended around 1840, and in 1917 one would have been hard put to find any Catholic university professors outside the Faculty of Theology. Did Heidegger know, one wonders, that this late Gothic niche, tucked away among the string of chapels ranged around the magnificent high altar of Hans Baldung Grien, also housed the earthly remains of his countryman Christoph Eliner, one-time professor of theology and rector of Freiburg University, who died in 1575, having endowed the scholarship grant enjoyed by the Messkirch schoolboy and student Martin Heidegger for more than five years? Eliner's memory is perpetuated in a wooden tablet embellished with a painting of the Resurrection which hangs opposite the altar of the university chapel. The altarpiece before which the bridal couple knelt (in all probability the work of Hans Holbein the Younger) is a triptych whose centre panel depicts the Adoration of the Shepherds and Magi, while the side panels portray the four great occidental Church Fathers: Ambrose, Jerome (who is also the patron saint of Freiburg University), Augustine and Gregory I. The chapel is spanned by a finely wrought eight-webbed net vault, crowned by a polychrome keystone in the form of a medallion cradled by the coats of arms of Austria, the City of Freiburg and the University. History breathes within these walls, audible to those that know.

A wartime wedding differed from normal weddings in another respect: it did not have to be entered in the marriage register of the parish in question, which in this instance was the Cathedral parish of Freiburg. The military clergy had their own registration practices, which permitted a good deal of latitude in the way marriages were recorded. 'Wartime marriage service without organ, bridal dress, wreaths or veils, coaches and horses, wedding breakfast or guests; conducted with the blessing of both sets of parents (conveyed by letter), but in their absence': this was how Krebs recorded the event on a page in his diary, which was laid out on the same lines as a form in a marriage register. The text of Krebs's marriage address has unfortunately not survived. The entry closes with the words 'Quod Deus benedicat!'

Although the wedding took place under a military religious dispensation, it was of course subject to the normal requirement of Catholic canon law that the children of confessionally mixed marriages must be baptized and brought up as Catholics. Relations between Heidegger and Krebs became noticeably, not to say pointedly, less close hereafter, which is hardly surprising in the case of a young married couple who are busy making a home for themselves and building up their own circle of friends. On top of that, Heidegger had embarked on a new line of religious and theological thought, which soon began to take clearer shape. From the letters written by Heinrich Ochsner we know that Heidegger was studying Schleiermacher in the summer of 1917, particularly the second of his 'Addresses on Religion', and was thereby led to consider the problem of religion in Schleiermacher. 'To raise the tone and say a rather special thank-you for our friendship, Heidegger delivered some observations on the problem of religion in Schleiermacher' (2 August 1917). A few days later Ochsner noted that the impact of Heidegger's talk had stayed with him all week.

Still assigned to his duties at the Postal Control Office and obliged under the terms of his lectureship to give lectures without pay, Heidegger had to come to terms with the failure of his Marburg candidacy – a further disappointment to add to the bitter experiences of 1916. Josef Geyser had finally taken up the chair of Christian philosophy at the start of the summer semester in 1917, and Heidegger's paid teaching appointment ended as of that moment. He did not form any closer ties with Geyser, and was content instead to remain dependent on Finke. At the same time his principal concern from now on was to develop closer relations with Husserl. Since 1916 he had been working hard to court the coy phenomenologist. Shortly before Christmas 1916, for example, Heidegger resorted to the tactic of an unannounced visit, only to miss the maestro, who had gone out for his regular short walk on the nearby Lorettoberg. An idyllic scene indeed. The teacher was normally accompanied on these occasions by the newly qualified doctor of philosophy Edith Stein, who had been appointed his personal assistant in October 1916, and was living not far from Husserl's home in a small boarding-house. Heidegger was invited to renew the visit – though he could 'perhaps give me some prior notice' – if his exacting duties (referring to his work at the Postal Control Office) permitted. 'If I can be of any assistance to you in your studies, I will certainly do what I can, if you so desire' (Letter of 10 December 1916). The tone is studiedly polite and unforthcoming.

So it was not that easy to get close to Husserl. Although he had been familiar with Husserl's phenomenology for years, Heidegger found it difficult to establish a personal rapport with him. Even when the inquiry came from Marburg Husserl made no attempt to get to know Heidegger better, although he did agree, at Heidegger's request, to make an appointment in October 1917. The language remains formal as before: 'I will be glad to assist you in your studies in so far as I am able' (24 September 1917).

But then the ice must have been broken. During the winter of 1917/18 the two philosophers became sufficiently close – and not just professionally – to discuss philosophy on an informal, personal basis. Similarly drawn to Husserl was Heidegger's friend Heinrich Ochsner, who sat in on Husserl's seminar on logic during the winter semester of 1917/18, and felt 'the presence of the divine'. It is a wonderful thing, he writes, 'the fundamental experience of all philosophy, that we only understand the world and the self through the absolute spirit of God' (Letter of 20 October 1917)[82].

Husserl was unquestionably the dominant philosophical figure to whom both friends turned with conviction. The Christian philosopher Josef Geyser was clearly out of the picture and of no further account. On the subject of Geyser's inaugural address, as Ochsner writes in his letter of 6 December 1917, 'the less said the better'. From the very clear accounts that Ochsner later gave from memory to his friend Bernhard Welte[83] it becomes apparent to what extent Heidegger was able to develop Husserl's phenomenological approach through fruitful dialogue with the older philosopher. Heidegger had understood right from the beginning that 'Husserl's approach, for all its significance, could not be a *prima philosophia*, because Husserl's "object" was the abstract object of theoretical scientific knowledge, and as such was a very derivative concept', whereas the materialized object that is constituted in the concrete forms of existence 'is something far earlier and more primordial'. Husserl, we are told, had ignored the problem of access to his phenomena, since for him the scientific existence of objects had been an unquestioned fact taken wholly for granted. Consequently, Heidegger had gone further than Husserl from the very beginning, radicalizing the latter's approach: for the question of the historical certainty of the existence of objects, and their accessibility, had assumed central importance for him.

On 17 January 1918 Private Heidegger of the territorial reserve had to report to barracks, for a short time at least, because his health was once more showing signs of strain. As Ochsner put it on 24 January 1918, Heidegger's health 'appears to be suffering greatly under the stress of his new duties . . . I am quietly hoping that this transfer will be revoked out of consideration for his health. It would be an immeasurable loss for me if he were to be assigned to the front.' Things had not quite reached that point, even if the defiant German High Command was obliged to mobilize all its resources in order to mount the big offensives on the Western front. But in the event he was transferred to the Heuberg area, with its adjoining troop training ground, for a period of military training: back to the land of his birth, in other words.

At the end of 1917 Husserl was regretting that he could no longer see Heidegger and enjoy the pleasures of philosophical discussion. But his own departure for the Black Forest – 'I fervently hope for a period of quiet contemplation'– and Heidegger's induction into the military ruled out any chance of a meeting. The cheerful greeting despatched to Freiburg from the Heuberg training ground by Private Heidegger, serving with the fourth company of No.113 Reserve Battalion to which he had been assigned on 28 February 1918, received a benignly paternal response from the patriotic Husserl (28 March 1918): clearly there were no worries about Private Heidegger's constitution being equal to the strain. The fact that Heidegger was obliged to put philosophy aside altogether for the moment was no bad thing. At a later date – 'hopefully the war won't last too much longer now after the splendid victories in the West' – he would be able to 'return with renewed vigour to the difficult problems', and Husserl would do his best to help him 'pick up the threads' of the philosophical debate again. Repeatedly Husserl uses the distinctive term *symphilosophein*, meaning 'to philosophize together, with mutual pleasure.'

And so the correspondence between the two scholars continued throughout the spring, summer and autumn of 1918. By April Heidegger was back in Freiburg, until he received orders on 8 July to report to 'Frontwetterwarte 414', an army meteorological observer unit from Württemberg, then stationed in Berlin-Charlottenburg, for basic training in meteorology.

From Berlin Heidegger wrote to his patron Husserl and described his impressions of the university and the intellectual life of the city. These impressions are mirrored in Husserl's letter of 10 September 1918, an extended monologue celebrating in fulsome terms Heidegger's youth, vigour and innocence, and extolling his 'clarity of vision, clarity of heart and clear sense of purpose', together with his forthright language. He thanks him for the gift of his letters: 'To be young like you! What a joy and a real tonic it is to share in your youth through your letters.' The tone is one of fatherly concern and gratitude.

When this letter was written Heidegger was already serving with his meteorological observer unit on the Western front, in the sector assigned to the 1st Army. 'Frontwetterwarte 414' was under the operational command of the 3rd Army's meteorological observer corps, and was stationed in the Ardennes, not far from Sedan. Its main task in the second battle of the Marne (which began on 15 July 1918) was to cover the left flank of the 1st Army as it advanced towards Rheims. These meteorological units had been set up to provide advance weather information in support of poison gas attacks. Detailed guidelines for the operational deployment of these units were given in an order issued by the Ministry of War on 25 August 1918.

Heidegger remained in the war zone for two months, until the last days of October. On 5 November 1918, only a matter of days before the November revolution, he was promoted to the rank of lance-corporal, and on 16 November he was discharged by the 10th Air Reserve battalion and sent home to Freiburg. Here the November revolution turned out to be a rather muted affair, in which the workers' and soldiers' councils played no central role.

The break with 'the *system* of Catholicism'

In the period prior to these events Heidegger's thinking, under the influence of Husserl, had started to undergo a fundamental change – for whatever specific reasons. That change culminated in a decision of momentous significance for the rest of his life, a decision that perhaps he never quite got over: he abandoned the faith of his birth – sweeping as that assertion may initially appear. Quoted below is the text of the letter of farewell written by Heidegger to his clerical friend Krebs on 9 January 1919, which was published a few years ago in a little-known journal.[84] A fuller interpretation of this document will follow in due course.

> Dear Professor,
>
> The past two years, in which I have sought to clarify my basic philosophical position, putting aside every special academic assignment in order to do so, have led me to conclusions for which, had I been constrained by extra-philosophical allegiances, I could not have guaranteed the necessary independence of conviction and doctrine.
>
> Epistemological insights applied to the theory of historical knowledge have made the *system* of Catholicism problematic and unacceptable for me – but not Christianity *per se* or metaphysics, the latter albeit in a new sense.
>
> I believe I have felt too keenly – more so, perhaps, than its official historians – what values are enshrined in medieval Catholicism, and we are still a long way removed from any true assessment or interpretation. I think that

> my phenomenological studies in religion, which will draw heavily on the Middle Ages, will do more than any argument to demonstrate that in modifying my fundamental position I have not allowed myself to sacrifice objectivity of judgement, or the high regard in which I hold the Catholic tradition, to the peevish and intemperate diatribes of an apostate.
>
> That being so, I shall continue to seek out the company of Catholic scholars who are aware of problems and capable of empathizing with different points of view.
>
> It therefore means a very great deal to me – and I want to thank you most warmly for this – that I do not have to forsake the precious gift of your friendship. My wife (who has informed you correctly) and myself are anxious to maintain our very special relationship with you. It is hard to live the life of a philosopher; the inner truthfulness towards oneself and those for whom one is supposed to be a teacher demands sacrifices and struggles that the academic toiler can never know.
>
> I believe that I have an inner calling for philosophy, and that by answering the call through research and teaching I am doing everything in my power to further the spiritual life of man – that and *only* that – thereby justifying my life and work in the sight of God. Your deeply grateful friend, Martin Heidegger.
>
> My wife sends her warmest regards.

Many things here recall earlier statements: the exploration of medieval philosophy, for example (mentioned in the letters to the Freiburg cathedral chapter written between 1913 and 1915), and the distinction he draws between the true philosopher and the 'academic toiler' (by which he means the academic historian of philosophy who typically spends his time editing scholastic texts – an activity that had been urged upon Heidegger himself a few years earlier). What is crucial, however, is the fact that Heidegger was no longer able to remain within the Catholic Church, within the *system* of Catholicism, which he terms 'an extra-philosophical allegiance'. But he did remain within the fold of Christianity, i.e., within the tradition of the New Testament and perhaps also of the patristic literature, though he does not elaborate further on this. Also significant is the reference to his future work on the phenomenology of religion, which clearly bears the stamp of Husserl. In sum – and let it be said without equivocation – this was an informal declaration to a friend of his intention to quit the family of the Catholic Church, the church visible of Christ Jesus, as it is defined in Roman canon law. With his theological training Heidegger knew only too well that Catholic ecclesiology had evolved a visible church as an institution with a hierarchy and a ministry, founded on the authority of dogma and papal infallibility, and that consequently, an individual Christian existence (with theistic leanings, perhaps) that placed itself outside this fellowship broke *ipso facto* with the society of the Church. It is no coincidence that Heidegger uses the phrase 'the peevish and intemperate diatribes of an apostate' in order to distance himself quite explicitly from behaviour of *that* sort: but he is well aware that he is an apostate nonetheless, albeit not the kind who sets out to foul the nest, but one who is determined to maintain his 'objectivity of judgement' and the 'high regard' in which he holds 'the Catholic tradition'.

The letter is written by one who already stands on the other side, outside the Catholic camp, and who will seek dialogue only with the understanding few within the camp who are willing to compromise. He ends with a theatrical flourish, justifying himself before God like a latter-day Martin Luther: the gesture is a trifle overplayed.

When Krebs received this letter at the beginning of 1919 it did not exactly come as a bolt from the blue. On the contrary: he was well prepared, since (as Heidegger mentions in the letter) his wife had 'already informed' him. Heidegger's official letter of farewell is a significant biographical document that demands to be read and understood in a wider context. The reference to his wife having 'informed' Krebs alludes to the following sequence of events: on 23 December 1918, the day before Christmas Eve – the weather was appalling – Elfride Heidegger called on the professor of dogmatics to inform him that she and her husband, who were expecting their first child, were no longer able to fulfil the undertaking they had given at their Catholic wedding and baptism to have their own children baptized and brought up in the Catholic faith. Here we might recall the Catholicizing inclinations of Elfride Heidegger, which Krebs had viewed with cautious scepticism in the spring of 1917. Such a step, he told her, should be carefully weighed, and was not to be rushed into. Engelbert Krebs was in the habit of couching important diary entries in the form of a dialogue or in direct speech, as we have already seen; so his accounts were clearly written while the conversations were fresh in his mind:

> My husband has lost his religious faith, and I have failed to find mine. His faith was undermined by doubts even when we got married. But I myself insisted on a Catholic wedding, and hoped with his help to find faith. We have spent a lot of time reading, talking, thinking and praying together, and the result is that we have both ended up thinking along Protestant lines, i.e. with no fixed dogmatic ties, believing in a personal God, praying to Him in the spirit of Christ, but outside any Protestant or Catholic orthodoxy. Under the circumstances we would consider it dishonest to have our child baptized in the Catholic faith. But I thought it my duty to tell you this first.

For Krebs it was a bitter disappointment to learn that 'my friend and his young wife' were intent on leaving the Church, and the diary entry continues: 'Having qualified to lecture on Catholic philosophy, Heidegger will get himself into a lot of trouble now for changing sides. He is growing away from Catholic thinking, going the same way that I saw Bühler going. How much of the responsibility for that rests with me, God alone knows. I was too naïvely trusting when I offered to help with the wedding.' It later appears that the form of church wedding conducted by Krebs when he deputized for the official military chaplain was not without its problems; at the very least, it seems, there were procedural irregularities. To the apostate Karl Bühler Krebs added the names K. Marbe, A. Messer, Horten and Verweyen – all of them one-time Catholic philosophers.

It was not until the autumn of 1919 that Heidegger and Krebs had a proper talk. They met in Messkirch, the home of the philosopher, on 15 September. On his way to Beuron Professor Krebs had stopped off in Messkirch to visit an old friend of his who was a local curate. Heidegger happened to be vacationing with his parents. 'After lunch, a little before 3 o'clock, we set off on a walk with our colleague

Heidegger . . . in the direction of Heudorf.' The curate soon took his leave of them, and 'now I had a chance to spend several hours in theological discussion with my friend and junior who had abandoned the faith – a discussion that he himself initiated.' The brisk walk up on the Heuberg plateau, through a landscape bathed in the colours of autumn, brought them to the ridge above Leibertingen, where their ways parted. 'I set off at a trot across country, down across fields and scree-covered slopes, and was in Beuron by six.'

A little later Heidegger played a distinctly passive role when he and Husserl were invited to spend the evening with Krebs on 16 January 1920. They talked at great length 'of philosophy and theology', stimulated no doubt by the contents of Krebs's well-stocked wine cellar. 'Heidegger hardly said a word, but Husserl more than made up for his silence.' Husserl, who had admitted to being 'pretty much irreligious himself, because totally committed to the scientific method', had listened reverently 'as I spoke to him of the lofty spiritual values, the joys and the profound emotional enrichment that Catholics derive from their faith'. They also discussed the problem of theology as the handmaiden of the ministry, which provided Krebs with a welcome opportunity to explain the Vatican doctrine of faith to the distinguished phenomenologist. Husserl had apparently listened intently, and then remarked: 'It's neat, very neat – and logical, too!' At that point, though, the old objection surfaced once again: 'Scientific study is bound to lose its freedom if people are in constant fear of being censured by a learned committee.' This was ground that Heidegger had covered long ago – hence the fact that he remained silent and withdrawn throughout the evening's little gathering. Did he think that Husserl was also talking about him when he remarked ironically to Krebs: 'There you are, dining on the fat of the land in your palace, while we poor beggars stand hungry at the door!'? At all events, Krebs accepted the parallel 'wholeheartedly'.

One more occasion that brought old familiar faces together again occurred in the summer of 1920, when Ernst Laslowski, a friend of both Krebs and Heidegger, took his doctoral examination under Finke. The reunion was again hosted by Krebs.

But thereafter their friendship soon faded into formal observance. It was as much as they could do to get together for a drink on the eve of Heidegger's departure for Marburg in 1923, where he had at last succeeded in obtaining a professorship. Their relationship was running out, like the sand in an hourglass. Krebs followed Heidegger's subsequent meteoric rise with a keen eye and a sorrowful heart, but the only further contact he had with him was in a purely official capacity (as when they met in the summer semester of 1933, while Krebs was serving as dean of the Faculty of Theology under Heidegger's rectorship). Unfortunately Krebs's diaries have not survived beyond 31 December 1932, so that we have no record of Heidegger's political involvement from the pen of this meticulous and scrupulous observer, of whose friendship Heidegger was careful to expunge all traces. The theologian Krebs, a man of upright character and unswerving convictions, soon got into trouble with the Nazi regime after 1933. In 1936 he was dismissed from his chair and forced into early retirement. He retreated into himself, living in lonely isolation until his death in 1950. He was rehabilitated after the Second World War, but he never recovered from his depression. A few years after the lapse of their friendly relations, once so close, Engelbert Krebs returned to the subject of Heidegger following an encounter with Edith Stein, who had come to see the theologian on a matter of special concern. The long diary entry includes an interesting, searching, almost prophetic comparison between Martin Heidegger and Edith Stein:

> On Friday, 11 April 1930 I received a visit from Dr Edith Stein of Speyer, Husserl's most outstanding pupil and a collaborator on the *Phenomenological Yearbook*. A Jew of Silesian extraction, she had been a student of Husserl's at Göttingen and from there had accompanied him to Freiburg. A friend of Frau Conrad-Martius, the other prominent woman on Husserl's team of collaborators, she soon shared the latter's leanings towards Catholicism. As a result of more profound study and prayer, she was converted in the early 1920s. Her conversion took place in the home, or at least the parish, of Frau Conrad-Martius, who stood godmother at her baptism, though not yet a Catholic herself. Today Frau Conrad-Martius is still not a Catholic, while Edith Stein continues to penetrate ever more deeply into the rich stores of our faith, and is

currently working on a German edition of the *Quaestiones de veritate* of St Thomas Aquinas. Last year she visited Husserl on his seventieth birthday, and went on afterwards to Heidegger's house in the company of Heidegger and a small party of Husserl's former pupils. She found him much changed from before, full of plans and work projects.

How very differently their destinies have turned out! Edith Stein won early recognition in the philosophical world. But she became small and humble, and a Catholic, and immersed herself in quiet study at the Dominican girls' school in Speyer. Heidegger started out as a Catholic philosopher, but then lost his faith and left the Church, and became a famous celebrity, the centre of attention among professional philosophers today.

As with Edith Stein, so it was also with Dietrich von Hildenbrand in Munich. Benedico te Pater, quia haec magnis et potentibus abscondisti, parvulis autem manifestasti. Sic Pater placuit tibi![85]

In seeking to illuminate this central episode in Heidegger's life, the great philosophical 'turn' or change of direction, it may be that we can never really get to the bottom of it. But before moving on, let us pause to formulate one or two further thoughts on an ideological change that went far beyond confessional allegiances.

In his doctoral dissertation and habilitation thesis Heidegger had still been concerned with the renewal of the epistemological foundations of theology. In his habilitation thesis on Duns Scotus he compared the 'two-dimensional existence' of modern man, whom he saw exposed to the danger of 'growing uncertainty and total disorientation', with the attitude of medieval man, conscious of his transcendental ties. A philosophy that eschewed all metaphysical or theological direction was simply not a viable possibility in his view. The acknowledgement of metaphysics and theology self-evidently implied the retention of ethical standards.

Philosophy cannot dispense permanently with its true perspective on the world, namely metaphysics. What this means

> for the theory of truth is the challenge of an ultimate metaphysical-theological interpretation of consciousness. It is here that value most specifically resides, in so far as it is a meaning-full and meaning-substantiating living act, which we have not remotely understood when it is neutralized in the concept of a blind biological factuality.

These statements sit perfectly well with the letters Heidegger wrote to the Freiburg cathedral chapter between 1913 and 1915 in connection with the Schaezler Foundation grant in memory of Thomas Aquinas. It was not a matter of making the right tactical moves: Heidegger was wholly and sincerely committed to the categories of thought of traditional metaphysics.

Within a few years Heidegger's attitude to the Catholic Christian tradition had undergone a fundamental and radical change, as can be seen from his letter of farewell to Krebs of 9 January 1919. 'Epistemological insights applied to the theory of historical knowledge': Heidegger's rejection of Christian philosophy began as a rejection of Catholicism; it was the great confessional turning point on an intellectual road at the end of which ethical and theological questions were deliberately not asked. This time of political upheaval and revolution in the aftermath of the First World War marked the first stage of the journey after the turning point. But what explanation can we find for the fact that he now began to distance himself from the philosophical and theological tradition to which he had pledged his life-long allegiance as late as 1915, promising to harness its intellectual and spiritual potential to 'the future struggle for the Christian-Catholic ideal'?

At this point we must consider, however briefly, the role of Protestantism in Heidegger's early life and career. Protestant thinkers, principal among them Friedrich Schleiermacher, opened up new perspectives for him, most notably the theology of Martin Luther, to which he was increasingly drawn. His response to Schleiermacher, to which we alluded briefly above, is a particularly characteristic example. The second of Schleiermacher's *Addresses on Religion*, on which Heidegger meditated as a junior lecturer in the summer of 1917, rejects the systematic approach of theological and philosophical thought. According to Schleiermacher, the essence of religion is neither thought nor action, but the contemplation of the universe and the communication of the sense of the infinite. He draws a rigid distinction between religion and philosophy, since metaphysics and philosophy are the corrupters of religion. Man should not use religion as if it were a mere instrument, for religion is given to him as a means or possibility of contemplation that is not some kind of accidental adjunct, but an essential part of his being:

> Morality starts from the consciousness of freedom, whose realm she seeks to extend throughout the universe, subjugating all to her alone. Religion draws breath wherever freedom has become the natural order once again; it comprehends man beyond the play of his peculiar powers and personality, and views him from the perspective where he must be what he is, whether he desires it or not (German text in *Schleiermachers Werke*, Vol. 4, 1968 reprint, p.241).

The move towards an existential view of religion that is already apparent here shaped Heidegger's study of Luther for some time to come, well into the early Marburg years. But interpretation of these matters is best left to those who are more competent to judge. This is not the business of the historian, whose real concerns lie elsewhere: typically with primary sources, contemporary documents that need to be studied in parallel.

In all this we must not overlook the active role played by Husserl, who had a considerable influence in matters of academic policy. For him it was self-evident that true philosophy had to be practised without preconditions, unfettered by any kind of confessional ties. As for the notion of a 'Catholic' science, he thought it a nonsense. In his work Husserl preserved a neutrality towards religion, in much the same way as a mathematician. On a personal level he did believe in God, but he never allowed this religious element to influence his thought (this at least is how Heinrich Ochsner summed it up in conversation with his friend Bernhard Welte).

Husserl's thinking on this point is sharply illuminated by the violent disputes that broke out in the Faculty of Philosophy at Freiburg University in 1924, when it came to appointing a successor to the vacant chair of Christian philosophy, following Josef Geyser's move to Munich. Husserl wanted to remove the confessional restrictions on the chair, which of course immediately brought the historian Heinrich Finke into the fray – Heidegger's patron from earlier years. Husserl's comment that 'the Catholic internationale had been accommodated to a very large extent during the war', but now the time had come to dismantle these structures, stung Finke into the following outburst: 'This is the kind of thing we have to listen to from an Austrian Jew. I've never in my life been an anti-Semite; but today I find it hard not to think along anti-Semitic lines.' But Finke found little support for his position. Only one member of the Faculty lamented the fact that Husserl had sunk so low, falling back on the old battle-cries of the *Kulturkampf* (see Josef Sauer's diary for 24 January 1924). In 1924 Husserl would have been only too glad to recall his protégé Heidegger to occupy the vacant chair. But the former Catholic aspirant, now the fully-emancipated philosopher, had just recently been appointed to a post in Marburg. From Malvine Husserl (the two philosophers' wives were very close at this time) Elfride Heidegger learned that there had been indescribable scenes of 'quarrelling and bitching' among Faculty members. The details would have to wait until they could meet and talk. 'It would have been good if we could have replaced G[eyser] with your husband' (19 February 1924).

But we are anticipating events that occurred much later. On 7 January 1919, two days before Heidegger's letter of farewell to Krebs, Husserl submitted a request to the ministry in Karlsruhe that Heidegger be appointed to the staff of the Philosophy department (I) on a fixed annual salary. Husserl explained that he relied on Heidegger to introduce beginners to philosophical phenomenology, while in return the latter needed some form of financial support. There was also a danger that Heidegger might be driven by the deterioriation in his financial circumstances to take a better-paid job outside academia, causing the university to lose one of its most promising young scholars. To lend added weight to his request, Husserl pointed out that Heidegger had been shortlisted for the Marburg post in 1917.

In order fully to understand what was at stake, one needs to know that in those days permanent posts for university assistants and lecturers were as yet unheard of. The only exceptions were the faculties of medicine, which established a handful of permanent positions in the interests of patient care and training for specialists. So Husserl was breaking new ground with this application. His request was granted in so far as Heidegger was given a paid teaching post. But Husserl refused to let the matter rest and kept in touch with the secretary for higher education at the ministry in Karlsruhe. In March 1919 he supplied him with details of Heidegger's straitened circumstances – straitened because the financial support from his wife's parents that Heidegger had enjoyed up until the end of the war had dried up when the family lost their fortune. 'I do not need to remind you that we are talking here about support for a valuable member of staff, who shows every sign of extraordinary promise' (letter of 22 April 1919). In the autumn of that year Husserl took up the cudgels again, pointing out that here was someone 'from the humblest family background' striving to make his way in the world, and that this was a perfect opportunity 'to practise the fine principle upheld by the democrats: "Give merit a chance!"' (13 September 1919). A year later Husserl finally succeeded in obtaining a regular teaching post for Heidegger as an *Assistent*. It was a post specially created for Heidegger, and therefore not a permanent addition to the strength of Husserl's department. Nevertheless, Heidegger's livelihood was now assured until he received the invitation from Marburg in 1922 – and accepted it in 1923.

So the years 1918/19 also saw the institutional 'turn' or change of direction, whereby Heidegger, who was still formally attached to the department of Christian philosophy (Philosophy II), now openly switched, in effect, to a different philosophical discipline. He became Husserl's collaborator, and perhaps his disciple – though the question of Heidegger's discipleship is still open to debate.

So is there a connection between Heidegger's statement of his position in the letter to Krebs, and the financial security provided for him as a result of Husserl's efforts? It is a question that cannot be answered. But one thing is certain. Husserl, a Lutheran by profession, was not unhappy to see his pupil convert from the Catholic faith to Protestantism, even if he was not minded to assist the process directly. Clear proof of this can be found in the long letter (already published in other contexts) written by Husserl to the Marburg historian of religion, Rudolf Otto on 5 March 1919, during the very period that interests us here.[86]

The ostensible purpose of this letter was to furnish a reference and recommendation for Heinrich Ochsner, whom we have already encountered as a figure of some moment in Heidegger's life. A deeply religious man, still struggling to complete his studies, Ochsner had hopes of obtaining a position on the staff of the theologian Rudolf Otto at Marburg.[87] In the course of his letter Husserl broadens his focus to take a more general view of the confessional issue, which again was a critical factor in the case (the Marburg syndrome once more).

The go-between who provided the link with Otto in Marburg was the Protestant curate Wilhelm Peter Max Katz (1886–1962). He had been given special charge of the diaspora in Riegel, just north of Freiburg, and was thus responsible for the Protestant faithful living in that predominantly Catholic region. Included within his remit was the birthplace of Heinrich Ochsner, Kenzingen im Breisgau. It is clear that Peter Katz was involved in Ochsner's conversion – or planned conversion. In the event Katz, a non-Aryan, emigrated to England after 1939, where he obtained his doctorate and became one of the foremost living specialists on the Septuagint.[88]

We need to know this in order to place the following passage from Husserl's letter to Rudolf Otto in its proper context:

> Herr Oxner [sic], like his older friend Dr Heidegger, originally studied philosophy under Rickert. Gradually, and not without much inward resistance, they both grew more receptive to my teaching; we also became closer in personal terms. During this same period both men were undergoing radical changes in their basic religious beliefs. Both are men of deep religious awareness: H's interest is predominantly theoretical-philosophical, while O's is more conventionally religious.

Later Husserl goes on:

> The Reverend Katz has undoubtedly told you that I will most gladly assist any efforts aimed at furthering O's career. But I must insist that my name is kept out of it. I must do nothing to jeopardize the tranquil pursuit of my labours in Freiburg.

> My philosophical labours do seem to have a remarkably revolutionary effect: Protestants become Catholics, and Catholics turn Protestant! But my intention is neither to catholicize nor to protestantize; all I want to do is to teach young people to think with radical honesty, to think in a way that does not lose sight of those primary concepts that determine meaning and are the essential basis of all rational thought, by obscuring and doing violence to them through verbal constructs and intellectual sleight of hand. In an ultra-Catholic city like Freiburg I do not want to be portrayed as a corrupter of youth, a proselytizer, an enemy of the Catholic Church. I am none of these things. I had no influence whatsoever on the decision of Heidegger and Oxner to convert to Protestantism, even though I am bound to welcome it as a free Christian (if I may call myself thus, as one whose eyes are fixed on an ideal goal of religious longing, which he sees as a never-ending personal challenge) and as an 'undogmatic Protestant'. For the rest I delight in influencing all truth-loving persons, be they Catholics, Protestants or Jews.

It is clear from this that Husserl viewed the conversion of Heidegger and Ochsner as a *fait accompli*, and that consequently Heidegger was seen in Husserl's circle as a Protestant Christian. Yet Heidegger regarded himself as a member of the Catholic church all his life, regardless of the strictures of canon law. The general tenor of this letter to Otto speaks for itself. We must strive to attain the freedom of a true Christian believer, rather than submitting to the extra-philosophical allegiance demanded by the system of dogmatic Catholicism.

But Husserl was not quite as objective as all this suggests. When he was informed in 1921 of the imminent conversion of his long-standing pupil and selfless collaborator Edith Stein, he reacted a good deal more sharply: 'What you tell me about Fräulein Stein has saddened me; she has not written to me directly. Conversions are all the rage at present, I'm afraid – a sure sign of people's spiritual poverty' (Letter to the Polish philosopher Roman Ingarden, dated 25 November 1921).

Be that as it may, it certainly did Heidegger's academic career no harm to have cast off his chains and cauterized the stigma of being a Catholic philosopher. It was superfluous ballast, the last thing he needed on his impending rise to fame: it would only drag him down into the stagnant lowlands of Catholic, or even scholastic, thought categories. A long and involved process of development appeared to have reached its successful conclusion. For the rest, the 'liberated' Heidegger could now enjoy a new circle of friends. Of central – one might even say fateful – importance were the friendly relations that now evolved between him and Wilhelm Szilasi. It is worth pausing here to sketch in a brief portrait of this man.[89]

A close contemporary of Heidegger (he was born on 19 January 1889 in Budapest), Szilasi was a political émigré with a colourful but rather mysterious past; many questions still remain unanswered. The son of the Jewish philologist Moritz Szilasi (1845–1905), he studied classical philology and philosophy at the University of Budapest from 1906 to 1910, concluding his studies with a dissertation on Plato's dialogues under the direction of Alexander Bernat (published in a revised version in 1910). At this time Szilasi had already come under the influence of Georg von Lukács, four years his senior. He likewise formed a close association early on with the poet and literary critic Mihály Babits. This was followed, in 1910/11, by a period of more intensive philosophical study in Paris and Berlin, where he was exposed to the influence of Emil Lask. In fact it was Lask who subsequently pointed the young Hungarian in the direction of Martin Heidegger. In 1911 Szilasi obtained a deputy teaching post at a Budapest grammar school, but the crucial event in his life at this time was his marriage to the daughter of a wealthy family (his father-in-law was the industrialist Hermann Rosenberg), which laid the foundation for a life of independence in the face of changing political fortunes.

Szilasi was a member of the legendary 'Sunday circle' established by Lukács in the summer of 1915 in Budapest, modelled on the group which gathered around Max Weber in Heidelberg, where of course Lukács himself had studied. Leading members included Karl Mannheim, Charles de Tolnay and Arnold Hauser. Szilasi played a marginal role, though significant enough to earn him a secure position in the educational system in the spring of 1919, when Lukács was appointed people's commissar for education under Béla Kun's Soviet Republic. By his own account Szilasi was given a full professorship at Budapest University, but more probably it was a professorial post within the restructured course system designed to integrate the functions of grammar schools and universities. In any event, Szilasi could not remain in Hungary after the fall of Kun's Republic, and in the autumn of 1919 he transferred his domicile to Freiburg, where he cultivated his relationship with Husserl, pursued his philosophical studies, and also dabbled in the natural sciences (principally chemistry), though it seems he never formally completed his studies.

So Szilasi's friendship with Heidegger came about via the circle around Husserl. When Szilasi settled in the summer of 1922 in Feldafing on the Starnbergersee, his house became a convivial meeting place for his philosopher friends: Martin Heidegger, Karl Löwith, Edmund Husserl and many others enjoyed his hospitality in a delightful rural setting not far from Munich, which was also a convenient stopping-off place en route to Austria. The somewhat fraught relationship between Heidegger and Szilasi is one that we shall encounter frequently hereafter.

But to revert to the great 'turn' in Heidegger's confessional and philosophical development: a stigma is not something that can simply be effaced. The mark remains, compelling its bearer to relive the conflict, even when it appears to have been successfully resolved. Karl Löwith has given telling expression to this underlying existential trait in Heidegger's personality: 'A Jesuit by education, he became a Protestant through indignation; a scholastic dogmatician by training, he became an existential pragmatist through experience; a theologian by tradition, he became an atheist in his research, a renegade to his tradition cloaked in the mantle of its historian.'[90]

The wound refused to heal; the thorn in his flesh remained virulent. How else are we to understand the passage in that letter of 1935, where a disenchanted Heidegger, gradually coming down to earth again after his soaring flights of politico-philosophical fantasy, takes up the old struggle with the faith of his birth? Seen from this perspective, many a reported remark of Heidegger's on the subject of the Catholic Church, on the purpose (if any) and justification of university chairs in Christian philosophy, delivered in a tone of ill-tempered polemic, not to say malice, begins to look more like an attempt to cover up or paint over an area of diseased tissue. Underneath, the sores and wounds continued to burn and smart, little soothed, one suspects, by the sticking-plaster of existential thought. The issue of his Catholic origins, of the faith of his birth, remained open and entirely unresolved. When the occasion arose the issue would always flare up again. And they still stand, the sombre words that Heidegger wrote in 1935, when he described 'the faith of my birth' as 'a thorn in the flesh'.

The second thorn in Heidegger's flesh in 1935 was 'the failure of the rectorship'. With this he sounded a theme that became a kind of *leitmotiv*, never to be silenced; it would be taken up again after 1945 with dark, menacing chords in which dissonant notes could soon be heard. The theme has been sustained until the present day, set down in a score that is not easy to decipher. But before we embark on that story, let us first take a brief look at Heidegger's years in Marburg.

The Marburg years
1923–1928

At the beginning of 1922 a further opportunity arose for Heidegger to obtain a professorial post in philosophy at Marburg, after his earlier prospects, in 1917 and 1920, had come to nought – as we have already seen.[91] Paul Natorp, the head of the Marburg School of Philosophy, was nearing retirement and was putting out feelers in various places, including Freiburg, for possible successors. He was 'inclined in principle' to give serious consideration to Heidegger, but to date, as Natorp noted, he had produced nothing to establish his credentials in phenomenology: 'I also worry a little that he may be better at absorbing and passing on the ideas and inspirations of others – which is extremely useful, of course – than at producing original creative work of his own.' This acute and telling assessment occurs in a letter to his colleague Husserl of 29 January 1922.[92] But Heidegger's fatherly friend leapt once again to his pupil's defence, emphasizing the originality of his thought, his teaching ability with beginners and advanced pupils alike, and above all the serious nature of Heidegger's commitment to the phenomenology of religion – though as long as he was in Freiburg, of course, Heidegger the one-time 'Catholic' could not devote himself to the really central issue, which was '*Luther*' (Husserl's italics). For these reasons a move to Marburg would probably be enormously important for Heidegger's future development. There he could act as an important link between philosophy and Protestant theology. He had a thorough knowledge of Protestant theology in all its forms, and a just appreciation of its unique value. What is more, Heidegger's appointment could well turn out to be a great asset to the University of Marburg.[93]

Husserl's line of argument is familiar enough by now. Heidegger is presented as the new-born Protestant emerging from the tutelage of Catholicism. The received image was already well established. But Natorp's initial response came as a disappointment: his successor had already been designated in the person of Nicolai Hartmann, associate professor at Marburg. That meant, of course, that they needed a replacement for Hartmann . . . It was a small consolation. But nothing had been decided as yet, not least because Heidegger had published so little. In the summer of 1922 Göttingen also indicated an interest in the young Freiburg lecturer: Husserl's old chair was now vacant again. Once more Husserl sang the praises of his pupil, who was just then completing a lengthy study on Aristotle that was due to appear the following year in Husserl's yearbook. This Aristotle manuscript – which has not yet come to light – has loomed large in the writings of historians of philosophy, though its relevance to our present inquiry is somewhat less. At all events, Heidegger was working feverishly on Aristotle at this time, particularly since the Marburg faculty was considering his name for the associate professorship in September 1922. They had also sent their scouts down south to Freiburg, the most celebrated of whom (in later years) was the student Hans-Georg Gadamer.

On 22 September 1922 Paul Natorp informed Husserl that they were 'going to take a fresh look at Heidegger' in Marburg, not only in the light of Husserl's glowing testimonial, 'but also in the light of what I have been told about his latest work, including reports from former Marburg men who have heard him lecture in Freiburg recently.' But here again he commented on how little 'he has published to date'.[94]

Heidegger, who in the meantime appeared to be developing the beginnings of a friendship with Karl Jaspers, wrote to Jaspers on 22 November 1922: having heard via Husserl that Natorp wanted some concrete indications about his future work plans, he had sat down for three weeks, summarized his own work, added a brief introduction, dictated the whole lot ('60 pages') and got Husserl to send a copy to Göttingen and a copy to Marburg. Reading on, we learn that while the department at Göttingen thought well of him, he had no reason to hope for anything. The reaction from Marburg had been more positive, and Natorp had written to tell him that he was 'very highly placed' on the list of candidates. Heidegger observed pessimistically that this probably meant 'the famous second place', and surmised that Richard Kroner would lead the field because he 'is my senior' and because he 'has all that paper to his name'. For his part, he could only look upon second place as a humiliation. All he wanted was an end to it all, 'one way or the other'. 'The endless dance they lead you, the half-prospects, the praise and flattery, etc. – you end up in a terrible state.'[95]

Heidegger's negative expectations were to be disappointed. By the end of October 1922 Natorp had written to Husserl to say that he and Nicolai Hartmann had 'read Heidegger's summary with the greatest interest, and found in it everything that was to be expected on the strength of your earlier communications . . . Above all it displays a remarkable originality, depth and intellectual rigour . . . I hope we shall at least be able to put him on the list of candidates, and high enough for him to receive the most serious consideration.'[96] The final list was drawn up by the Faculty of Philosophy at Marburg during December 1922, with Heidegger's name at the top. He was offered the position on 18 June 1923: 'the associate professorship with the status and rights of a full professor', as he wrote to Jaspers on 19 June. A few weeks later he filled in some of the background for Jaspers: Richard Kroner, his chief competitor, had been placed no higher than third, and had gone around complaining to all and sundry; he had even gone to Berlin (to the ministry), and turned up in person in Marburg to offer himself for the job, saying that if he were appointed he would even sit in on Hartmann's lectures. 'I certainly shan't be doing that', Heidegger informed Jaspers, 'but I shall give him hell by the *manner* of my presence; a whole combat patrol of sixteen is coming along with me, some of them the usual fellow travellers, but there are a few good, competent men among them.' The tone is military, and suggests a kind of logistical thinking. And indeed, the same letter contains some fundamental observations from Heidegger on the philosophy of the future, which he plans to construct (he says) in a fighting alliance with Jaspers. Heidelberg, of course, was the place they could *really* get things done. As it was, he would have to rely on an unseen meeting of minds in order to do anything in Marburg, which did not even have a decent library.

Heidegger never took to Marburg. He never felt at ease 'in that foggy hole', and was constantly complaining about the 'stuffy, stifling atmosphere that envelops one again'; apart from his term-time teaching duties, there was nothing to keep him in Marburg. The message of the letters to Jaspers during the Marburg years is absolutely clear. His real abode, his home in fact, was the mountain hut in Todtnauberg, to which he was invariably drawn as soon as the term ended in Marburg. He always looked forward 'to the bracing air of the mountains – this soft, flabby stuff down here ruins your health in the long run.' Chopping wood, 'followed by more writing'. He would have spent the whole winter up at the hut if he could, staying on to write. 'I have no desire to spend my time with university professors. The local country folk are far more agreeable – and indeed more interesting.' *Being and Time* was written up at the hut, where life 'appears to the mind as something pure, simple and immense' (letters of 23 September 1925, 24 April and 4 October 1926).

Nonetheless Heidegger did find an outstanding discussion partner at Marburg in the person of Rudolf Bultmann, the distinguished exegete and systematic theologian, whose seminar on St Paul he attended in that first winter semester of 1923/24. Husserl's forecast – that Heidegger would act as a link between philosophy and Protestant theology – was already coming true. Heidegger also lived up to the reputation that had preceded him, namely that he was a Protestant who had converted from Catholicism. Bultmann regarded him as *the* leading expert on Luther, as he wrote in a letter to his friend Hans von Soden on 23 December 1923, describing his first, very impressive experiences of the newly appointed Heidegger. The latter (he writes) is not only thoroughly versed in scholasticism and Luther, but is also well acquainted with the modern theology of such practitioners as Friedrich Gogarten and Karl Barth.[97] Heidegger gives an account of the ambience at Marburg in a letter to Jaspers of 13 June 1924, when he was in his second semester at the university: 'It's a beautiful day outside; at the university there is nothing happening, no stimulus at all. The only real human being here is the theologian Bultmann, whom I see every week.' According to Hermann Mörchen, Heidegger sometimes remarked that his years at Marburg had been the happiest of his life.[98] Doubtless he idealized the past in his memory, speaking from the bitter experience of the harder times that followed. But the fact is that Heidegger felt isolated within his specialist field, despite the success he enjoyed – even among the students. He had settled in and found his feet, he wrote to Jaspers in May 1925, but 'a unified teaching approach based on a consistent level of achievement is simply not possible': the kind of philosophy being taught by his faculty colleague Jaensch was too primitive even for primary school teachers. And now Nicolai Hartmann was on the point of leaving them. He would have done better to accept the offer from Japan that was officially made to him in the spring of 1924: three years' paid research at an institute founded by the Japanese nobility and Japanese high finance 'for the study of European culture', with particular reference to the humanities. The teaching load would have been very light, and he would have been helping to edit a quarterly journal, all in return for a fabulous salary (17,000 marks per annum: an unheard-of sum of money for a German university teacher in those days). One could save enough money to build a house, commented Heidegger, when he was trying to make the invitation sound appealing to Jaspers in June 1924. But first of all he needed to publish the Aristotle manuscript. Obviously his horizons would be broadened, he would be able to work in peace, and arrangements had even been made for him to teach at the University of Tokyo. But he wasn't entirely sure if he needed a protracted study trip of this kind – or indeed if it was right for him to take it on.

But in 1925, when the internal tug-of-war to select Hartmann's successor got under way, when the intrigue and manoeuvring in Marburg and at the ministry in Berlin were at their height, and Heidegger was looking for a full professorship on a higher salary, he may well have thought back wistfully to the rich fleshpots of Japan, given the provincial narrowness and petty-mindedness that he saw all around him. Husserl was again approached by Marburg, this time in the person of Ernst Jaensch, who asked him for a comparative assessment of two of his former pupils, Heidegger and Dietrich Mahnke. In his reply of 30 June 1925 Husserl expressed his high opinion of Dietrich Mahnke, whom he regarded as eminently well qualified for the associate professorship (i.e. to succeed Heidegger): the Faculty would be well served by this 'man of sterling character and scholarly distinction'. But he would be less than honest (he went on) if he did not state quite clearly

> that Heidegger, in my view, is unequivocally the preferred choice for the permanent professorship. By that I mean: not just preferred to Dr Mahnke, but preferred to *any* candidate you might have in mind for the post. Among the younger generation of philosophical talents I have not met anyone who exhibits such fresh and boundless originality, or who is so wholly devoted to philosophy to the exclusion of all worldly interests. The unique character of his teaching, which speaks to the whole man in his listener and commands by the sheer force of his philosophical earnestness, must be well known among his colleagues in Marburg. To my mind Heidegger is without doubt the most important figure among the rising generation of philosophers. Unless some singularly unhappy stroke of chance or fate intervenes to prevent it, he is predestined to be a philosopher of great stature, a leader far beyond the confusions and frailties of the present age. Just how much of originality he has to say, having kept his own counsel for years until he could arrive at a definitive and cogent statement of his position, will appear from his forthcoming series of publications.[99]

Such lofty sentiment – bordering on the excessive – was entirely characteristic of Husserl's response to Heidegger after 1918, deeply convinced as he was of the young philosopher's greatness.

In its recommendation of 5 August 1925 the Faculty was no less fulsome than Husserl: Heidegger, they noted, had established himself as a scholar and teacher 'of the first rank'. Although the book on Aristotle had not yet been published, the final manuscript, completed 'after careful and thorough revision', had been ready for some time and was 'about to be published' (in fact it never appeared). The report then refers to the second main work that was soon to appear – named here as 'Time and Being' – 'which shows us yet another side of Heidegger, namely the creative, independent thinker. This work offers nothing less than a fresh appraisal of the ultimate and fundamental ontological questions, representing a synthesis of the phenomenological approach (freed here for the first time from all subjectivism) with an analysis of the great traditional legacy of ancient, medieval and modern metaphysics.' In contrast to the older representatives of the phenomenological method, who in a sense had only prepared the ground, Heidegger's central concern with the fundamental philosophical problems was clearly apparent. There was nothing of comparable merit in contemporary philosophy to set alongside his thought: Heidegger was a worthy successor to the chair on which Natorp had so decisively set his stamp.[100] In effect the Faculty was saying that it wanted to keep Heidegger at Marburg, doubtless in the knowledge that Husserl's chair at Freiburg would soon become vacant.

Meanwhile the ministry in Berlin was making difficulties, sending back the list of candidates; and although the proofs of *Being and Time* were submitted in support of Heidegger's candidacy, more than two years passed before he was offered the permanent post at Marburg in October 1927 (the associate professorship was given to Mahnke at the same time). In his correspondence with Jaspers Heidegger referred repeatedly to the background intrigues behind the delaying tactics; the theologian Otto was working hard against him, he noted, while Jaensch was playing his own game, determined to ensure that 'only mediocrities, nobody who posed any kind of threat', should be allowed in at Marburg. When the list came back from Berlin for the second time in December 1926, Heidegger vented his frustration in a letter to Husserl, who consoled him by recounting his own similar experiences when he was at Göttingen. He pointed out that at least the Faculty at Marburg was firmly behind him, and furthermore:

> ... how fortunate that you are about to publish the work through which you have grown to be what you are, and with which, as you must surely know, you have begun to realize your own true being as a philosopher. From that beginning you will grow to new and greater stature. Nobody has more faith in you than I – faith, too, that no ill feelings will confuse or divert you from the work that is purely a consequence of the talent entrusted to you, conferred upon you at birth.[101]

We will do well to remember these words later, when we come to consider the alienation between Heidegger and Husserl – an alienation so absolute that it seems beyond human reason.

Being and Time appeared in the spring of 1927. The great breakthrough, correctly predicted by Husserl, had finally come. His eventual appointment to the chair at Marburg in October 1927 appears as a rather empty gesture, not least because it was only a matter of weeks before moves were afoot in Freiburg, at Husserl's instigation, to offer Heidegger the chair there. The ripening seed, nurtured so long, had at last been harvested and gathered in. Even before Heidegger accepted the offer from Freiburg, the plot of land above Freiburg-Zähringen had been purchased, and the builders worked throughout the summer of 1928 to get the house ready for occupation by the winter. The exile had returned.

Heidegger's style of public speaking was characterized by 'the cold logic of syllogistic reasoning', as one of his Marburg students notes in his *Philosophical Autobiography*:

> But on two occasions I witnessed outbursts of intensely felt emotion, which may well have underlain even the seemingly cool and analytical parts of his discourse: once during the speech he made on the death of Scheler, and again during his farewell address when he was leaving Marburg. On these occasions his thoughts sprang from emotions so intense and spontaneous that the tears welled from his eyes and choked his voice.[102]

PART THREE

'The failure of the rectorship'

'National Socialism: Germany's preordained path'

Karl Löwith, a pupil of Heidegger's from the Marburg years, whose habilitation Heidegger had approved prior to his departure from Marburg, was a Jew who drifted and travelled for a while before finding temporary asylum in Italy in 1934, where he was able to study in Rome with the aid of a Rockefeller grant. It was here, on 2 April 1936, that he met his old teacher, who was giving a lecture that day at the Istituto Italiano di Studi Germanici – the now-famous lecture on 'Hölderlin and the essential nature of poetry'.[103] The following day Löwith sent a postcard to Karl Jaspers, written in his characteristically minuscule hand, giving his immediate impressions: it was an elegant and beautifully-constructed discourse, 'but what the essential nature of this poetry has to do with the swastika is hard to see. For him the answer is presumably: "To choose is to be guilty, no matter what one chooses". He closed with "Bread and Wine", verse 7: "I know not what to do or say in the meantime, and what is the use of poets (= philosophers) in an impoverished age?".'[104] These lines from Hölderlin are later echoed in the title of Löwith's 1953 memoir: *Heidegger – a philosopher for an impoverished age*.

With acute insight Löwith put his finger on the darkly conflicting elements in Heidegger's personality, the unsolved enigma: how are we to reconcile the man and the works? Is it possible to separate Martin Heidegger the philosopher from Heidegger the man of political action? And if so, how and to what extent? At the heart of Heidegger's personality structure lies an ambivalence that defies comprehension.

The swastika, the Party emblem, ostentatiously worn in the

buttonhole as a profession of allegiance (and this in Fascist Italy): what has this got to do with Hölderlin's poetry, asks Löwith (and his question is essentially rhetorical, implying that Hölderlin's poetry and the symbol of National Socialism have nothing in common). It was 'hard to see' indeed.

Löwith has left us a detailed account of his meeting with Heidegger and Heidegger's family in Rome in the spring of 1936. They were no strangers, after all: they had been friends for many years, at least since 1919, well before their teacher-pupil relationship began. There were so many bonds between them: the holidays spent together at Szilasi's villa on the Starnbergersee, for instance. And Löwith had often looked after Heidegger's children. But now the political divide had become sharply apparent: Frau Heidegger welcomed Löwith, a Jew, 'with polite but formal reserve'. What particularly struck him was the way Heidegger wore his party badge throughout: 'It obviously hadn't occurred to him that the swastika was out of place when he was spending the day with me'. In the course of conversation Löwith made their positions clear: it was his belief that Heidegger's support for National Socialism lay in the very nature of his philosophy.

> Heidegger readily agreed with me, and explained that his concept of 'historicity' furnished the basis for his political 'service'. He also left me in no doubt about his faith in Hitler. There were only two things he had underestimated: the vitality of the Christian churches and the obstacles in the way of union with Austria. He remained convinced that National Socialism was Germany's preordained path. It was just a question of 'seeing it through to the end'.[105]

Löwith wrote these words in 1940 when the memory was still fresh in his mind, a memory sharpened by his own bitter experiences: sad at heart that his old teacher had got himself so disastrously involved. His diagnosis is clear enough: Heidegger's philosophical undertaking is intimately bound up with National Socialism. Heidegger himself accepted this analysis, substantiating and interpreting it through a form of historicism that emerges at many points (of course) throughout his writings.

On 8 April 1936 Heidegger gave a second lecture in Rome

(unpublished as yet) under the title 'Europe and German philosophy'. The setting was the Bibliotheca Hertziana in the Kaiser-Wilhelm-Institut; Löwith was not invited, because Jews were 'unwelcome' there.

Heidegger published the Hölderlin lecture in the December 1936 number of the journal *Das Innere Reich* – an esoteric organ in which the unseen Germany contained within German intellectual life was to be given a voice: subservient in a special, subtle way to the *Führer* Adolf Hitler, the miracle worker chosen by Providence, in whom, 'inspired by a gracious omnipotence, the knowledge of the eternal riches of the German mind' was at work (thus the editorial introduction to the first issue). Anyone who contributed to this journal had not in any sense embarked on a course of 'inward emigration', but belonged rather to that group of 'characters', viewed with faint amusement or open contempt by the more down-to-earth National Socialists, who were grudgingly allowed a fool's licence in these pages. Hence the short shrift given to Heidegger's Hölderlin piece in the review columns of *Wille und Macht*, the '*Führer* organ for National Socialist youth' published by the *Reichsjugendführer*, Baldur von Schirach. The reviewer was a certain Dr Willi Fr. Könitzer, who, as Martin Heidegger wrote to his publisher's reader, had been running around Marburg as a Social Democrat as late as the summer of 1933, but was now a big noise on the staff of the *Völkischer Beobachter*. He had heard this from an old SA leader, who was familiar with the scene in Marburg.[106]

Be that as it may, Heidegger had long since classed himself as a political failure: 'the failure of the rectorship' was the phrase he used to Jaspers, who was bound to understand him, knowing as he did the aims with which Heidegger had entered the political and ideological arena in 1933. But his party piece had failed to draw the crowds, the show was over – and now the public was applauding others. And yet – of this there was no doubt at all in Heidegger's mind – to him alone had been vouchsafed the quasi-mystical vision of the true essence of National Socialism, 'the inward truth and greatness' of the movement; and this knowledge he could never disavow, never, as long as he lived.

But let us move on now towards 1933, which was to be the critical year of destiny for Heidegger. If we accept the testimony of René

Schickele, the writer who spanned two cultures in the borderlands of the Upper Rhine valley, who noted in his diary for August 1932 that the word in Freiburg University circles was that 'Heidegger now consorts exclusively with National Socialists' – and Schickele made a note to look into this, since it seemed scarcely credible – then his first actual encounter – leaving aside the similarities in thinking – with people who thought of themselves as National Socialists must have come at an early date.[107] This would presumably accord with the veiled and sibylline references made by Heidegger to various correspondents in 1930, after he had turned down the offer of the Berlin chair – as in the letter to his colleague Julius Stenzel at Kiel, an expert in ancient philosophy: a clear small voice within him (he writes) is telling him that he must save himself in the coming years for something far more important (17 August 1930).[108]

Later on, after Heidegger had ostentatiously joined the NSDAP on 1 May, the 'Day of National Labour', the National Socialist daily *Der Alemanne* was jubilant, claiming that the formal adoption of membership was simply the public expression of a position that Heidegger had long held in private. Latterly, it was said, no National Socialist had been turned away from Heidegger's door, and he had loyally supported the movement from within. Whatever the truth of the matter, this period remains shrouded in a fog of mystery, for the time being at least. One view is contained in the dry summary prepared by the denazification commission installed at the University of Freiburg by the French military government in 1945, based on statements made by Heidegger himself:

> Prior to the revolution of 1933 the philosopher Martin Heidegger lived in a totally unpolitical intellectual world, but maintained friendly contacts (in part through his sons) with the youth movement of the day and with certain literary spokesmen for Germany's youth – such as Ernst Jünger – who were heralding the end of the bourgeois-capitalist age and the dawning of a new German socialism. He looked to the National Socialist revolution to bring about a spiritual renewal of German life on a national-ethnic basis, and at the same time, in common with large sections of the German

> intelligentsia, a healing of social differences and the salvation of Western culture from the dangers of Communism. He had no clear grasp of the parliamentary-political processes that led up to the seizure of power by the National Socialists; but he believed in the historical mission of Hitler to bring about the spiritual and intellectual transformation that he himself envisaged.

This account, distilled from a series of interviews, furnishes a mirror in which many intellectuals, and especially university teachers, were able to recognize themselves, at least in broad outline.[109] It paints a picture of innocence coupled with political naïvety, plus a stiff dose of anti-democratic sentiment: a man, above all, who was unpolitical, who lived only in the world of the intellect, possibly sympathizing with the intellectual leaders of the conservative revolution (Ernst Jünger *et al.*), but with no formal ties to any organization or institution.

The botanist Friedrich Oehlkers, a member of the denazification commission, was utterly convinced that Heidegger had never been a Nazi in the conventional sense. In a letter to his friend Karl Jaspers of 15 December 1945, in which he asked for a reference or report on Heidegger, Oehlkers examines the problem from many different angles. He saw a tragic destiny at work in Heidegger, who had been 'utterly unpolitical'; 'the special brand of National Socialism he had concocted for himself' had nothing whatever to do with reality. He had performed his duties as rector 'in this self-created vacuum', doing 'untold damage to the University', until one day 'he suddenly noticed the wreckage lying all around him'. Only now was he beginning to understand how this had come about. The picture he presents is that of a politically naïve philosopher who did not know what he was doing – an innocent, essentially, who had got caught up in something against his will. Oehlkers portrays Heidegger against the dark background of Frau Heidegger's doings. She was a National Socialist activist whose record now made her husband's position even worse. She had earned the hatred of the local community (Freiburg-Zähringen) by 'brutally mistreating the women of Zähringen involved in the digging of entrenchments' in the autumn of 1944, and had not scrupled to 'send sick and pregnant women to dig entrenchments'. But the conduct of Heidegger's wife did not fall within the commission's remit.[110]

In due course we shall come across evidence that does not quite square with this picture of blue-eyed innocence. Contacts must already have been established with student groups – or, to be more precise, with Nazi cells within the student body – not only in Freiburg, but also in Berlin. In other words, when the brown tide swept through Germany Heidegger was ready and waiting for the revolution. Indeed, this was essential to his whole concept of historicity.

Yet this is precisely what he denied so emphatically, from his initial declarations before the denazification commission in July 1945 to the various statements and summaries and the numerous reworkings of his apologia, which was published for the first time in 1983 under the title *Facts and Thoughts* – appropriately enough in conjunction with the reprint of Heidegger's rectorship address of 27 May 1933, 'The Self-Affirmation of the German University'. In the course of the debate that took place in 1947/48 between Heidegger and Herbert Marcuse, Heidegger's pupil at Freiburg from 1928 to 1932 (an episode to which we shall return later), Heidegger gave the following account of his position and conduct in 1933:

> With regard to 1933, I looked to National Socialism to bring about a complete spiritual renewal of life, the healing of social differences and the salvation of Western culture from the dangers of Communism. These thoughts were expressed in my rectorship address (did you read the *complete* text, I wonder?), in a lecture on 'The essential nature of science' and in two speeches I made before the lecturers and students of the university here. Plus an election proclamation of some 25/30 lines that was published in the local student newspaper. There are one or two sentences in it that I now consider misguided. But that's all.[111]

Painstakingly listed in this fashion – ticked off on the fingers of one hand, as it were – it seems to amount to precious little: 'That's all.' But important questions are ignored, and the most important question of all is not even asked.

In fact Martin Heidegger, rector of the University of Freiburg from 22 April 1933 until his resignation on 23 April 1934 (or until the acceptance of his letter of resignation by Baden's minister for education on 27 April 1934) has repeatedly been the subject of controversy on account of his rectorship year. At the same time he has courted contention and kept himself in the public eye, from the famous *Spiegel* interview of 1966 (not published, at his own request, until immediately after his death at the end of May 1976) to the apologia *Facts and Thoughts*, likewise published posthumously in 1983. The latter text was promptly translated into English and French, in order that the authorized version might be spread throughout the world and a consistent account be given of that period in which the philosopher, abandoning his aloofness from public life, sought to intervene directly and decisively in the course of events – indeed, believed himself called to do the bidding of Being itself.

But what purpose is served by the continual harping on this theme, which many believe has been worked to death in the last few decades, and squeezed to the very last drop? Have not the party lines long since been drawn, ever since the days immediately following the national catastrophe of 1945, when the author of *Being and Time* was called to account? And is not anyone who dares to voice doubts now, when the philosopher's legacy has been made known and the final word spoken, immediately identified with the hostile camp? Are we not to be guided by Heidegger's own dictum in *Facts and Thoughts*?: 'To those and those alone who take pleasure in focusing on what they see as the shortcomings of my rectorship, let me say this: in themselves these things are of as little account as the fruitless rooting around in past efforts and actions, which are so utterly insignificant within the planetary will to power that they cannot even be termed minuscule.'

Does not this dictum serve as a verdict on anyone who, with all due critical care, fails to descry the paths that one individual walked in difficult times? Not just anyone, it must be said, but an intellectual and spiritual leader, whose word was meet to be heard, and was heard, and whose early going-forth was a summons and an inspiration, an example for others to follow.

How Heidegger became rector

It is not with the aim of finding fault with the philosopher and former rector of Freiburg University that we propose to retrace those paths now, armed with the equipment of the historian, who is used to searching for clues and traces, who uncovers traces that have been deliberately obliterated so that the paths appear to run more straight and true, with fewer twists and turns, than was in fact the case: paths that peter out into nothing – or lead towards the abyss. *Sine ira et studio*: dispassionately, but with due purpose and commitment. The historian must examine many things before he reaches the solid ground on which his edifice may be raised up plumb and true. First he will weigh and weigh again what the subject of his investigations has said in his own reflections, and will look at this in relation to what is revealed by the available sources. So let us begin by allowing Heidegger to speak for himself in the opening sentences of his apologia *Facts and Thoughts*:

> In April 1933 I was elected rector by unanimous vote of the University's plenary council. My predecessor in that post, von Möllendorff, had been obliged to resign after a brief period in office at the direction of the minister. Von Möllendorff himself, with whom I had discussed the matter of his successor on a number of occasions, wanted me to take over the rectorship. Similarly the previous rector, Sauer, tried to persuade me to take on the position in the interests of the University. On the morning of the election I was still not sure about it and wanted to withdraw my candidacy. I had no connections with the relevant government or Party officials; I was not a Party member myself, nor had I ever

> been politically active in any way. So I could not be certain of being heard in the corridors of power with regard to what I saw as a necessity and a task to be accomplished. I was equally uncertain as to how far the University would accept the challenge of rediscovering and redefining its own true identity – a challenge I had already outlined publicly in my inaugural address back in the summer of 1929.

It is a consistent and plausible account, which links together actual events and psychological motivation. Without going into great detail, which is neither possible nor desirable – and the same applies to other episodes in the course of the narrative – it must be said that the actual course of events was *not* as described here. It was not the case that the previous rector von Möllendorff, Professor of Anatomy, who held office only for a few days after his election, was forced to resign at the direction of the minister. He relinquished the position voluntarily, as a committed democrat and republican who refused to be a party to the *Gleichschaltung* that was about to be imposed on the University, or to the measures discriminating against Jewish academic staff. Nor was it true that the colleagues mentioned here by Heidegger had pressed him in this way to take over the rectorship. The real reasons lie deeper. In the letter to Karl Jaspers of 3 April 1933 Heidegger wrote: 'I was still hoping for some definite news about the plans to reorganize the universities.' Baeumler was keeping quiet; in his brief letter he sounded annoyed. Krieck in Frankfurt was not giving anything away either. And Karlsruhe was sitting tight. A meeting of the Working Party on the Philosophical Faculties had been arranged for 6 April; Freiburg was sending Schadewaldt as its delegate, but the identity of Heidelberg's man was not known in Freiburg. Perhaps the Berlin representatives were the most likely source of information on this occasion. A working party set up in Frankfurt under the direction of Krieck was also getting nowhere. Freiburg's rector (Dean Sauer), with whom he had spoken, was simply appalled at the ineptitude of the Conference of German Rectors ... Here is a comprehensive summary of the situation, analysed by one who was awaiting new developments with impatience, eager to play a part in shaping their course. It is in this light that we are to understand the closing sentences of the letter: 'Although a lot of things remain obscure and questionable, I feel more and more that we are emerging into a new reality, and that the old era has run its course. Everything depends on whether we can bring philosophy to bear in the right place and help it to do its work.'

The mention of Alfred Baeumler and Ernst Krieck introduces two key figures with whom Heidegger worked very closely indeed, at least in the initial phase of the 'national revolution'. It would not be long before relations with both became very strained in their different ways, to the point of outright hostility in the case of Krieck; but nobody could have foreseen this at the start of the summer semester of 1933, when all three were keen to play their part, working together to achieve certain goals within the National Socialist scheme of things. Caution is indicated, therefore, where this early and intensive collaboration is passed over in silence, and a touched-up portrait has been placed in the gallery of history. At the time there were any number of distinguished competitors jockeying for position at the starting line, hoping to get off to the fastest start and come in first. Unfortunately they had no clear view of the course, and the event turned out to be anything but a sprint; it was more like a long-distance obstacle race, littered with traps and nasty surprises. For the moment, however, everyone was riding high on a mood of buoyant optimism.

In fact it was the young Freiburg classicist, Wolfgang Schadewaldt, who twice went to see the incumbent rector Sauer before his period of office ended in mid-April – an unusual move in extraordinary times – to urge that the rectorship should go to Heidegger rather than von Möllendorff, the rector-designate chosen back in December 1932 – given the special political circumstances that now obtained. All this came as a great surprise to Sauer, who simply could not see the point of this extraordinary request. Moreover he did not think Heidegger was the right man to exercise this important function at the present critical time, given that he had no experience of university government.[112]

Schadewaldt's first visit took place on Good Friday (14 April), the day before the transfer of the rectorship. It was an unusual time for a Catholic theologian to call:

> Then Schadewaldt arrived and stayed until 1.30 p.m. He talked about the whole question of *Gleichschaltung* at the University, and wondered if we shouldn't put Heidegger in as rector. I pointed out that he was hardly the man for the administrative and business side of things, which would be a good deal more difficult now than it had ever been in the past ... I reminded him that Möllendorff was still there, and was probably the man best qualified for the job.

This diary entry – a key text for resolving the question of who actually pointed Heidegger in the direction of the rectorship – needs a word or two of explanation. Schadewaldt was a full professor of classical philology ranking alongside his older (Jewish) colleague Eduard Fraenkel, who had now been retired in accordance with the recently revised legal position in Baden. A devoted supporter of the new regime, Schadewaldt was clearly one of the prime movers behind plans to oust von Möllendorff – an unacceptable choice from a National Socialist point of view – in favour of Heidegger, and hasten the process of *Gleichschaltung* within the organs of university self-government by the removal of non-Aryans from the University Senate and the replacement of any non-Aryan deans of faculty.

The incumbent rector Sauer, a theologian and gentleman of the old school, had no idea at this time that a cadre of Nazi professors or close sympathizers had begun to form at the University. In the first few days of April the new Nazi secretary for higher education at the Ministry of Home Affairs in Karlsruhe, Eugen Fehrle, later professor of folklore studies at the University of Heidelberg, came to Freiburg on a fact-finding visit. He not only had talks with the official representatives of the University (i.e. the incumbent rector Josef Sauer and the rector-designate Wilhelm von Möllendorff) but also met with a small group of Nazi professors to discuss the way forward for the Party at the University. One of the professors present referred to these discussions in a report dated 9 April 1933, from which the following extract is taken:

> To take the first point raised at our recent discussion, concerning the alliance of National Socialist university teachers, we have ascertained that Professor Heidegger has already entered into negotiations with the Prussian Ministry of Education. He enjoys our full confidence, and we would therefore ask you to regard him for the present as our spokesman here at the University of Freiburg. Professor Heidegger is not a Party member, and he thinks it would be more practical to remain so for the time being in order to preserve a freer hand *vis-à-vis* his other colleagues whose position is either unclear still or openly hostile. He is quite prepared, however, to join the Party when and if this should be deemed expedient on other grounds. But I would particularly welcome it if you were able to establish direct contact with Professor Heidegger, who is fully apprised of all the points that concern us. He is at your disposal in the coming days, but I should say that there is a meeting in Frankfurt on the 25th which he could usefully attend as the spokesman for our University.

The report goes on to consider further tactical measures, and there is some discussion of Möllendorff's rectorship, since the latter is said to be 'an avowed democrat'.[113] So Heidegger did *not* end up in the rectorship through a series of chance events, dutifully shouldering this burden at the urging of the worthy Sauer and von Möllendorff. There was instead an internal conspiracy, plotted behind the scenes by a small clique of Nazi sympathizers, while the on-stage action followed a carefully prepared script. The new rector, von Möllendorff, called his first meeting of the Senate on 18 April, the Tuesday after Easter. The day began ominously with the publication of the following article in the National Socialist newspaper *Der Alemanne*:

> Professor von Möllendorff has been elected to the rectorship of the University. From this position of leadership he will

be expected to work for and contribute to the cultural reconstruction of Germany. It is evident that the work of reconstruction can only be successful if all those in positions of responsibility set about their task with fierce determination and energy. But that task is doomed to failure if men with old-fashioned scruples and liberal views oppose the process of *Gleichschaltung*, or even labour actively against it. There is a risk that active opposition of this kind is to be expected from Professor von Möllendorff, at least in staff-related matters. After all, if the rector of Freiburg University can take up the cause of a mayor whose connections with the University were surely tenuous at best, and whose possible dismissal does not in any way touch upon the rector's own area of responsibility, what will it be like when he has to deal with decisions that *do* fall within his remit as rector? In our opinion the appointment of a man who holds such views is entirely out of keeping with the spirit and aims of the national revolution. Neither can we imagine how a common ground of trust could ever be established between Professor von Möllendorff and the student body, which is overwhelmingly National Socialist in its sympathies. But even if these differences could be overcome by honest hard work, the resistance to official government policy in Baden, and indeed in Berlin, would lead to the kind of friction that is best avoided in the interests of steady progress. Unnecessary time and energy would also have to be spent in easing strained relations. The whole point of *Gleichschaltung* is to avoid this kind of thing: men with a shared aim and purpose are supposed to pool their resources and work together for the attainment of the common goal. There must be no more pulling in different directions, no more dissipation of resources. Nobody who wishes to play his part should be excluded: all the more reason, therefore, to ensure that the common enterprise is not hindered or disrupted by unnecessary opposition. We urge Professor von Möllendorff to seize the opportunity – and not to stand in the way of the reorganization of our university system.

This was no isolated incident, but part of a concerted campaign as dictated by the script. This attack in *Der Alemanne*, which was widely viewed as the unofficial voice of the Party, had the desired effect. Events unfolded rapidly over the next two days, to the point where von Möllendorff called an extraordinary meeting of the Senate for 20 April (Hitler's birthday!), at which the Senate and the rector moved to resign and convened a plenary assembly for the following day, 'for the purpose of electing a new rector and Senate'. The last obstacle to *Gleichschaltung* at the University of Freiburg had been removed.

As far as von Möllendorff was aware, therefore, the process of choosing his successor was compressed into these few dramatic hours. What he did not know was that the drama had been carefully scripted, and that Heidegger's entrance as rector-to-be had been prepared well in advance.

And so Heidegger was elected to the rectorship on 21 April 1933 by an almost unanimous vote – though admittedly a number of university professors who were eligible to vote were not present. At several points in the official minutes of the meeting we can sense the profound shock wave that had passed through the ranks of Freiburg's senior academics: for the first time colleagues who were otherwise eligible to vote had been excluded on racial grounds – thirteen out of ninety-three, to be precise. This situation was a consequence of the new legal position created on his own authority by Robert Wagner, *Gauleiter* and Reich Commissar for the *Land* and *Gau* of Baden, who had issued the appropriate decree. Non-Aryan professors at Baden's universities were suspended until further notice, and had to resign any academic offices they held (including deanships and senatorial seats). Internally a good deal of concern and disquiet was voiced, and there were many private expressions of sympathy: but academic bodies and committees were 'purged' in accordance with the decree against the Jews, and the governing bodies of universities, institutes and clinics 'enforced' the Baden decree 'without incident'. This shows how deep-rooted was people's faith in the state, and how few questions were asked about the legality of such rulings, which ignored basic human rights and opened the door to brutality and inhumanity. This readiness to conform, to swing into line behind the totalitarian regime, was extremely widespread, not to say universal.

The new law was established on the foundations of Hitler's Enabling Act – and was therefore still within the framework of the Weimar constitution.

The minutes of the meeting on 21 April 1933 do show that the dean of the Faculty of Theology, Engelbert Krebs, Heidegger's one-time friend, made a statement expressing sympathy for his dismissed colleagues: a handsome gesture, but for internal consumption only. Given the state of majority opinion, the University of Freiburg felt it could no longer make a public show of solidarity with its dismissed Jewish colleagues. The sails had been hauled in, the ship was rolling at anchor in a quiet inlet, hoping to sit out the storm that was raging out at sea. But Heidegger, as the new rector, had other ideas: he wanted to make ready the great ship of the university, and put out to sea at the head of the armada, braving the tempest. The local Nazi party newspaper, *Der Alemanne*, rejoiced at the news of Heidegger's election. This it saw as part of the general process of *Gleichschaltung*, whereby the academic staff at Freiburg University had declared their willingness 'to put their shoulder to the wheel in the cause of national and social revolution'. The student body, it went on, had pledged its 'allegiance and co-operation' to the new rector 'as the *Führer* of the University'. The report concluded: 'The way has thus been paved for the University, whose organization and teaching work with young scholars must be in harmony with the guiding principles of the State, to pursue its development along the one true path.'

What programme did the *Führer*-rector Heidegger carry with him as he embarked on this new stage in his life's journey, swept along by the great national awakening, and stepping out into the limelight under the gaze of a public that was by no means confined to the University of Freiburg? To answer that question we would need to examine in detail the programmatic rectorship address given by Heidegger on 27 April 1933 – a task that has already been performed with great diligence by scholars working in this field, who have placed the address in the context of Heidegger's earlier thought and dissected what he had to say about the theory and concept of the state in an attempt to arrive at a definition of Heidegger's own political philosophy.[114] But this is not our particular concern here. As rector, the philosopher acknowledged his commitment to the intellectual leadership of Freiburg University, and at the same time defined the essential character of the German university *per se*, which only 'attains to clarity, status and power if the leaders, first and foremost and at all times, are themselves *led* – led on by the inexorability of that intellectual and spiritual mission which forces the destiny of the German nation into the mould of its history'. The grounding of leadership and allegiance in this essential character could only be achieved through the self-affirmation of the German university, interpreted as the will to its own essential character. 'The will to the essential character of the German university', we read, 'is the will to science and learning as [it is] the will to the historic spiritual and intellectual mission of the German nation.' For the philosopher, however, the guardian of Being, this means that the intellectual leaders of the nation, the academic staff of the universities, must really advance 'into the most forward outpost exposed to the constant danger of world-uncertainty [Weltungewissheit]', since the spiritual and intellectual world is the nation's only guarantee of greatness. This spiritual world of the nation, however, is the power of preserving at the deepest level those forces that are rooted in the earth and its own blood, as the power of the innermost excitation and most far-reaching convulsion of its being-there [*Dasein*]. Starting from these basic principles, couched in the language of his philosophy, in the language of Being, Heidegger unfolded his programme in detail: a programme, much debated since, that came to nothing.

The curious yearning for hardness and rigour

Placing the rectorship address in its martial context

Heidegger's rectorship address – 'a call to arms, an intellectual summons, a resolute and urgent stepping-into-line with the times', to quote the perceptive summary of R. Harder, writing in the journal for classical philology (*Gnomon*) in 1933 – also (indeed specifically) contained very concrete directions for action: the concept of freedom for German students, he claimed, was now being restored to its true meaning, out of which the future 'loyalty and service of the German student body' would evolve. The new rector identifies three sorts of service: labour service, military service and the service of knowledge. It need hardly be emphasized that all this is of course predicated on the specific concept of 'truth' developed by Heidegger, which is to say: 'The second loyalty is to the honour and destiny of the nation within the community of nations. It demands a willingness, firmly grounded in knowledge and ability and sharpened by discipline, to give of one's utmost. This loyalty will in future embrace and permeate the whole of student existence in the shape of military service.' 'Military service', of course, had a familiar enough ring in the German universities of the day. The militarization of the German nation was an urgent imperative fuelled by the shame of the Versailles Treaty, and much of the impetus for that was to come from the student generation. Premilitary training organizations of various kinds – not just the *Stahlhelm* – had proliferated, with strong support from the students' duelling corps. 'Military service' was already a national slogan when it was hijacked by the National Socialists for their own ends in 1933.

It was in the martial figure of Hermann Göring, veteran air force captain of the First World War, the last commander of the Richthofen Fighter Squadron, holder of the *Pour le Mérite*, long established as one of the leading lights of the Hitler movement – president of the Reichstag since the summer of 1932, Reich minister and acting Minister of the Interior for Prussia (in charge of the police) since 30 January 1933 – that Heidegger apparently saw his ideal of new German manhood embodied. Why else would he have taken along a copy of Martin Harry Sommerfeldt's book *Hermann Göring: Ein Lebensbild* as a thank-you present for the family of his old art historian friend Hans Jantzen, whom he visited two days before the crucial Reichstag elections of 5 March 1933? He inscribed the book (a copy of the 3rd edition (Berlin 1933), which had just appeared that February[115]) with a dedication: 'To my dear friends the Jantzen family, as a memento of 3 March 1933 in Frankfurt am Main. Martin Heiddegger.'[116]

Göring's swashbuckling deeds of derring-do are extolled in purple journalistic prose, with particular emphasis on the stylization of his struggle at Hitler's side – his 'most faithful paladin'. We are given a definition of National Socialism, and are told the goals of the movement in ringing phrases. The nimble Sommerfeldt, appointed Göring's press spokesman (for the Prussian Ministry of the Interior) in early February 1933,[117] promptly updated his eulogy. New brushstrokes were added to the picture: now we get to see Göring as 'Hercules', cleansing the Augean stables. 'Yes indeed: the pace is going to be fast and furious this year, because Hermann Göring has forgotten nothing!' Germany was about to find out 'that Hermann Göring did not become Minister of the Interior in order to "dispense justice", but in order to carry out a political programme; a political programme that is wholly and unequivocally Prussian, German, and unreservedly national.' How very true! The terrible consequences of that political programme were displayed for all to see during the weeks before the Reichstag elections: the Reichstag fire and emergency decrees, the hunting down of political opponents, the mobilization of the mob on the streets. 'Göring will use this power, ruthlessly, unflinchingly – until the successful outcome proves him right before the German nation, Europe, the world, and the tribunal of History': so writes Sommerfeldt in the closing paragraphs of his tract, putting the passage in spaced type so that it cannot possibly be missed. This journalistic rag-bag was the book Heidegger chose to bring with him as a present. Perhaps he had not actually read it – as he claimed not to have read Hitler's *Mein Kampf.* We must assume otherwise, seeing that Sommerfeldt's *Hermann Göring* was an important source of material for Heidegger's rectorship address, his proclamations during the autumn of 1933, and the statements he issued in his own defence in 1945 – even down to the linguistic nuances. What a come-down! And what about Heidegger's later contribution to the *Festschrift* for Jantzen, published in 1951 on the scholar's seventieth birthday – an essay entitled 'Logos', which examines the Heraclitus fragment B50? Where does that fit in to the picture?

But it did not stop at heroic declarations. Practical training was soon introduced, and student SA units at Freiburg University, together with other groups whose precise extent and origins are difficult to determine, received their first paramilitary training in the summer semester of 1933 under the direction of Dr Georg Stieler, associate professor of philosophy and pedagogy. He had formerly been a career officer on active service, ending the war with the rank of naval commander, and subsequently serving with enthusiasm in the *Stahlhelm.* The tall figure of the philosopher (he stood a full 6 feet 7 inches) could be seen exercising with the informally constituted student companies in the claypits of a brickworks at the foot of the Schönberg just outside Freiburg; in the absence of real weapons the students were issued with dummy rifles made of wood. For the time being such training activities had to be conducted in secret. Recruits received their 'call-up' orders by word of mouth. However, all these training exercises were cleared in advance with the appropriate authorities in the *Reichswehr.* An eye-witness, who was at that time a junior lecturer in medicine and who had himself fought in the First World War and come home seriously wounded, told me how he had been prompted by curiosity to go and watch these rather childish war games. Suddenly the rector's official car drew up, and Heidegger jumped out. Professor Stieler, as the 'commanding officer', stood to attention in front of the great philosopher, who was short in stature, and made his 'report' in the correct military manner, as if the rector was the commander-in-chief of his military forces. It was a touching, and at the same time slightly farcical scene, given the considerable difference in the two men's physical size.

Such antics must have been viewed by Heidegger as a confirmation of what he had said in his rectorship address: that the student body was 'on the march', in search of leaders. And to make sure that this assertion stuck in people's minds, the rector had had the words of the *Horst-Wessel Lied* printed on the back of the programme notes that accompanied the inauguration ceremony – a suitably martial text, hammered home by a relentless marching beat, to be sung following the address by the leader (*Führer*) of the Freiburg student body. So the worthy professors raised their voices in song, picking their way through the unfamiliar words of this literary work, the official hymn of the movement:

> Raise high the flag, stand rank on rank together!
> Stormtroopers march with firm and valiant tread,
> Comrades gunned down by Red Front and reaction
> March on in spirit, swelling still our ranks.'

After the second ('Make straight the way for the brown battalions . . .') and third ('Now sounds the final call to arms . . .') verses, the first is repeated as verse 4, accompanied by the following direction: 'During the singing of the fourth verse the right hand shall be raised in a greeting to the dead' – though doubtless only to the dead of 'the movement', the 'comrades gunned down by Red Front and reaction'.

There had been a good deal of commotion in the days preceding the rectorship address. On 23 May 1933 Rector Heidegger had issued a memorandum (No. 5193) on the order of events at this inauguration ceremony, which was something quite new and therefore without precedents. The raising of the hand and the demeanour to be adopted during the shout of 'Sieg Heil!' were among the points considered, and there was heated debate about whether the raising of the hand during the singing of the *Horst-Wessel Lied* was tantamount to a declaration of allegiance to the NSDAP.

The following day, in an effort to clarify the position, Heidegger published the official government communiqué received from Karlsruhe (memorandum No. 5288), which stated that the raising of the right hand during the singing of the national anthem and the *Horst-Wessel Lied* (verses 1 and 4) and during 'the loyal salute "Sieg Heil!"' was not in itself an avowal of Party membership. Rather it was the case that the gesture had become the national greeting of the German people, and was intended simply as an expression of the individual's allegiance to the present state and his inward identification with the new Germany. The rector did make one concession, doubtless in part out of consideration for the infirmity of some of the more senior academics present: 'After conferring with the leader of the student body I have decided to confine the raising of the hand to the fourth verse of the *Horst-Wessel Lied*.'

The significance of the ambience in which Heidegger's rectorship address was embedded – indeed rooted – cannot be overemphasized. Josef Sauer, who served as pro-rector alongside Heidegger, has described his impressions on the day: visually it was a very altered scene, with numerous Nazi uniforms and weather-beaten faces placed in the rows reserved for special guests – Party officials, for whom even the wives of elderly academic colleagues were turned out of their seats. As a marshal Sauer had tried to intervene, but to no avail. Heidegger had been almost incomprehensible.

Speaking at the dinner that followed the ceremony, which was attended also by the university rectors of Karlsruhe and Heidelberg, Heidegger had begun in a rather frosty and formal vein. Sauer had tried to relieve the atmosphere by introducing a note of humour into his remarks, whereupon Heidegger rose to speak again, his tone now warmer and more relaxed; this time he talked about their earlier relations, and the fact that he had once been Sauer's protégé.[118] Heidegger specifically noted this incident later in his apologia; it seems that the Ministry subsequently intimated that his after-dinner speech, with its explicit references to Sauer, had been something of a *faux pas*.[119]

At the time the militaristic and martial poses struck by the new rector were the target of much derision in Freiburg University circles. Among the academic staff were numerous highly-decorated First

World War veterans, who were well acquainted with the details of Heidegger's own undistinguished military past – men such as the economist Walter Eucken, one of Heidegger's most outspoken opponents inside the University, a former front-line officer who viewed Heidegger's military antics with ill-concealed contempt.

An unmilitary and indeed unheroic figure, Martin Heidegger was for one thing an admirer of Ernst Jünger – first and foremost, no doubt, as the brave holder of the award *Pour le Mérite*. Through such exemplary figures from the First World War, with whom he sought to identify, he was able to compensate for the great experience of front-line action that he himself never had. We have already seen how very little Heidegger's military service amounted to: it was hardly something to shout about. The philosopher Max Müller, an expert on Heidegger who also considers himself a disciple of his, believes that Heidegger's own unheroic past probably 'contributed to elevate the experience of the front-line soldier to the status of a heroic myth' in his thinking.[120] So it is not surprising that the later Heidegger either did not respond to official inquiries about his military past, or else gave a heavily doctored version of it; in 1927/28, for example, the curator of Marburg University asked Heidegger no less than five times to supply details of his military service for purposes of calculating his pension entitlement. The curator received no reply.

In the *Deutsches Führerlexikon* of 1934/35 Heidegger portrayed himself as a volunteer for war service who was subsequently discharged for health reasons. In fact Heidegger, who had not done military service because he was originally a theologian, was conscripted after the outbreak of war in the normal way and registered as a 'Landsturmmann' or territorial reservist (with no prior service). Since his military activities at the Freiburg postal censorship office were open to misunderstanding, it was always presented as a period of uninterrupted military service; but Heidegger was hardly ever quartered in barracks, and consequently enjoyed a privileged freedom. Heidegger also stated that towards the end of the war he was stationed near Verdun – an attempt to bask in the glory that attached to that name. This was his characteristic way of glossing over a period of uneventful service as an army meteorologist in the Ardennes.

But things did not stop at the military training programmes

organized by Professor Stieler. In the spirit of martial endeavour Heidegger asked Stieler to draft a code of honour for the soon-to-be-established university lecturers' association, which he submitted to the authorities in Karlsruhe and Berlin with his recommendations. It was based on the military officers' code of honour. The purpose of the exercise was to restore the sense of honour of which the nation had been stripped by its past political leaders, with the result that 'everywhere a brutal struggle for survival is going on without regard for one's fellows and colleagues'; the conviction that 'a process of cleansing and renewal' was now needed had to come from within the ranks of university lecturers:

> We lecturers seek to rise up and come to ourselves again. We seek to cleanse our ranks of inferior elements and thwart the forces of degeneracy in the future. By nurturing our sense of honour we seek to teach and instruct each other, thereby ensuring that there is no possibility of falling back into the old ways. Finally, and most importantly, we seek to nurture and develop that spirit of true comradeship and genuine socialism among ourselves which does not view one's colleagues simply as rivals in the struggle for survival . . .

This ideal of comradeship – fellow members of one class and one nation – was realized by 'the officer corps in its heyday' – whenever that might have been. 'Let us embrace Fichte's idea that we are a single whole that has grown inextricably together, in which no individual member regards the fate of another member as foreign and touching him not!' Such exalted justifications for the introduction of an honour code clearly reflect the confusion inherent in National Socialist ideology: the restoration of honour (perceived as the honour of an officer caste) was equated with the efforts of the body academic to cleanse itself of inferior or degenerate elements.[121]

Participation in field sports was a requirement for every student after May 1933 – to the no small irritation of many professors, who could muster little enthusiasm for this particular manifestation of 'the militarization of academic youth', which interfered with the regular teaching schedule. But Heidegger was conspicuously supportive of these militaristic field sports. To the students of Heidelberg

University, for example, he addressed the following exhortation in his notorious speech of 30 June 1933, directed against the professors: 'How can we speak of "wasting time" when it is a question of fighting for the state? It is not in working for the state that the danger lies: the danger lies in indifference and opposition.'[122]

A special paramilitary sports camp was quickly set up for this purpose. Since it could not be sited in Freiburg or its immediate vicinity because of the demilitarized zone, a suitable location was found not far from Löffingen, in the area known as the Baar, some fifty kilometres east of the university city. In the months of August and October successive contingents of three hundred students at a time attended the camp for three-week periods, receiving training from members of the *Reichswehr*, SA and SS. But this new political student class, these model SA students who were to be trained in the camps, were nothing like the student figures invoked by Heidegger in heroic style in his rectorship address: 'The students of Germany are on the march. And they are searching for the leaders through whom they will advance their own appointed claim to a well-founded, knowing truth, and place it in the clear light of the interpretative word and work.'

Heidegger had long since lost contact with reality, if indeed he was ever in touch with it in the first place. He was permanently confronted with a reality that at first he simply denied, on the principle that what cannot be is therefore not so. One example will serve to illustrate the point. Certain core elements within the student body at Freiburg who were organized into the student section of the SA, and who assembled that first summer for practical service at the hastily-erected sports camp near Löffingen up on the Baar, took part in an operation of a kind that was becoming increasingly common in those days. The aim was to 'soften up' a local politician from one of the centre parties who had incurred the displeasure of the National Socialists, which was done by stage-managing gatherings of outraged citizens to furnish a legal pretext for arresting the man. The 'outraged citizens' in question were the SA students from Freiburg University. A Freiburg professor of civil law, a war veteran and as good a nationalist as the next man, learned of these events and wrote to Rector Heidegger at the beginning of September, asking whether such behaviour was in keeping with the national honour that Hitler had in mind. Heidegger, clearly indignant, vouchsafed the distinguished member of the Faculty of Law and Political Science the following testy reply:

> I have noted your remarks about Löffingen. But you appear to be unaware of the dubious conduct of the local resident concerned, which apparently prompted these scenes of public disorder. Your complaint is based on the report of a gentleman who is unknown to me. Under the circumstances I am unable for the moment to reach any firm conclusions in this matter. While it is important to voice doubts and misgivings, I would in future be obliged, for the sake of the 'realization of the Third Reich', if some of the suggestions submitted to me were of a more constructive kind.

'Think positive' is Heidegger's message to his colleague: for the Third Reich waits upon its 'realization' . . . In the not too distant future it would be made abundantly clear to Rector Heidegger that the leadership no longer rested with him, but with the hierarchy of the SA student organization and its Party organs. But for the moment Heidegger cherished his illusions – 'like a boy who dreams and knows not what he does'? – or else gratified his 'lust for power'. Both are pictures we have seen before; neither is accurate. Heidegger knew exactly what he wanted to achieve, and had good reason to be optimistic that in alliance with persons of influence and key agencies – particularly those in Berlin – he would succeed in his aim.

The clue to the compensatory mechanism at work here lies in Heidegger's language (and we still await an in-depth study of his use of the words '*Kampf*' [= struggle, fight] and '*kämpferisch*', together with their compounds and derivatives[123]). But it is fair to say that few writers use these words with more frequency and variation than Heidegger. And if, with Heidegger, we take the essence of language (which continues to elude us, it must be said) to be 'that it is the house of the truth of Being' (*Letter on Humanism*), then we cannot overestimate the significance of his words. The rectorship address alone contains plenty of material: 'struggle' is central to his whole discourse, especially when we think of the contemporary political context in which the word 'struggle' flourished. Heidegger protested vigorously against such simple-minded and perhaps malicious interpretation, insisting that he be construed in the Heraclitean sense. Be that as it may, Heidegger laid himself open to misunderstanding – and was indeed widely misunderstood – on account of the ambiguity of his address.[124]

It certainly gives us pause when we hear what Heidegger wrote to his pupil Karl Ulmer at the eastern front, towards the end of the Russian campaign: that the only worthy kind of life for a German was a life at the front.[125] For Heidegger the word 'front' always signified the place of maximum exposure to danger. We need only examine the rectorship address for the range of military or quasi-military vocabulary it contains: 'danger', 'distress', 'hour of need', 'force', 'power', 'discipline', 'elite', 'last-ditch stand', 'remorseless clarity'. And when Heidegger defined 'life at the front' as the only worthy kind of life for a German, he was following exactly the line that he took in the rectorship address, where he spoke of 'the resolute determination of the German student body to bear the burden of Germany's destiny in its hour of need'. This sort of value judgement – and Heidegger applied himself assiduously to the work of passing judgement, despite the vehement efforts of his supporters to dissociate him from axiological enterprise – is related to his cartoon-like analysis of contemporary history, in which Europe, and at its centre 'our nation', the 'metaphysical nation', are caught 'in a giant pincer grip between Russia on the one side and America on the other'. Both powers (he opines) are exactly the same 'seen from a metaphysical perspective': 'the same soulless spectacle of technology run riot and ordinary people at the mercy of social organization without roots'.[126]

Germany, 'as the one in the middle', felt the squeeze most acutely. And then, after Stalingrad, comes a letter in this vein to his pupil Karl Ulmer – despite the fact that the fate of the German nation, this metaphysical nation, was already sealed: doomed to destruction, burdened with the crimes that the *Führer* had ordered, that same Leader who 'alone and of himself *is* the present and future German reality and its law.'[127] Heidegger never subsequently retracted this statement.

For his own part he had no desire to endure the worthy life at the front. The verbal swagger and the actual deed were worlds apart. Acting on Heidegger's authority, Heinrich Wiegand Petzet spread the legend that Heidegger was conscripted into the *Volkssturm* (Home Guard) in November 1944 'doubtless in the hope of getting rid of him once and for all: an act of perfidy that certainly failed of its purpose.'[128] But what is the true story behind the incident darkly hinted at here?

On 23 November 1944 Heidegger left Freiburg with a *Volkssturm* contingent, heading towards Breisach on the Rhine. They did not get very far, since Neu-Breisach had already been taken by French troops along with Strasbourg, making it impossible for the Germans to conduct further military operations across the Rhine. Heidegger was able to avoid further military duty, not least thanks to the strenuous efforts undertaken via the Party organization known as the *Reichsdozentenbund* (the Reich League of University Lecturers). Professor Eugen Fischer, the internationally renowned anatomist and eugenicist and former director of the Kaiser Wilhelm Institute in Berlin, who had retired to live in his native Freiburg, was a member of the leading circle around the head of the *Reichsdozentenbund*, Dr Gustav Adolf Scheel. On the very day that Heidegger left Freiburg Fischer sent a cable to Dr Scheel, who was also *Gauleiter* of Salzburg: 'To *Gauleiter* Salzburg. Acknowledge *Volkssturm* and present exigencies, but support Faculty petition for release military service Heidegger stationed Alsace, irreplaceable thinker unique asset nation and Party. Eugen Fischer.' This curiously garbled text is an interesting document, whose full meaning only emerges from the follow-up letter that Fischer sent. He and the Faculty of Philosophy fully understood that at the present time 'all our fighting strength and all our fighting resolve – and therefore the *Volkssturm* – must come before everything else.' The University, he went on, had relinquished many 'comrades' for service in the *Volkssturm*, in order to defend the Rhine front. But it was only right and proper 'to request exemption for certain individuals in rare and exceptional circumstances. And Heidegger is deserving of such exemption; though one may take exception to some of his remarks, there is no denying the value of his work.' The Faculty was not seeking to promote its own interests, he pointed out, but simply standing up for 'an intellectual leader and thinker of unique stature. We really do not have that many great philosophers – and even fewer with National Socialist sympathies.' Fischer, who had been a friend of Heidegger's for many years, and who had joined with Heidegger in the infamous Leipzig proclamation of 11 November 1933 (Fischer was rector of Berlin University at the time), was demanding that the philosopher Heidegger be treated in the same way as the great inventors, physicists and chemists who were excused from military service 'because they can be of more service elsewhere'.[129]

The case of Heidegger had been satisfactorily resolved in the meantime, making it unnecessary for the *Gauleiter* of Salzburg to intervene personally. (He would happily have done so, as hc told Eugen Fischer a little later.) Following the devastation of Freiburg on 27 November 1944, Heidegger withdrew to rescue his manuscripts – at a safe distance from the Rhine front, which others were welcome to defend. In the request for leave that he sent to the rector of Freiburg University on 16 December 1944 from Messkirch, he explained that the new military situation on the Upper Rhine made it necessary for him to remove the manuscripts to a place of safety, 'so that as far as possible they are out of any immediate danger'. It is in the nature of philosophical work, he writes, that it 'is more closely bound up with the person of its author'. Yet: 'In truth my work does not belong to me personally, but to the future of Germany, in whose service it stands. Its removal to a place of safekeeping may therefore claim to be a matter of some priority.' But what kind of future did Germany have now, when the nation had been exhorted to rise up – 'a nation is rising up, a storm is breaking loose' – when the order from the *Führer* of 18 October 1944 had called for all able-bodied males between the ages of sixteen and sixty to be enrolled in the German *Volkssturm*, whose task it was to defend 'the soil of the fatherland with every available weapon and means'?

Was Martin Heidegger reminded of the closing sentences of his rectorship address of 27 May 1933? 'But we shall only fully comprehend the majesty and glory of this brave beginning if we bear within us that profound and broad reflectiveness out of which the wisdom of the ancient Greeks uttered the saying: "All great things stand fast in the storm . . ."' (Plato, *Republic* 497d, 9) – a highly individual translation and interpretation of this sentence from Plato that has been fiercely debated ever since. But in the Advent weeks of 1944, as the year drew towards its close, there was no more talk of brave beginnings. All that was left was the promise of fulfilment in some distant future; but the reality of the present catastrophe was denied. We have already read the words that Heidegger wrote to Rudolf Stadelmann in Tübingen on 20 July 1945: 'Everyone now thinks of doom and downfall. But we Germans cannot go under because we have not yet arisen, and must persevere still through the night.'

Heidegger had retreated from the 'front' to the safety of his birthplace in Messkirch, entering into the idyllic surroundings of the Upper Danube valley – and into colloquy with Friedrich Hölderlin. He obtained a medical certificate on 8 February 1945 which recommended that he absent himself from his place of duty for a period of three months in the first instance. As the certificate pointed out, this effectively meant getting out of Freiburg. The Baden ministry of education, which had relocated to Annette von Droste-Hülshoff's home town of Meersburg on Lake Constance following the fall of Strasbourg, granted this petition on 16 March 1945 – one of its last official acts before the first French armoured units thrust into the south-west corner of Germany, where they met with little resistance.

When the French units reached the village of Hausen im Tal on 21 April, Heidegger had been in hiding for several days at 'the Lodge', as it was known, high above the Danube valley, together with the occupants: a forest warden with his wife and family, the Princess Margot von Sachsen-Meiningen and the Prince von Sachsen-Meiningen, who had served as a 'special commander' – that is, a uniformed officer attached to the *Wehrmacht*, but without combatant status, who helped to organize provisioning and supplies. It was a question of survival.

The net that the French military government cast over the Danube valley was not particularly tight. The controls were fairly lax and Heidegger did not experience any difficulties. There was no sense of total collapse. Not a shot had been fired in anger. A solitary tank round had been fired beneath Castle Wildenstein, more for practice or fun, but that was all. The chaos was unfolding elsewhere. But Heidegger's two sons were stationed on the eastern front. Nothing had been heard from them for some time, and the uncertainty about their fate was a great worry to their parents and young wives. The case attracted a good deal of sympathetic interest in view of Heidegger's international standing.

From the religious philosopher Enrico Castelli we learn how Heidegger broke through the carapace of hardness and rigour when the hand of friendship was extended.[130] Castelli travelled through the French zone of occupation in the summer of 1946, visiting Tübingen, which had survived the war intact, and Freiburg, which had been heavily bombed. From there he drove on up to Todtnauberg with a friend on 9 June 1946 to visit Heidegger in his mountain hut. He brought with him a message from the French religious philosopher Jacques Maritain, who was France's ambassador to the Vatican at the time (1945–1948): Maritain offered to intervene through official French and Vatican channels to try and get some news about Heidegger's sons. Frau Heidegger's answer was a polite refusal; she felt there was no point. Castelli surmised that she found it beneath her dignity to accept help from strangers. She also refused to divulge certain information she had about concentration camps. Heidegger himself remained silent. But as he was escorting his visitors on the long walk from the hut over to the mountain station of the Schauinsland cable railway, he said – 'speaking in hushed tones' – that he would welcome such an intervention. He then wrote the details, with his thanks, in the damp earth around the base of a felled tree. Heidegger was visibly moved.

But we have jumped ahead in the chronology of events again, and it is time to return now to the rectorship address of 27 May 1933. There is no doubt that among a whole series of addresses delivered by university rectors throughout the German Reich in May 1933 this one stands alone in the power of its rhetoric and the compelling force of its ideas: a speech that few of his listeners understood at the time, shaped and formed as it was by Heidegger's philosophy and language, which present difficulties enough. A speech that was probably liable – not to say bound – to be misunderstood and misinterpreted by the majority, against the background of this metalanguage: a speech, as Heidegger himself later wrote, that was 'spoken into the wind': not the wind that fills the sails and gets the ship under way, but the wind that blows nowhere, accomplishing nothing. This was no way to get the ship afloat; or was it that the sails were wrongly set, the rigging ill-matched to the wind? Had the philosopher who acclaimed the *Führer* Adolf Hitler, if not explicitly in this address then in many of his other pronouncements, in fact omitted to read his programme, and failed to take on board his political intentions, his political thinking? When questioned by a member of the denazification commission in 1945, Heidegger said that he had read Hitler's book *Mein Kampf* only in part 'because I found its contents repugnant' – to quote his exact words. Heidegger further stated in 1945 that he had realized by mid-June 1933 that political events were not taking the course he had anticipated, by which he presumably meant the course outlined in his letter to Karl Jaspers shortly before Christmas 1932: 'Will it be possible, I wonder, to create a firm foundation and a place for philosophy in the coming decades? Are there men coming who bear a distant dispensation within them?' This secularized Advent expectation, this hope of a secular Messiah, would find its fulfilment in the mission of the *Führer*.

But did not this pseudo-religious hope of salvation conceal a monstrous hubris, and was not Heidegger's thinking in its radically formulated and elemental simplicity merely a straightforward substitute for the Christian world-view he had thrown overboard? Was there not 'a necessary and absolute antithesis to all forms of Christianity' inherent in the '*philosophical* approach to the problem of existence' as taught in *Being and Time* – as Heidegger had put it in a letter to a reviewer of his book, Julius Stenzel, back in 1928:[131] morally neutral, designed only to illuminate the essential modality of existence (*Dasein*) as a 'being-in-the-world'? Had the advent of the *Führer* now brought the forest clearing of Being that much closer? Heidegger undoubtedly thought so. It was not his intention to develop an anthropology or a 'specific system of ethics'; his sole task was to do the bidding of Being by preparing 'a possible basis for a strict interpretation of Being in all its possible variations and regions', as he formulated it in 1928.

Had he, the hidden artisan, who with the hammer of his thought had chipped away the weaker stone and cut through to the hard bedrock in the essentiality of his questioning – had he now been given, in the person of the *Führer* Adolf Hitler, the new hammer that would open up a path to the lodes of Being itself? The metaphor is Heidegger's, used by him, albeit in a different context, in a letter to Stenzel at the end of 1929.

Is it frivolous, or even unseemly, to ask such questions? Nowhere, as far as I can see, have any of the professional philosophers ever offered a satisfactory comparative explanation of the proposition put by Heidegger as *Führer*-rector to the students of Freiburg University at the beginning of the winter semester 1933/34, drawing on the power of his internationally acclaimed thought:

> May you ceaselessly grow in the courage to sacrifice yourselves for the salvation of our nation's essential being and the increase of its innermost strength in its polity. Let not your being be ruled by doctrine or 'ideas'. The *Führer* himself and he alone *is* the German reality, present and future, and its law. Study to know: from now on all things demand decision, and all action responsibility. Heil Hitler!

To those who can decode the metalanguage it is readily apparent – assuming they are not dazzled or blinded – that the proclamations of Heidegger the rector were based on the fundamental arguments put forward by Heidegger the philosopher in 'On the Essence of Truth': 'Let not your being be ruled by doctrine' – referring presumably to ecclesiastical and theological dogmas – 'or ideas', referring presumably to what Heidegger sees as enslavement to Plato's theory of ideas and to the entire Western philosophical and theological tradition. If his students continued to pursue these chimeras, they would lose the essence of 'unhiddenness' (*Unverborgenheit*). Rather: 'The *Führer* himself and he alone *is* the German reality, present and future, and its law.' The verb 'to be', italicized by Heidegger for emphasis, contains within it the message of Being. Here the philosopher-rector – not just any academic figure who happens to be the leader and rector of the provincial university of Freiburg, but a true leader of German science and scholarship, which can only mean philosophy, Heidegger's philosophy – has succeeded in reducing his thought, by a monumental feat of compression and compaction, to a single valid formula. Here is the summons of Being in the forest clearing, the *aletheia*: the locus of truth. In the troubled times before Adolf Hitler, Being itself – and no longer just the entities of being in the world – was called into question for the first time. He who asked such questions was following in the footsteps of Heraclitus, 'the obscure' and yet 'the bright', inasmuch as he illumines the primordial essence of truth in its hidden beginnings.

Attempts have been made, and are made still, to put a different interpretation on Heidegger's words. 'Doctrine' is alleged to refer to the Nazi party program, while 'ideas' refers to Nazi ideology. In support of this view it has been pointed out that Heidegger ended his rectorship address with a quotation from Plato, '*ta . . . megala panta episphale*', which he translated as 'All great things stand fast in the storm'. But Heidegger knew Greek (we are reminded), and therefore knew very well what Plato had really said: 'All great things are destined to fall.' This is used to point out the subtle equivocation, the deliberate ambiguity of Heidegger's formulation. In reality the philosopher, it is alleged, sought to play off Hitler against the Party, which 'brought the wrath' of the Party down upon him, while Hitler himself knew nothing of all this. Such is the line taken, for example, by Walter Bröcker, professor of philosophy at Kiel and Heidegger's former assistant.[132] This of course is a very superficial analysis, which fails to explore the underlying layers. 'Doctrine' and 'ideas' must be seen as references to the traditional ballast of Western philosophy, which now, at long last, can be jettisoned, now and in the time to come, for 'the *Führer* himself, and he alone, *is* the German reality, present and future, and its law'. As for the Plato quotation, Heidegger was simply taking the basic meaning of the Greek ('destined to fall' or 'put to the proof') and adapting it to fit in with the rhetoric of 'struggle' that informs the rectorship address – indeed, using it to dramatic effect as a resounding final flourish.

Most recently Otto Pöggeler has interpreted Heidegger's dictum as an expression of his concrete political beliefs: 'It is not about doctrines and ideas, and therefore not about the National Socialist party programme or racial theories: it is about the chancellor of a national coalition putting himself above his party and thereby becoming the true Leader (*Führer*) of the new national awakening.'[133] But isn't this a rather contrived interpretation, a case of thinking around corners, as it were? No account is taken of the atmospheric affinity with the rhetoric of struggle so prevalent in late 1933. And who cared a fig for any 'national coalition', when so much power was already concentrated in the hands of the Reich Chancellor? Writing from a contemporary perspective, Heidegger's pupil Herbert Marcuse had no difficulty in relating these remarks to what is manifestly their true context, namely philosophy. In his 1934 essay entitled 'The struggle against liberalism in the totalitarian state' he analyses the two November proclamations and sets Heidegger's dictum against a quotation from Hegel:

> Hegel still cherished this belief: 'Whatever in life is true, great and godlike, is all of these things through the Idea ... Everything that holds human life together, that has value and validity, is intellectual and spiritual [*geistig*] in character, and this realm of the intellect and the spirit exists only through the consciousness of truth and right, through the apprehension of Ideas' (From Hegel's introductory address to his students at the start of his course of lectures in Berlin in 1818).

Marcuse was in no doubt that Heidegger, as an exponent of philosophical existentialism, had politicized his philosophy and bound man 'to the *Führer* and the movement that is utterly committed to him', as Heidegger had proclaimed to the students of Freiburg on 10 November 1933. In Marcuse's words, this amounted to 'an act of self-abasement on the part of existentialism that is without equal in the whole of intellectual history'.[134]

Martin Heidegger never retracted these remarks, or indeed any others, throughout the period of the Third Reich. For who could escape the visionary power? When did the Delphic Oracle ever retract one of its utterances? When has a god ever erred, dwelling in the place of Being, sending and conferring upon the nation the destiny of its essential character? If a nation denies its destiny it goes astray, remains in darkness – night begins to fall. But how can any guilt be ascribed to the thinker who has drawn nigh to the dwelling place of Being? Who calls him to account, makes him answer-able – the medium, of whom thought has taken possession?

Karl Jaspers clearly understood nothing of Heidegger's thought, since he repeatedly urged his one-time philosopher friend to confess his sins, like Augustine of old, or to emulate the conversion of Saul. He and others who took the same line failed to comprehend the essence of truth, thought in 1931 and recounted for the first time in Beuron, that place of early and enduring proximity – though not published until 1942: they failed to comprehend because, like Plato's cave-dwellers, they were dazzled by the light and a prey to mere appearances. They did not understand that the sacral dimension was to be revealed – could *only* be revealed – in the 'forest clearing' (*Lichtung*), in the self-disclosing disclosure of Being. Was not the salutation 'Heil!' an invocation of the sacral (*das Heilige*), of what was whole and healed (*das Heile*)? Heil Hitler! It was not a case of having the gall, but of having the right, when in 1947 Heidegger wrote in the *Letter on Humanism*: 'Perhaps the distinguishing feature of the present age lies in the fact that wholeness as a dimension of experience is closed to us. Perhaps this is the only evil.' And that in the wake of the evil that had been visited on mankind in the name of the whole-healing *Führer*!

And what about the essential unity of life and thought in Heidegger, the congruence of the man and the *oeuvre*? Here a somewhat sophistical distinction was drawn between the existential and the existent Heidegger, between one who analyses existence as such, and therefore speaks existentially, and one who exists by the power of decision, analysing the existential. But what happens then to the existent Heidegger and his system of values, which he acknowledges in a spirit of responsibility? Or is there also a sidling-away from responsibility here? Isn't there something almost shady about this? That question will probably never receive a satisfactory answer. The opposing positions are too hardened for that, the distinction between the possessors of the sacral and the secular too sharply drawn.

The story of how Rudolf Bultmann and Martin Heidegger met up again after 1945, having gone their separate ways, is both thought-provoking and amusing. The episode is narrated here in the words of Fischer-Barnicol:

> The two friends didn't meet again until after the war. One day Bultmann received a telephone call: 'It's Martin here'. He was so taken by surprise that he replied: 'Forgive me, but which Martin is that?' Heidegger then explained the reason for his visit: 'I should like to ask your pardon'. The joyful reunion took place. Their former intimacy, their delight in mutual understanding, were spontaneously restored. The day passed in a lively and animated exchange of ideas – just like the old days. Though times were hard and food was in short supply, they ate together and drank each other's health. The past was forgotten. If he had once been drawn to National Socialism for good reasons, they had soon turned to disillusionment. Nothing more stood between us. As we were saying goodbye, I came back to what he had said on the telephone: 'Now you'll have to write a retraction, like Augustine', I said, '. . . not least for the sake of the truth of your thought.' Heidegger's face became a stony mask. He left without another word . . . I suppose one must look for a psychological explanation (See Neske 1977, p.95f.).

Not at all. The explanation lies in two conflicting perceptions and definitions of truth. If the concept of truth is seen in terms of Heidegger's understanding of immanent being, then there is nothing to retract. If we take truth in its traditional sense, which Heidegger equates with simple 'rightness' (*rectitudo*), then Heidegger's withers are unwrung. To that extent, it was inevitable that all the accusations thrown at him from the time of the Freiburg denazification commission onwards would simply bounce off him. Heidegger remained unmoved. When he justified himself – which he did, as we have seen, even from beyond the grave – he stepped on to a kind of intermediate plane, where the distinction between the existent and the existential is blurred, not least in terms of the argument. But let us return to the scene of the action again, examining the events as they unfold and seeking to understand the larger picture.

In the weeks preceding the rectorship address Rector Heidegger was already establishing a new set of ground rules. He had no intention, for example, of convening the Senate to discuss the major problems associated with *Gleichschaltung*. As far as he was concerned, the University had already been 'co-ordinated' and 'brought into line' (*gleichgeschaltet*). It had, in other words, been made subservient to the 'leadership principle' (*Führerprinzip*), to which the only appropriate response was obedience, not corporate democracy – an outdated and crumbling edifice, no longer capable of supporting the weight of the new. For this reason he was quite indifferent to the extreme degree of ill-feeling he provoked, particularly in the Faculty of Medicine and the Faculty of Law and Political Science at the University, which in his view were still stuck in the old well-worn rut of faculty politics, and whose concept of science was consumed in the fiery heat of philosophy as the one true science. In those heady days of May 1933 the economist Walter Eucken (son of the philosopher and Nobel Prize winner Rudolf Eucken) and Wilhelm von Möllendorff were already the main focus of opposition to the new rector. Eucken made his feelings known to Pro-rector Sauer, who noted the incident in his diary: 'He said that Heidegger was acting as though he wanted to run the whole show himself, on the principle of the *Führer* system. He obviously saw himself as the natural philosopher and intellectual leader of the new movement – and as the only truly great thinker since Heraclitus.' Eucken had certainly put his finger on it, characterizing Heidegger's view of himself with remarkable accuracy and defining the extent of his intellectual-political ambitions. As far as the rector was concerned, of course, these were yesterday's men, the denizens of an outmoded world of mere appearances, disengagement and inauthenticity. Heidegger's real aim, made known to the world through his carefully staged entry into the NSDAP, appears from the letter he wrote to the secretary for higher education at the Karlsruhe ministry of education, Professor Fehrle, on 9 May 1933: 'My warmest thanks for your words of welcome upon my entry into the Party. We must now commit all our strength to conquering the world of educated men and scholars for the new national political spirit. It will be no easy passage of arms. Sieg Heil! Martin Heidegger.' Here is the rhetoric of struggle once more: 'commit all our strength', 'conquer', 'passage of arms', and of course 'Hail to victory!' (*Sieg Heil!*). So what Heidegger himself has to tell us about the circumstances and motivation of his entry into the Party can safely be disregarded. The fact of the matter is that he proceeded along strictly tactical lines, delaying his entry into the Party until such time as those colleagues 'whose position is either unclear still or openly hostile' no longer presented any obstacle. The rectorship election was over. The first of May, a national holiday, was a highly appropriate day for a public announcement of Party membership. The importance attached to this new national holiday, the 'day of national labour', by the newly installed rector appears from a circular dated 27 April 1933, in which all lecturers are asked to take part in the public demonstration – 'a dictate of the hour. The construction of a new intellectual and spiritual world for the German nation has now become the single most important task of the German universities. This is "national labour" of the highest kind.'[135]

Other incidents and remarks of a similar nature could readily be adduced, some of them even plainer in their meaning. Of a piece with such thinking, for example, is the text of a telegram Heidegger sent to the former *Reichskommissar* and newly appointed *Reichsstatthalter* (governor), Robert Wagner, in the first few days of May 1933: 'Delighted by your appointment as *Reichsstatthalter*, the Rector of the University of Freiburg im Breisgau greets the *Führer* of our native borderland with a "Sieg Heil!" from a brother-in-arms. Heidegger.' All right, many will say: but how many people in positions of authority in those heady days of revolution did *not* deliver themselves of similar verbal torrents! This applied even to the highest Church dignitaries, as Heidegger angrily pointed out in that 1950 draft letter to the editor of the *Süddeutsche Zeitung* that we have already had occasion to look at – a comment aimed primarily at his Messkirch compatriot and loyal patron, Archbishop Gröber. This was undoubtedly true. And let it be said again that it is not a matter of dragging a scholar before the tribunal of a later generation, but simply and solely of making clear the line of defence and drawing attention to the problem of responsibility in a philosopher of worldwide renown whose apologia is offered for worldwide consumption. In this we should be guided by the sound scholastic principle that states: 'contra factum non valet argumentum' – '(mere) argument shall not prevail against solid fact.'

This telegram to *Reichsstatthalter* Wagner was addressed to the man responsible for the Baden decree against the Jews, to which, as has already been indicated, a number of professors at Freiburg University fell victim, including several retired professors against whom further discrimination was quite pointless. Among the non-Aryan *emeriti* was Heidegger's own predecessor and teacher (however we choose to assess the teacher-pupil relationship), Edmund Husserl, a thoroughly patriotic German, whose two sons had volunteered for the student batallion at Langemarck. One son, Wolfgang, who held the rank of Lieutenant, was killed in action in 1916. The other, Gerhart, ended the war as a highly decorated officer, returning home from the front seriously wounded. In 1933, as a professor in Kiel, he too was suspended from his position, and later dismissed in accordance with the infamous Reich law for the re-establishment of a permanent civil service.

Edmund Husserl and Martin Heidegger

A personal and political profile

This seems to me the right moment to examine in more detail the relationship between Husserl and Heidegger: for I have the impression that this is an issue of considerable concern to many people, particularly since there are so many opinions, rumours and conjectures on the subject – few of them inspired by an academic interest in the history of philosophy. The year 1933 may be viewed as a kind of mirror or lens, which serves to focus both the years immediately preceding and the period that followed. We will begin with a statement made by Heidegger in the autumn of 1945 to the chairman of the denazification commission, concerning his conduct towards Husserl:

> The allegation that as rector I banned Husserl from the University and the library is a particularly vile calumny. I never ceased to look upon Husserl with gratitude and respect as my teacher and mentor. It is true that my philosophical studies moved away from his position in many respects, with the result that Husserl himself attacked me publicly in 1931 in his great speech in the Berlin Sportpalast.[136] So the ties of friendship had begun to slacken long before 1933. When the first law against the Jews was passed in 1933, which deeply shocked both me and many others who were favourably disposed towards the National Socialist movement, my wife sent a bouquet of flowers to Frau Husserl, together with a letter – in both our names – that expressed our undiminished respect and gratitude, and condemned the harsh measures

> against the Jews. When a later edition of *Being and Time* was in preparation, my publisher wrote to tell me that it could only be published if the formal dedication to Husserl was dropped. I agreed to this, on condition that the substantive dedication to Husserl on page 38 of the text was retained – which is what happened. When Husserl died I was ill in bed. When I had recovered, however, I did not write to Frau Husserl; and in that I was undoubtedly remiss. The reason for this omission was the bitter sense of shame I felt about what was now being done to the Jews – far beyond the scope of that first law – and which one was powerless to prevent.

It is interesting to note Heidegger's use of language here: 'one' was powerless to prevent it . . . But the truth is revealed when we compare this with the telegram sent by Heidegger in early May 1933 to *Gauleiter* and *Reichskommissar/Reichsstatthalter* Wagner, who alone was responsible for the Baden decree against the Jews.

In the *Spiegel* interview of 1966 Martin Heidegger gave a more detailed account of this episode. This handful of sentences, in which central issues are discussed alongside more trivial matters, covers an extended period of time. To deal with the more trivial matter first: the allegation that Heidegger banned Husserl from the library and university premises is indeed false – 'a particularly vile calumny', as he put it in 1945. This version of events has been reiterated so often that it has become a kind of topos. The historian and publicist Golo Mann, for example, in his *Thoughts and Reminiscences*, tells the story of how he and his dissertation supervisor, Karl Jaspers, fell out in 1963. In an effort to avert the impending breach, Golo Mann sent Jaspers a bouquet of flowers with a note asking if he could come and visit him. The philosopher's response was brusque: 'He said he did not want to see me. My flowers, he said, reminded him of the ones Martin Heidegger sent to his teacher Edmund Husserl when Heidegger was rector of Freiburg University, on the day he banned Husserl from the university library.' To which Golo Mann rejoins: 'This insult I could only take to mean the end of our friendship of more than thirty years' standing.'[137]

The allegation that Heidegger banned Husserl from the university and the library during his rectorship has been repeated over and over again, even in serious scholarly publications. Even Golo Mann does not for a moment question the authenticity of the story; he simply takes exception to the parallel drawn by Jaspers, which he regards as an insult. So what is the real story behind the infamous library ban and the bouquet of flowers?

Let us be quite clear about one thing: as rector and head of department Heidegger did not issue a ban of any kind on the use of the university library or the departmental library. This oft-repeated charge is without foundation.[138]

The question of the bouquet of flowers and a letter is rather more complex. Here we have two separate strands of tradition to draw on, so to speak: on the one hand we have the direct, contemporary evidence of Husserl's wife Malvine, who recounts the incident in a letter of 2 May 1933[139]; on the other hand we have Martin Heidegger's own version of events, represented in the first instance by the very early account compiled at his instigation by Alfred (Frédéric) de Torwanicki, which appeared under the title 'Visite à Martin Heidegger' in Sartre's journal *Les Temps Modernes* (1945/46) in conjunction with Heidegger's first attempts to defend his actions after 1945; and subsequently reiterated in Heidegger's own words in the *Spiegel* interview of 23 September 1966, which was published after his death on 31 May 1976.

In a private letter dated 2 May 1933 Malvine Husserl states that she has received a letter from Elfride Heidegger 'that upset us a good deal. She feels compelled to tell us (in these difficult weeks), speaking on his [Heidegger's] behalf as well, how they still remember with undiminished gratitude everything that has gone before, etc.' A translation of part of this same letter from Elfride Heidegger can be found in de Torwanicki's article (taken from a copy that the Heidegger family had kept, and no doubt still has in its possession, the original having been destroyed by fire during military action in the port of Antwerp in 1940, along with many other Husserl family papers).

This letter from the Heideggers reached the Husserls on 1 May 1933, a Monday, having been sent on to them in Orselina near Locarno, where the couple had travelled on Saturday 29 April 1933. The letter, sent to accompany the flowers – the first reference to the flowers appears in de Torwanicki, who says they were sent, 'd'une manière officielle', on 23 April 1933) – probably arrived at their home, Lorettostrasse 40 in Freiburg, on the day of their departure, so was very likely written the day before on 28 April (the *Spiegel* interview talks of May). The flowers that came with the letter had no doubt withered by the time the Husserls returned from Orselina in mid-May. At all events, they make no mention of flowers.

From the above-mentioned documents we can reconstruct the general tenor, and in places the actual text, of Elfride Heidegger's letter as follows: although her husband had gone his separate way in philosophy, he would never forget what he, as Husserl's former pupil, owed to his mentor. Nor would she, Elfride Heidegger, ever forget the kindness and friendship Frau Husserl had shown them in the difficult years after the First World War. It had pained her that she had not acknowledged this debt of gratitude in recent years, because of the misunderstandings fomented between their two families by others. She had been shocked to read that their son, Gerhart Husserl, had been suspended, and she hoped that it was just a temporary measure, the work of some over-zealous bureaucrat in the general confusion of the times – like the similar abuses that occurred during the revolutionary weeks of 1918. After all, the Husserl family had been staunch patriots during the First World War. Such was the gist of her letter.

Anyone familiar with the warm and friendly relations between the two families, which had grown steadily closer since 1918 until the tone was one of easy intimacy, particularly between the two wives, will be taken aback by the formal, almost stilted style of this letter, which also omits any reference to the important fact of Edmund Husserl's own suspension from office. Instead the writer chooses to focus on the action taken against Husserl's son Gerhart, who was teaching in Kiel as a professor of law – only to make light of it as an error on the part of some jumped-up official: 'the old story', as Malvine Husserl notes of this passage.

The early personal intimacy had grown steadily more distant after Heidegger's appointment to the chair at Freiburg in 1928, which Husserl had pushed through with single-minded determination, until by 1930 all real contact between the two families had effectively ceased. This has been established beyond a doubt by a number of documents published since.

So what prompted this letter now, complete with its reference to the patriotic credentials of the Husserl family, who – heaven knows – had paid their dues in blood during the First World War? Let me explain. In Baden – ever the model province – on 6 April 1933, the day before the infamous Reich law 'on the re-establishment of a permanent civil service' (directed primarily against Jewish officials) came into effect, the then *Reichskommissar* and *Gauleiter* of Baden, Robert Wagner, issued a special decree (on what legal basis, nobody knows) whereby all civil servants 'of non-Aryan origin', irrespective of religious denomination, were to be suspended from office – including, absurdly, those who were already living in retirement. As far as Freiburg University was concerned, this last dispensation affected the two emeritus professors Edmund Husserl and Otto Lenel (the famous teacher of Roman law). At the same time the organs of self-government and administration at Baden's universities were to be made *judenfrei*: all Jewish deans and senate members were to be dismissed from their posts.

At first there was great confusion at Freiburg University over the significance of this decree, which many felt should not really be taken seriously. But the ministry of education insisted on its implementation. And so on 14 April 1933 Edmund Husserl was formally notified in his turn of his enforced 'leave of absence', which created an enormous stir, especially beyond the borders of the German Reich. A few weeks later Husserl wrote that he regarded this suspension as the supreme affront of his life. 'I fancy I was not the meanest among Germans (of the old style and of the old school), and that my house was a place of true national sentiment, as evinced by *all* my children, who volunteered for service in the field or (in Elli's case) in a military hospital during the war.'

It was partly because of this intolerable action that von Möllendorff, the rector originally elected for the academic year 1933/34, had resigned his rectorship after just a few days. Having succeeded him, Heidegger now had to administer decrees of this kind. On 28 April

1933 the *Reichskommissar*'s decree of 6 April was rescinded because it had been superseded by the national Reich law already referred to – and in fact constituted an infringement of that law. Rector Heidegger knew about this on 28 April, whereas those who were directly affected, including Husserl, did not learn of it until later: it was only when he returned from Orselina in mid-May that Husserl found the letter of notification (that the decree had been rescinded) waiting for him.

However, on 28 April, the day his wife wrote to the Husserls, Heidegger knew that he would be joining the NSDAP on 1 May, the 'Day of National Labour', thereby making his commitment spectacularly public. The date and time of his formal entry into the Party had already been discussed at the beginning of April by the cadre of National Socialist professors referred to above, who had chosen Heidegger as their spokesman – though the final decision rested with the ministry in Karlsruhe.

Such were the events surrounding the bouquet of flowers and the letter from Elfride Heidegger. That the Husserls were quite taken aback is apparent from a letter that Edmund Husserl wrote to his pupil Dietrich Mahnke from Orselina on 4 May. The events of the last few months and weeks, he wrote, had struck at 'the deepest roots' of his existence. Of all his former pupils Heidegger had disappointed him the most, not least by the way he 'broke off all relations with me (and very soon after his appointment)', despite the years of trust Husserl had placed in him. 'The perfect ending to this supposed bosom friendship between philosophers was his very public, and very theatrical, entry into the National Socialist Party on 1 May.'

Anyone familiar with Heidegger's version of the alienation between himself and Husserl, as offered for public consumption in the *Spiegel* interview, will now be able to read between the lines. The only surprising thing is that Malvine Husserl bothered to reply at all. Heidegger's attribution of blame to the Husserls (claiming that it was Frau Husserl's letter in response to their own that ended their relationship) is a facile evasion; it may be true in a strictly formal sense, but that is all. So how did the story about Heidegger banning Husserl from the library come about? I simply don't know.

For his part, Karl Jaspers knew perfectly well that Heidegger had not behaved in the manner alleged. When Hannah Arendt sent Jaspers the English version of her essay 'What is Existenz Philosophy?' [*sic*] in the summer of 1946, he corrected the story repeated there, to the effect that Heidegger had forbidden his 'mentor and friend' Husserl to set foot in the Faculty because he was 'a Jew': 'What you say about Heidegger does not quite square with the facts. I suspect that the letter to Husserl was simply the letter that every rector in those days was required to write to persons dismissed by the regime' (Letter of 9 June 1946).

It is possible that Jaspers then came across the essay by Alfred de Torwanicki, which is the only place, so far as I know, where the story about the flowers is recorded prior to the *Spiegel* interview. Did Jaspers later weave together different strands from different sources? After all, 'stories' as such can become a constituent part of reality – perhaps mirroring the truth in a deeper sense.

The flowers sent to Frau Husserl by Frau Heidegger make sense only in relation to the offensive treatment meted out to Edmund Husserl in his retirement by the Baden decree against the Jews of April 1933, and the subsequent withdrawal of the decree on 28 April 1933. Whatever we choose to make of the motives, the fact remains that the Heidegger family, after more than fifteen years of friendship, wished to signal a degree of solidarity in the face of changing fortunes and estrangement. As for Heidegger's explanation of the decision to drop the dedication from the fifth edition of *Being and Time* in 1941, there is little that one can add to that. He need not have complied, after all. The fourth edition, which appeared in 1935, still contains the formal dedication at the beginning:

Dedicated to Edmund Husserl
in respect and friendship.
Todtnauberg, in the Black Forest, 8 April 1926.

Edmund Husserl had celebrated his sixty-seventh birthday on 8 April 1926. When the fourth edition appeared the infamous 'Reich citizenship law', approved at the Nuremberg 'Party Conference for Freedom' (*Reichsparteitag der Freiheit*) on 13 September 1935, had not yet been invoked against the distinguished phenomenologist. By the end of 1935, when the new law had been fully implemented, Husserl had become a 'non-person' within the university community: in the mealy-mouthed legal jargon of the day the professor, 'having been released from his official responsibilities', was notified of the withdrawal of his teaching rights. The Freiburg University authorities forwarded the order to Husserl, as they were obliged to do. There was no accompanying letter, not even a gesture of regret or solidarity: doubtless it was deemed unwise in the prevailing political climate. From the summer semester of 1936 the University no longer included Husserl's name on the lecture list. The venerable institution was thus absolved from any obligation to commemorate the death of Edmund Husserl in April 1938.[140]

Heidegger associated himself with this expunging of Husserl's memory, or at least he submitted to the order. His action sent a clear message to others, even though Heidegger himself repeatedly pointed out that the passage on page 38 of the first edition of *Being and Time*, in which he explained and justified the formal dedication to Husserl at the beginning of the book, remained unchanged in the later editions: 'If the present study goes some way towards an elucidation of "the things themselves", the writer's thanks are largely due to *E. Husserl*, who by a combination of forceful personal direction and the most generous indulgence acquainted the writer during his apprenticeship years at Freiburg with the most diverse areas of phenomenological research.' Heidegger subsequently recited this passage verbatim in the *Spiegel* interview.

The human failure he displayed during the long years of Husserl's lingering illness and at the time of his death and burial is bad enough, and nothing can excuse it. How very different, by contrast, was the solicitous concern that Edith Stein showed for Husserl's well-being during the final phase of his life, from the enclosure of the Carmelite convent in Cologne.

This brings us to the real heart of the matter. Earlier we recounted in some detail the growing rapprochement between Heidegger and Husserl, particularly in 1917, noting how Husserl took thought for his material needs as well – like a father, in fact; what hopes he cherished for Heidegger, and how he worked to secure Heidegger's appointment to the chair at Marburg. We know from many contemporary accounts that Husserl saw Heidegger as his disciple, who alone was capable of carrying on his philosophy and raising it to new heights. He made available the pages of his *Jahrbuch für Phänomenologie und phänomenologische Forschung* for Heidegger's principal work, *Being and Time* (which appeared in 1927 as Volume VIII of the journal, as well as coming out simultaneously in a separate edition), and he was in no doubt that Heidegger would succeed him in the chair at Freiburg: in 1928 Heidegger's was the first and only name on the list of candidates. As the outgoing professor Husserl personally drafted and formulated the Faculty's recommendation – a most unusual procedure – and on 21 January 1928, immediately after the decisive meeting of the appointments board, Husserl wrote in triumph to Heidegger in Marburg: 'My dear friend! The board's decision: *unico loco*. Not a word to anyone, of course.' Not long afterwards, Edmund Husserl informed him that his draft recommendation for submission to the ministry had been approved. Everything had 'worked out very well' for Heidegger. In short, Husserl was passionately concerned to ensure that Heidegger and Heidegger alone would be appointed to the chair at Freiburg.

Husserl was warned about Heidegger often enough: 'Heidegger's phenomenology is said to be totally different from mine; his university lectures and his book are not so much a continuation of my scholarly work as a series of open or covert attacks aimed essentially at discrediting it. When I put this to Heidegger in a friendly kind of way, he just laughed and said "Nonsense!"'. Writing here to Alexander Pfänder on the feast of Epiphany in 1931, Husserl takes bitter stock of a long period of hopeful collaboration, academic friendship and indeed fatherly feelings. Heidegger is seen as a whole chapter, so to speak, in Husserl's life-story. He had cast himself as Husserl's devoted pupil and future colleague, sharing common ground with Husserl's constitutive phenomenology in all essentials of method and approach. 'The growing impression I had of this young man's extraordinary talent, his absolute devotion to philosophy, his enormous intellectual energy, led me in the end to a rapturous assessment of his future significance for a scientific phenomenology along my own lines.' Ultimately the only explanation

Husserl could offer for his childlike trust was 'blind infatuation'. This was why Martin Heidegger was the only candidate considered for the chair. But disillusionment soon set in after Heidegger's appointment in 1928: 'After he took up his post our regular contacts lasted for about two months; then it all came quietly to an end. He simply kept out of the way, avoiding any opportunity for philosophical discussion, which he obviously regarded as unnecessary, unwelcome and unsettling. I see him once every few months, less often than my other colleagues.' Husserl goes on to say that he has only recently made a close study of *Being and Time* and Heidegger's later writings:

> I came to the sad conclusion that as a philosopher I have absolutely nothing in common with this Heideggerian profundity, this inspired disregard for scientific rigour – that Heidegger's overt and hidden critique is based on a gross misunderstanding, and that he is embarked on the development of a systematic philosophy of the kind that I have always thought it my life's work to render permanently impossible. Everyone else realized this a long time ago – everyone except me. I have not concealed my conclusions from Heidegger. I am not passing judgement on his character, which has become totally incomprehensible to me. For almost a decade he was my closest friend, but that is all over, of course; you cannot be friends with someone you do not understand. This turnaround in my scientific judgement of the man and in my personal relations with him has been one of the bitterest blows of my life.[141]

Conflict between the generations, the problems of the teacher–pupil relationship, opposing schools of thought and much more besides: this is nothing new in the history of scholarship. Likewise the deep disappointment, indeed the hurt, suffered by a teacher. Rather more problematic are the human aspects revealed here. They do nothing to explain Heidegger's behaviour after 1933, when he no longer had any particular occasion to consort openly with Husserl, by then politically outlawed as a Jew. The ageing phenomenologist could have had no inkling that for years now his 'pupil' Heidegger had been speaking and writing of the master, his methods and his 'scientific' philosophy with biting sarcasm – among close friends, that is. When and if we ever have free access to Heidegger's correspondence, the testimonies to this will be legion. Here is one example, taken from a letter to Jaspers dated 14 July 1923, in which he acknowledged Jaspers' congratulations on his appointment to Marburg:

> There's a lot of idolatry to be rooted out. The various medicine men among today's philosophers must have their miserable and ghastly trade unmasked once and for all – and we must do it during their lifetime, so that they don't get the idea they have ushered in the kingdom of God on earth. I expect you've heard that Husserl has been offered a post in Berlin; he's acting worse than a junior lecturer who regards a full professorship as the nearest thing to paradise. It's hard to make out what is going on, but at the moment he sees himself in the role of *praeceptor Germaniae*. Husserl has completely lost his marbles – if he was ever 'all there' to start with, which I've begun to doubt more and more lately; he trots around from pillar to post, uttering banalities to make one cringe. He's obsessed with his mission as the 'founder of phenomenology'. Nobody knows what that is – anyone who spends a semester here soon gets the picture. He is beginning to realize that people are not going along with him any more; he says it's too difficult, of course, obviously nobody understands a 'mathematical theory of ethics' (that's the latest thing!) – even someone who is further on than Heidegger. And now he's telling people that it's because I had to start giving lectures of my own straight away that I couldn't go to his, otherwise I *would* be further on than I am ... And *that's* the man who wants to save the world in Berlin! This kind of atmosphere wears you down even if you stay completely out of it.

Thumbnail sketches in this vein, dashed off with careless abandon, occur frequently in the letters to Jaspers. Heidegger wrote this kind of thing during the period when he was doing everything he possibly could to win Karl Jaspers' friendship (in the same letter he said:

'it is time for our friendship to take concrete shape'), and indeed to develop it into a 'fighting alliance': 'the growing certainty of a fighting alliance in which both "sides" are equally sure of themselves' (in a letter of 19 November 1922), which will serve as a point of departure for the total transformation of philosophical thought at the universities. 'And the more organic, concrete and discreet the manner of its overthrow, the more enduring and certain will that overthrow be' (14 July 1923). Husserl had been a stepping-stone on the way to the top; he had served his purpose, now he could be discarded. Jaspers was next in line. But how did Jaspers characterize Heidegger's brand of friendship in his memoirs? 'He appeared to be a friend who betrayed one in one's absence, but who was unforgettably close in moments that were of no consequence in themselves' – this observed with the trained eye of the psychiatrist. He goes on:

> Only a friendship in which there is no holding back, no hidden reserves, in which dependability in simple matters of right and wrong is the norm, in which there is loyalty in thought and word and deed, can acquire a solidarity that is proof against the slings and arrows of public life. The fact that such a friendship did not grow up between us is not a matter for reproach on either side. It has fostered an atmosphere full of unresolved and ambiguous possibility.[142]

Husserl would likewise have found it hard to imagine that *Being and Time*, the printed sheets of which were lying on his desk in the autumn of 1926, was written 'against' him – and quite consciously so: 'If the treatise is written "against" anyone it is written against Husserl, who realized it straight away, but chose from the beginning to concentrate on the positive side' (letter to Jaspers, 26 December 1926). Be that as it may, this was the emancipation of a brilliant pupil from the spell of an imposing, dominating teacher, a clearing kick into the vast open spaces of thought. We are brought back to the world of practical, everyday reality, the world of human relationships, in which matters had been greatly complicated by the prominent political stance taken by Heidegger as rector. Let us return to our account of the situation during the spring months of 1933.

Among the few who now remained loyal to the Jew Edmund Husserl was the philosopher Dietrich Mahnke in Marburg, who for some years had been Heidegger's closest colleague at that university. In the difficult weeks of April and May 1933 he gave much succour to the elderly Husserl couple when it was a matter of looking for a possible opening at another Prussian university for their son Gerhart Husserl, who had been suspended from his post at Kiel. For the Christian-Albrechts University at Kiel was being groomed as a model academy for the elite of the 'movement'. Non-Aryan professors and those who were politically unreliable no longer had any place there. We know from the reminiscences of Gerhard Leibholz (*'Als es umschlug an den deutschen Universitäten'*: 'The change of climate in the German universities') that the University of Göttingen fought tooth and nail against the possibility of having Gerhart Husserl 'shunted off' to Göttingen.[143] So it is hardly surprising that the Husserls were touched by gratitude for these small marks of loyalty and friendship: 'Allow me to tell you first of all that we are both deeply touched by your heartfelt understanding of our situation. The only thing that can sustain one today is the loyalty of those precious few who openly declare their support for those who have been sorely afflicted'. So wrote Malvine Husserl to Dietrich Mahnke on 18 May 1933, ending her letter with these words: 'On top of everything else we have also had some bitter personal experiences. One of the most distasteful has been our mutual friend (or rather enemy) H. There's a lot more I could tell you – but only by word of mouth.'

Shortly before this, in the long letter of 4 May 1933 that was cited earlier, Edmund Husserl had once again taken bitter stock of his situation (following the earlier letter to Alexander Pfänder written in January 1931). In a fundamental review, looking at his own work in philosophy and the development of his pupils, he writes. 'Others, however, have turned out to be a great personal disappointment – the most recent, and the most hurtful, being Heidegger; what hurt me most was the fact that I had put my trust, for reasons I no longer fully understand myself, not just in his talent, but also in his character.' And now, on 1 May, Heidegger joined the Party in a manner that Husserl found 'very public, and very theatrical':

> Before this he broke off all relations with me (and very soon after his appointment) and in recent years has allowed his anti-Semitism to come increasingly to the fore, even in his dealings with his group of devoted Jewish students and his Faculty colleagues. That was hard to swallow. What was also hard to take was the way Heidegger and the other proponents of 'Existenz' philosophy – largely derived from caricatured versions of the ideas contained in my writings, lectures and personal teachings – twisted the radical scientific purport of my life's work into its very opposite, damning that work by praising it fulsomely as something that had been entirely superseded, something that it was quite unnecessary to study any more . . . But the events of the last few weeks and months have struck at the deepest roots of my existence.

And Husserl the patriot, whose family had paid the price in blood in the First World War, and for whom the public commemoration of Langemarck was not some nationalistic abstraction but a time of painful mourning for the loss of a son, wrote these prophetic words: 'The future alone will judge which was the true Germany in 1933, and who were the true Germans – those who subscribe to the more or less materialistic-mythical racial prejudices of the day, or those Germans pure in heart and mind, heirs to the great Germans of the past whose tradition they revere and perpetuate.' Husserl was isolated, an outcast from the community, 'excluded from the beloved collective of the nation': a difficult and testing time. 'For the moment I am, so to speak, a refugee from Germany.' Here in the Ticino he was free to 'be a German again', instead of being branded as a 'Jewish intellectual'.

In the light of such remarks, Heidegger's attempt to justify his behaviour towards his predecessor in the chair of Philosophy (I) pales into insignificance – a halting catalogue of feeble excuses, scraped together from the wreckage that littered his path. And the pathetically small number of Faculty members who attended Edmund Husserl's funeral in April 1938 demonstrates how very few dared to show their last respects to a man who had been outlawed by Hitler's Germany.

With what tender concern, by contrast, did the Carmelite nun Edith Stein, under her religious name of Sister Teresa Benedicta of the Cross, sustain her revered master Edmund Husserl in his affliction – deeply inspired, of course, by her monastic conviction and calling. As a Jew she understood what it was like for a Jewish professor to have to live out his last days in 1937/38. And for Husserl it would undoubtedly have been worse still had it not been for another of his loyal pupils, Dr Adelgundis Jaegerschmid, a Benedictine nun from the convent of St Lioba in Freiburg-Günterstal, who also became a spiritual companion to him and a support for his wife. Edith Stein wrote to Sister Adelgundis from the Carmelite convent in Cologne on 15 May 1938: 'I don't know anything about the funeral. There was no mention of it in the notice. I wonder how the University reacted? And what about Heidegger?' Her dark suspicions proved well founded. Edith Stein and Dr Adelgundis Jaegerschmid had been friends since the time they studied together under Husserl at the end of the First World War, and had remained close over the years.

The rector's dilemma: caught between Scylla and Charybdis

Heidegger's anti-Semitism, raised as an issue by Husserl, is substantiated by the latter's observation that it had 'come increasingly to the fore in recent years', even in his dealings with his own Jewish students and Jewish Faculty colleagues. Here we are stepping on to controversial, not to say shaky, ground. Whether, as has been maintained,[144] it is 'foolish' to inquire into allegations of anti-Semitism, on the grounds that this only muddies the waters and makes it impossible to define the place of politics in Heidegger's thought, is a point of view on which I forbear to comment. To my mind the issue is by no means unimportant, given the catastrophic consequences of the anti-Semitism that was an integral part of National Socialism. And anyone who cast Hitler in the role of the great leader figure sent by Being was bound to take account, at the very least, of the man's appalling anti-Semitism.

One thing seems certain: if Heidegger subscribed to anti-Semitism, it was certainly not on the basis of the crude racial ideology embodied in Hitler's *Mein Kampf*, Rosenberg's world-view or Streicher's antics. Heidegger was too cultivated a man for that. How else could he have maintained a special relationship with the Jew Hannah Arendt over such an extended period of time, in flagrant breach of bourgeois convention? How else could he have brought in the Jew Werner Brock from Göttingen to be his personal assistant, because he thought highly of his abilities? Although he could not keep his assistant when he became rector, he did at least help him to make a new start in England. The available sources do not present a consistent picture. On the one hand we have to take seriously the statement from Husserl, who had access to inside information and was a contemporary witness; likewise the vigorous assertions, both spoken and written, of Frau Edith Eucken-Erdsiek, another contemporary witness who has testified to the anti-Semitism of Heidegger and his family. On the other hand, as we shall see later, Heidegger did stand up for threatened Jewish colleagues when he was rector, albeit more out of concern for the possible political repercussions. His claim in the foreword to *Facts and Thoughts* that he banned the display of the 'Jewish poster' within the University is less than specific. Exactly what kind of poster it was remains unclear, but Heidegger was probably referring to the poster sent to him by the leaders of the German student body on 12 April in connection with the forthcoming campaign 'Against the un-German spirit', for which preparations had been under way since the beginning of April. This was a large white poster measuring 47.5 cm by 70 cm, on which the twelve theses of the German student body were set out in bright red Gothic script under the heading 'Against the un-German spirit'. Thesis No. 7, for example, read as follows:

> We regard all Jews as foreigners, and we intend to take our national identity seriously. We therefore exhort the censorship authorities to allow Jewish works to appear only in Hebrew. If they appear in German, they must be clearly labelled as translations. We want tough action against the abuse of German or Gothic script: German script is only for the use of Germans. The un-German spirit must be rooted out from all our public libraries.[145]

The precise significance of Heidegger's ban – which was quite possibly justified on aesthetic grounds alone – must remain an open question for the moment. As long as the rectorship papers in the Freiburg University archive are not fully catalogued and access to them is blocked by data protection legislation designed to safeguard the privacy of the individual, it is impossible to establish the facts of the case. When and if these obstacles are removed, it should also be possible to determine how much truth there is in the accusation levelled by the denazification commission in 1945 that Heidegger shielded those responsible for looting the Jewish student fraternity building (meaning the students); or in the related charge that he sanctioned the display of another poster encouraging students to inform on their fellows.

Heidegger linked his ban on the display of the 'Jewish poster', pronounced on his second day in office, with the emergence of early differences between himself as rector and the leaders of the National Socialist students. The dispute, he says, was referred to the highest levels in Berlin and became something of a trial of strength. These claims are difficult to understand, given his very close contacts with the leadership of the German Student Union, which was radically anti-Semitic and anti-Marxist in complexion, and whose leader Gerhard Krüger, who was on friendly, not to say intimate terms with Heidegger, had been quietly planning the campaign 'Against the un-German spirit' behind the scenes. It culminated in the nationwide book-burning on the night of 10 May 1933. In Freiburg too the bonfires burned that night on the square outside the University Library. Rector Heidegger had done nothing to prevent this act of barbarism; there was nothing he could do. The Italian philosopher Ernesto Grassi, who was living in Freiburg at the time (we shall encounter him again later) was a member of the circle around Heidegger, and a close friend of Wilhelm Szilasi. Recording his memories of Szilasi later, he writes: 'Then the years from thirty-three onwards, suddenly bursting upon us with destructive force: Heidegger's rectorship, his inaugural address; the burning of Jewish and Marxist books and the writings of "corrupting" science under his rectorship. The fires were burning in front of the University Library' (Quoted in Grassi, 1970). I have spoken to a number of eye-witnesses who have corroborated this account. The only problem was that it rained on the evening of 10 May 1933 – unpropitiously for the *auto-da-fé*. This conflicts with Heidegger's unequivocal claim, published in the *Spiegel* interview, that he had banned the projected book-burning.

At any rate, a few weeks later he sent copies of his rectorship address to the leadership of the German Student Union, to be distributed (according to Heidegger's accompanying letter) to Gerhard Krüger, Georg Plötner, head of the Central Office for Political Science, Andreas Feickert, who subsequently became Reich leader of the German Student Union in 1934, and Hanskarl Leistritz, head of the 'Academic Scientific News Service' and the Central Office of Information and Advertising. Heidegger must have known that it was

Leistritz who directed the lighting of the bonfires in Berlin, at the infamous burning of the books on the Opernplatz. It was in all the newspapers the next day.

I myself know of only one incident where there is firm evidence that Heidegger was prepared to use the jargon – and specifically the anti-Semitic jargon – of National Socialism to smear a political opponent. Some time in 1934/35 Marianne Weber, the widow of Max Weber, sent Jaspers a copy of a report on Eduard Baumgarten that Heidegger had drawn up on 16 December 1933 for the attention of the League of National Socialist University Lecturers in Göttingen. The background, in brief, was as follows: in 1931 Baumgarten had applied for the position of Heidegger's assistant, but Heidegger had passed him over in favour of the Jewish philosopher Werner Brock, making arrangements for Brock's habilitation qualification, obtained in Göttingen, to be transferred to Freiburg. The effect of the report in question was to block Baumgarten's academic career for the time being. According to the copy made by Baumgarten from the files of the Göttingen League of National Socialist University Lecturers, Heidegger's report contained the following remarks:

> Here at any rate [i.e. in Freiburg] Baumgarten was anything but a National Socialist. In terms of family background and intellectual sympathies his roots lie in the Heidelberg circle of liberal-democratic intellectuals around Max Weber. Having failed to secure an appointment with me, he established close contact with the Jew Fraenkel, who used to teach in Göttingen and has now been dismissed from here. Through him he managed to find a place in Göttingen ... It is too early to reach a final verdict on him, of course. He could still develop. But a decent probationary period needs to elapse before he can be permitted to join any organization of the National Socialist Party.[146]

The Eduard Fraenkel mentioned here was a full professor of classical philology at the University of Freiburg up until the summer semester of 1933. Given the history of the documentary record, the view has been expressed, and is reiterated still by certain writers, that this report is a forgery, prepared with the malicious intention of bringing the philosopher Heidegger into disrepute: a typical manifestation of the 'desperate search for documents of any kind, driven by hatred of the great thinker'.[147] Certainly Jaspers was profoundly shaken; indeed, this revelation in 1934/35 finally opened his eyes to Heidegger, whose political activities to date he had continued to attribute to pure naïvety. He confronted Heidegger with this in his first letter of 1949: 'It was not only the way you broke off all contact after 1933, but also, and chiefly, your report on Baumgarten, a copy of which I saw in 1934. That moment was one of the most distressing experiences of my life. My own feelings of dismay were inseparable from the objective significance of the incident.' It was doubtless because he did not want to start off on the wrong foot that Jaspers did not press for an explanation from Heidegger of the phrases 'the Jew Fraenkel' and 'the circle of intellectuals around Max Weber', of which Jaspers himself was of course a member. Jaspers had already recounted the Baumgarten affair at length in his 1945 report, the contents of which were known to Heidegger, and the case had been discussed by the denazification commission: so the authenticity of the documentary evidence cannot be doubted. If there had been a forgery, or even a serious distortion of the original report, Heidegger would have set the record straight. The fact remains, therefore, that in making his report to the League of National Socialist University Lecturers the rector had used Baumgarten's 'Jewish affiliation' as an argument against him; in other words, he had made a verbal obeisance to anti-Semitism. How many politically coloured reports Heidegger addressed to party agencies in the course of the following years is not known. But we shall have occasion to look again at the practice of political vetting when we come to consider the cases of the two Catholic lecturers in philosophy, Max Müller and Gustav Siewerth.

Commentators have continued to puzzle over the vexed question of Heidegger's anti-Semitism. A few years ago, in a piece that appeared in the arts pages of the *Frankfurter Allgemeine Zeitung* under the rubric 'Faktum' (2 January 1984), it was alleged that Heidegger had sent a copy of his rectorship address, personally inscribed, to his fellow philosopher Richard Kroner in Kiel, knowing full well that Kroner was a Jew. 'What prompted Heidegger, normally so cautious, to send his political speech to the Jew Richard Kroner?' asks the writer, having regard to the fact that Kroner 'was in a particularly vulnerable position? Was he confused, outraged, saddened? Did he feel provoked?' The *FAZ*'s contributor sums up the problem thus: 'If Heidegger did not regard Kroner as a Jew in Hitler's sense, what was it they had in common that would account for the dedication of that speech?' Richard Kroner, who was forced to leave Kiel, then being groomed as the elite academy of National Socialism, to be shunted off to Frankfurt like Gerhart Husserl, had been Heidegger's rival in 1923 for the associate professorship at Marburg, but was only placed third on the short-list of candidates. To find the answers to such questions it may be helpful to look at what Heidegger wrote to Jaspers in July 1923, characterizing his defeated rival Richard Kroner after his acceptance of the Marburg appointment: 'I have never met such a pitiful human being. Now he's going around looking for sympathy like some old woman. The kindest thing they could do would be to withdraw his lecturing qualifications forthwith.'

How remarkably prophetic! Heidegger's meeting with his old pupil Herbert Marcuse in 1947 and the ensuing correspondence show Heidegger's mentality in a revealing light. In his letter of 28 August 1947, in which he addresses the problem of how Heidegger's life relates to his thought, Marcuse dwelt in particular on the profound dismay felt by Jews. It is possible for a philosopher to deceive himself in matters of politics – in which case he will openly acknowledge his error. 'But he cannot deceive himself about a regime that murdered millions of Jews simply *because* they were Jews, a regime that made terror the norm and perverted everything ever associated with the words "mind" and "intellect", "freedom" and "truth", into its bloody opposite.' Marcuse told Heidegger he would be sending him a food parcel. His friends had advised against it, accusing him of 'helping a man who identified himself with a regime that sent millions of my fellow believers to the gas chamber'.

Heidegger received the parcel, but arranged for it all to be distributed to former pupils 'who were not Party members or otherwise connected with the Party in any way'. That should reassure Marcuse's friends, observed Heidegger in his reply of 20 January 1948. As for the taint of genocide – and Heidegger conceded that his strictures were justified – he 'can only add that if you substitute "East Germans" for

"Jews", then exactly the same charge can be levelled against one of the Allies, the only difference being that the international public knows very well what has been going on since 1945, whereas the Nazis' bloody reign of terror was actually kept secret from the German people.' The *Historikerstreit*, the celebrated revisionist controversy of 1986, is effectively anticipated here by Heidegger.

But let us get back to the sequence of events as they unfolded during the rectorship. In the early weeks, indeed the early days, of his period of office the new rector displayed a remarkable degree of activity. On 24 April, immediately after taking over the rectorship, he got in touch with the leadership of the German Student Union in Berlin, which had now been brought fully into line with the new regime, and proposed a central training seminar for the political leaders of the student union. It is clear from the correspondence that both parties were well acquainted. In other words, there was already a long-standing connection between Heidegger and the officials of the German Student Union, which since 1930 or so had been under the control of the National Socialist League of Students. This suggestion was well received in Berlin, and Plötner, whom Heidegger evidently knew very well, went to work at once to organize the first Reich conference of 'student officers for science' for 10/11 June 1933. The most important outcome of this first training seminar for the student political leadership was the project for an 'academic summer camp' (*Wissenschaftslager*) – the brainchild of Heidegger himself, which he proposed to set up that October in Todtnauberg. The small group of speakers who were considered for the June conference likewise points to a community of political purpose; the two names mentioned in conjunction with Martin Heidegger were Alfred Baeumler and Ernst Krieck. Both names, as we have already seen, cropped up at the beginning of April in Heidegger's letter to Jaspers. There must have been an understanding between them based on shared political views, which lasted until their one-time alliance swiftly turned to mutual hatred and antagonism; but that did not happen until the end of the year or the early months of 1934.

Suffice it to say in passing that Heidegger and Baeumler had known each other for a long time through their common philosophical interests, primarily in the field of Nietzsche studies. As a member of the Fighting League for German Culture, Baeumler had been associated, even before Hitler's seizure of power, with Alfred Rosenberg, who had made him head of the Department of Science, and in the summer semester of 1933 he was appointed to a chair in 'political education' at the University of Berlin – one of the first such 'political' chairs to be established during the Third Reich. Under Rosenberg, Baeumler's career rose rapidly to new heights, while his philosophical standards sank steadily lower, until in the end he was little more than Rosenberg's court jester – as when, for example, writing in the *Völkischer Beobachter* in October 1942, he celebrated the fact that *The Myth of the Twentieth Century* had now sold more than a million copies. It was, incidentally, Baeumler's first lecture in Berlin on 10 May 1933 that gave the signal for the wholesale burning of books that took place that night in the capital of the Reich. Baeumler marched in person at the head of the leading cohort of the torchlight procession, and was one of the first to throw his torch into the waiting bonfire of books.

Ernst Krieck, who hailed from the region known as the Markgräflerland in the extreme south-west corner of Baden, had worked his way up the academic ladder from humble beginnings as a primary school teacher: a dogged old campaigner who was now reaping his full reward. The association between him and Heidegger was probably more a matter of academic politics, since Krieck was elected rector of Frankfurt University in 1933. In terms of their scholarly work, certainly, they had absolutely nothing in common.

In order to attain his political, academic, indeed historic goals, Heidegger had to work, not to say fight, in a variety of different arenas. His own university was only ever a base, a point of departure, and an occasional refuge – at least to begin with. Even before his formal installation as rector he had already begun to stake out the territory that he planned to occupy. Some considerable stir, not to say indignation, was caused among the few people in the know in Freiburg when it emerged that Heidegger had sent the following telegram to Adolf Hitler on 20 May 1933: 'I respectfully request postponement of the planned reception for the Board of the Association of German Universities until such time as the much needed realignment of the

Association in accordance with the aims of *Gleichschaltung* has been accomplished.'

With this the new rector had unequivocally stepped up on to the national stage, which he no doubt saw as his proper field of action – though there is not a word about this in the apologia published in 1983. To sketch in the background briefly: the University Association in those days – in contrast to its post-1945 successor – was the corporate union of all German universities, whose principal purpose was to represent the interests of university teachers as a social and professional class. In effect it was an organ of the Conference of German University Rectors. Heidegger planned to replace this dual structure, not least because it had overtones of a parliamentary system, with a single, integrated Conference of Rectors, modelled on the principle of totalitarian leadership (*Führerprinzip*).

The agitation in Freiburg was provoked principally by the reference to *Gleichschaltung*, whose meaning then, in the early summer of 1933, was clear enough: the realignment of all institutions, all areas of life, in conformity with the principles of the totalitarian state and totalitarian society and the new power structures of the centralized National Socialist regime. This telegram weighed heavily against Heidegger in 1945, and in November of that year he submitted the following explanation to the chairman of the denazification commission – furnishing further evidence of the way he conducted his defence:

> Although the telegram mentions 'Gleichschaltung', I was using the term in the same way that I used the term 'National Socialism'. It was not, and never had been, my intention to impose Party doctrine on the University; on the contrary, I wanted to bring about a transformation in thinking both within National Socialism and with regard to it. It is untrue to claim that National Socialism and the Party had no intellectual plans for the universities or for science and learning: they had them only too clearly, citing Nietzsche as their authority, who taught that 'truth' does not have any content or substance of its own, but is merely an instrument of the will to power, i.e. a mere 'idea', a totally subjective

> concept. What was and is so grotesque about it, of course, is that this 'politicized' science and learning is essentially in line with the teachings of Marxism and Communism on the 'idea' and 'ideology'. It was against this that my rectorship address of 23 May, given three days after I had sent the telegram [Heidegger confuses the 23rd with the 27th of May], was clearly and explicitly directed.

Anyone who compares the telegram of 20 May 1933, the rectorship address of 27 May 1933 and Heidegger's other public proclamations, particularly during the autumn of 1933, with the explanation offered by him in 1945 will have no difficulty in seeing that Heidegger was interpreting here with the benefit of hindsight. At the time the only issue was *Gleichschaltung*, and what *that* meant had been abundantly clear since March 1933: every day and everywhere, existing political and social structures were being broken up and replaced by new ones. Heidegger neatly turned the tables by applying the concept of *Gleichschaltung* to the concept of National Socialism itself, which he claimed to have understood in his own special way – a version of National Socialism, as critics pointed out (this we know from other statements by Heidegger): a National Socialism that was yet not founded on the world-view of National Socialism. Among other elements underlying the view of events presented in 1945 is the hidden confrontation with Alfred Baeumler and his interpretation of Nietzsche, which in a sense became the official Party line. For Baeumler held one of the most influential posts in terms of shaping academic policy: he was head of the 'Department of Science of the Commissioner appointed by the *Führer* to oversee the intellectual training and education of the NSDAP' (so clumsily were official functions defined under a totalitarian regime). But this confrontation with Baeumler the philosopher did not occur until much later, when Heidegger's utopian goals had finally receded beyond his reach, and he was thrown back on his previous existence as a plain professor of philosophy at Freiburg University.

In order fully to understand the version of events given by Heidegger in 1945, we would really need to make a close reading of the Nietzsche lecture he gave in the winter semester of 1936/37

('Nietzsche: The Will to Power as Art': not published in full until 1985), in which he takes polemical issue with the Nietzsche interpretation put forward by Baeumler and Jaspers. This is not possible here, but such a reading would show that in his efforts to exculpate himself in 1945 Heidegger simply lumped together Baeumler, the 'political' interpreter of Nietzsche and the mentor of Alfred Rosenberg, with Jaspers, the 'unpolitical' interpreter of Nietzsche and his one-time fellow philosopher and friend, producing an amalgam of contradictory and antithetical elements. But isn't that the philosophical method in a nutshell? Such a reading would further show that Heidegger poured the essence of his philosophy of Being into statements of this kind made in 1945, just as he had fought with Baeumler in 1936/37 after Baeumler failed to take on board Heidegger's interpretation of Heraclitus. But that would take us too far afield, into areas that lie outside our professional expertise.

In 1933 at all events – and that is all that concerns us here – Heidegger had no such intentions. On the contrary, he was intent on co-operation with both Baeumler and Krieck; indeed, Heidegger was determined to use the Conference of University Rectors, scheduled to take place in Berlin on 8 June 1933, to elbow out the University Association, which had since undergone *Gleichschaltung*. But a majority of the rectors supported the Association, and Heidegger was forced to rely on a few old comrades-in-arms, specifically Krieck, who was joined by the rector of Göttingen, the Germanist Friedrich Neumann, and the rector of Kiel, Lothar Wolf, who was a natural scientist (Kiel, as mentioned more than once before, was intended to be the showcase university of National Socialism). This quartet, having been defeated, left the Conference of Rectors under protest; their departure excited little notice. In the weeks that followed, when (as we saw earlier) Heidegger collaborated closely with the then leadership of the German Student Union in conjunction with Baeumler, a lively correspondence developed between Heidegger and his opposite number at Göttingen. From this it becomes clear that this inner circle, this 'gang of four', plotted together behind the scenes to overturn the majority at the Conference of University Rectors and push through their own ideas on educational policy, nourished by the spirit of National Socialism.

A meeting was scheduled for the end of the summer semester to compare notes and plan future strategy. It is interesting to note that Heidegger and Krieck were evidently still collaborating very closely in the summer of 1933. And when Heidegger was again offered an appointment at the University of Berlin at the beginning of September 1933 ('this appointment would be associated with a special political mission', as Heidegger explained in a letter to his ministry of education) he combined the visit to Berlin with a long-planned meeting of the 'gang of four': 'On Wednesday morning [6 September 1933] I shall be travelling to Bad Homburg to attend a special meeting with my three friends and colleagues, the rectors of Kiel, Göttingen and Frankfurt', he writes in the same letter. So the contact was kept up, and the 'special political mission' associated with the Berlin offer opened up a whole new horizon, to which we shall return later.

It is worth noting that alongside his official duties Heidegger found time to give lectures in Heidelberg and Kiel (end of June/ beginning of July), which earned him a reputation as a particularly radical spokesman for the movement. These activities were in line with his long-term objective, which was to become one of the intellectual leaders of the movement in terms of academic and scientific policy-making – perhaps even its leader *tout court*. At any rate, the rector of Freiburg played a central role in the reshaping of Baden's university constitution, presumably in close collaboration with Krieck, who as a native of Baden had his sights set on a chair at Heidelberg, and for that reason alone was anxious to get on with the job in hand.

A provisional new university constitution for Baden came into force on 21 August 1933, whereby the rector was appointed 'leader' (*Führer*) of the university by the minister of education with effect from 1 October 1933; the university itself was given no say in the matter, not even a right of nomination. No limit was set on the period of office. Baden, ever true to its reputation as a model province, had pushed ahead of the field, eager to set the standard for others to follow. The rector himself appointed the deans as 'leaders' of the faculties; in future the university would be run in strict accordance with the 'leadership principle' (*Führerprinzip*). The remaining *Länder* of the German Reich, principal among them

Prussia, Bavaria and Saxony, adopted an initial policy of wait-and-see. 'This was Heidegger's doing. "*Finis universitatum*": the end of the universities', wrote the pro-rector of Freiburg University, Josef Sauer, in his diary entry for 22 August 1933. Nominally Heidegger's deputy, Sauer had long since been pushed aside. 'And we're in this mess because of that fool Heidegger, whom we elected to the rectorship in order to bring about a new intellectual flowering of the universities. What an irony! For the moment we can do nothing except hope that the other German universities, particularly in Prussia, will not take this step into the abyss, even though they are being urged in no uncertain terms to do so; that will soon put paid to Baden's little aberration' – a prediction that turned out to be thoroughly mistaken.

The precise extent of Heidegger's contribution cannot be established in the absence of documentary substantiation. But when the denazification commission observed in the autumn of 1945 that 'he played a very active part in reshaping the university constitution in line with the new "*Führerprinzip*"', Heidegger could not deny it. In his apologia published in 1983 he claimed that he proposed the change of constitution in order to clear the way for the appointment of deans who would guarantee that 'the essential character of the faculties and the unity of the university were preserved'. Heidegger himself, as expected, was appointed the first 'Leader-Rector' of Freiburg University with effect from 1 October 1933 – undoubtedly in the certain belief of ministry and appointee alike that he would be in office for many years to come. The new university constitution, for which Heidegger bore his full share of responsibility, was the expression and product of his thought and conduct, and the preamble to the Karlsruhe proclamation breathes his spirit and speaks his language. The total renewal of the German university system, he declared, could only be accomplished if the university reform programme was implemented across the board throughout the Reich – which meant that the powers hitherto vested in the university senate would pass to the rector, the deans would hold their posts in trust from him, and the guiding spirit of the rector would pervade the entire university community. The appointment of Heidegger was intended to herald a new era, for only now was it possible to break free from the ossified, dead forms of the old university system whose day was done.

Heidegger's philosophical conversation partner, Karl Jaspers, also took a positive view of the constitutional reforms, welcoming this manifestation of the aristocratic principle. From his many years of university experience he could 'not but find the new constitution right and proper. The regret I feel that a great era in university history, whose end has long been in sight, is now ending in this striking and sudden fashion, is the pain of a pious respect that I am not ashamed to own', wrote Jaspers to Heidegger on 23 August 1933, thanking him for his rectorship address, which had just been published by the Korn-Verlag in Breslau – a publishing house that specialized in the more militant brand of National Socialist literature (including the works of Moeller van den Bruck). Jaspers broadly welcomed Heidegger's inauguration address of 27 May: 'Your bold move in beginning with early Greek culture struck me once again like a new and yet self-evident truth. Here you are in agreement with Nietzsche – but with this difference: there is reason to hope that you will one day put into effect what you say through your work of philosophical interpretation. And that lends real substance and credibility to your speech.' Jaspers praises Heidegger's style and pregnancy of diction, ranking this speech above all the other rectorship addresses of that summer semester. He goes on:

> My confidence in your philosophical enterprise, which has been renewed and strengthened since the spring and the conversations we had then, is not shaken by certain characteristics of this speech that are very much of their time, by something in it that strikes me as a little forced, or by sentences that do seem to me to have a hollow ring to them. All in all I am just glad that someone is able to say these things and reach out to the genuine frontiers and origins.

So Heidegger is compared with Nietzsche – but a new Nietzsche who will actually put his philosophy into practice. Praise indeed! Heidegger did not reply to this letter. In fact it was two years before he took up the correspondence again, writing to Jaspers (in a letter we referred to in an earlier chapter) from the isolation and solitude of his new condition. As for his opinion of Jaspers' Nietzsche interpretation, we know that from his Nietzsche lecture of 1936/37, where he discusses the doctrine of eternal recurrence. Jaspers (he says)

> . . . sees that here is one of Nietzsche's crucially important ideas. But Jaspers fails to associate this idea with the basic question of Western philosophy, despite the talk of Being, and consequently fails to establish any real connection with the doctrine of the will to power. The reason for this seemingly puzzling stance is this: for Jaspers, to put it bluntly, philosophy is an impossibility. At bottom it is an 'illusion', whose purpose is the moral illumination of the human personality. Philosophical ideas lack the power of truth, either a truth of their own or *the* truth of essential knowledge. Because Jaspers, in his heart of hearts, no longer takes philosophical knowledge seriously, he has ceased to ask any real questions. Philosophy is reduced to the moralizing psychology of human existence. With such an attitude he cannot hope, however hard he tries, to achieve any kind of critical understanding of Nietzsche's philosophy (*Complete Works*, Vol.43, 1985, p.26).

To think that it had come to this, between philosophers who had once been friends! But then, Jaspers had already been dismissed from his academic post by this time, on the grounds that he was married to a Jew.

While the new university constitution for Baden was being finalized, Heidegger was still negotiating with Berlin on the appointment he had been offered. A powerful lobby at Freiburg University was urging the ministry in Karlsruhe to hang on to Heidegger: 'We ask the Ministry to do everything in its power to retain this distinguished scientific figure and leader for the University'. Heidegger's departure could have a disruptive effect on the changes that had already been set in train. He had thrown himself into his duties with all his strength and with the full force of his personality, building up a fund of experience, personal connections and trust. This appeal was signed at the end of September 1933 by leading members of the five faculties – proof that Heidegger's rectorship was viewed in a very positive light, at least within senior professorial circles. The signatories stated that their views were 'shared by very many members of our University'.

By this time Heidegger was no longer negotiating for the Berlin post. He remained in Freiburg – in the provinces. In a radio talk entitled 'Why we chose to remain in the provinces' he explained his motives, listed the factors that influenced his decision and painted a portrait of his state of mind at the time. In 1930, when Heidegger was first offered a position in Berlin, the question for him had been whether it made sense to squander 'the limited powers of a solitary individual' on an 'unnatural monster' like the University of Berlin. 'Big-city life only serves to create a sense of excitement and animation: the outward appearance of alertness. Even the best intentions are stifled by sensation and show – the scourge of all philosophy' (letter to Julius Stenzel, 17 August 1930). Three years later Heidegger's aversion to city life will have changed but little; indeed, his radio talk lifts that aversion on to a poetic plane.

But what about the 'special political mission' that was associated with the Berlin appointment? What exactly was this? Here we can only surmise, since the documentary evidence, if it exists, is not open to inspection – one of the costs of a divided Germany. Probably what was envisaged was some kind of collaboration with the 'political' professor, Alfred Baeumler, but this time at a new, national political level, as evidenced by the arrangements for the Berlin training seminar of June 1933. We have a reference written by Baeumler for Heidegger, dated 22 September 1933; the identity of the addressee is not known, but it was probably some non-university agency, since the actual offer of the position had been made to Heidegger long before. This report focuses on the academic importance of Heidegger, and sets out to provide information of a fairly general kind. Baeumler describes Heidegger as 'the most important figure in philosophy since Dilthey'. He had revolutionized current thinking in philosophical research both in systematic and in historical terms. Contemporary philosophical thought had entered a new phase with the publication of *Being and Time*. Every philosophical undertaking now had to take account of this book, either for or against. 'Furthermore, Heidegger's influence on contemporary philosophy in systematic terms, not only here in Germany, is quite incalculable.' Heidegger had an unerring instinct for the really crucial issues, particularly in Greek philosophy. 'If his approach to historical problems is sometimes a little wilful, he is simply exercising the prerogative of the philosophical genius.' Baeumler's report gives the impression of having been written for a Party agency that had a special interest in the Berlin appointment because of the 'special political mission' associated with it.

For his part, Heidegger informed the academic staff at Freiburg of his decision to decline the Berlin offer at the beginning of October, concluding his somewhat non-committal but cryptic statement with these words:

> I shall not go to Berlin, but shall seek instead, here at our University, to turn the opportunities created by the provisional new constitutional arrangements in Baden into a genuine and tried reality, in order to pave the way for the co-ordinated development of the new national constitution for all German universities. At the request of the government authorities in Berlin I shall continue to keep in very close touch with the work that is going on there.

Reading between the lines, we can see that the role Heidegger played in piloting through the new university constitution for Baden and pioneering the new-style rectorship modelled on the *Führerprinzip* was intended to make him an expert in these matters at the national level. The formulation used by Heidegger in his letter of 30 September 1933, thanking the ministry in Karlsruhe for his appointment as rector, is simply phrased differently, without actually telling us much more. 'At the request of the Prussian Ministry of Education I have placed myself at their disposal for further consultations in the future; likewise in the event that they decide to embark on a full-scale implementation of the National Socialist programme of university reform.'

But how definite were these arrangements, one wonders, given that events were still in such a state of flux? Was it not a programme built on high hopes and expectations, now overtaken – and long since, perhaps – by reality? And was Heidegger the only one who failed to realize that he was being used as a figurehead, for purely tactical reasons, by the government authorities in Berlin, Party organizations and others, who had chosen men of a different stamp to fill the real positions of power: old Party faithfuls, veterans of the struggle, who suited the crude ideology of National Socialism? Had not Heidegger been forced to observe, in the closing days of September, how brutally the leaders of the German Student Union, the pioneers of those heady early days, who still retained something of their idealistic fervour, were ousted by others? The struggle for power was now on, and open trench warfare had erupted throughout the Polycratic single-party system.

But if it is true, as Heidegger claimed before the denazification commission in 1945, that he had realized by the summer semester of 1933 that political events were moving in a direction he had not anticipated, then the declarations, both public and private, made by the *Führer*-rector in the autumn of 1933 become all the more puzzling. We have already come across his proclamation to the students of Freiburg on 3 November 1933, followed a week later by another announcement in the student newspaper calling on students to vote in the elections of 12 November, the so-called *Reichstag* elections. But the rhetoric of these utterances pales into insignificance beside the declaration that Heidegger made before the world as a prominent signatory to the Leipzig proclamation of 11 November. The vote on 12 November was essentially a referendum on Hitler's policies: the election of a single-party slate to the *Reichstag*, coupled with endorsement of Germany's withdrawal from the League of Nations. The rector of Freiburg University interwove major elements of his rectorship address, indeed of his philosophy, with the practical politics of Hitler to create an amalgam of ontological domestic and foreign politics – possibly his worst publicly recorded aberration, which compromised all his subsequent philosophical endeavours. But *was* it an aberration?

So far as I know, there is not a passage anywhere in Heidegger's work, or in the correspondence or official records, where he retracts a single word of this proclamation or of other statements like it. Heidegger stood behind what he had said, because it was a premise of his thinking that he was not liable to error. That was a risk to which others were exposed, those who did not listen to him, the prophet of Being. Ideally we ought to quote the Leipzig proclamation in full, in order to get a sense of the wider context. But a few of the more memorable passages will have to suffice:

> We have renounced the idolization of groundless and powerless thinking. We see the demise of the philosophy that was subservient to it. Of this we are certain: that the clear hardness and workmanlike assurance of an unyielding, simple questioning of the essence of Being is now coming back. The primordial courage either to grow or to perish in the encounter with the entities of being [*Seiendes*] is the innermost motive of the inquiry conducted by an ethnic and national [*völkisch*] science . . . And so we declare, we who are in future to be entrusted with the safekeeping of the will-to-knowledge of our nation, that the National Socialist revolution is not simply the assumption of a power already present in the State by another party that has grown large enough to wield it: this revolution brings with it *the total transformation of our German being*. From now on all things demand decision, and all action responsibility. Of this we are certain: when the will to answer for oneself becomes the law by which nations coexist, then every nation can and must become the instructor of every other nation in the riches and power of all the great deeds and works of human existence. The act of choice that the German nation now has to perform – *that event in itself*, regardless of the outcome – is the strongest possible manifestation of the new German reality of the National Socialist State . . . the *Führer* has brought this will to its full awakening throughout the nation, and forged it into a *single* mighty resolve. No man can stand aside on the day when that will is made manifest. Heil Hitler!' (Schneeberger 1962, Document No. 132).

The dominant pseudo-religious rhetoric will sound more than a little familiar to those acquainted with the formulas of recantation and affirmation of faith used in early Christian baptism. Philosophy *ante Heidegger* was mere idolatry: 'We have renounced.' But now that the National Socialist State has been instituted in accordance with the 'primordial imperative of all Being to save and preserve its own essence', the doctrines of the one true philosophy can be invoked, and the new faith openly professed: 'And so we declare' – 'we' meaning the initiated, those who know – '*that the National Socialist revolution is the total transformation of our German being.*' So Heidegger cannot have become disenchanted by the late autumn of 1933: far from it. Not even when the new leadership structure was imposed on the Conference of University Rectors in November 1933 – and Heidegger found himself excluded.

But all this was just talk. If Heidegger was going to prove himself, it had to be within the framework of his own university, and following his appointment as *Führer*-rector he was intent, as we have seen, on turning the opportunities created by reform 'into a genuine and tried reality'. *Hic Rhodus, hic salta*! But this was stony ground indeed, hard to work, with little prospect of any harvest. The lofty flights of Heidegger's speeches, addresses and proclamations were in stark contrast to the toil in the sweat of his brow on the unyielding earth.

Before we turn to Heidegger's rather slight, not to say meagre, attempts to bring about a reform of his university in line with the dictates of the National Socialist revolution, let us say a word or two about Heidegger's handling of staff matters in relation to the notorious Reich law for the re-establishment of a permanent civil service, which had been promulgated on 7 April 1933. We have already encountered this piece of legislation in an earlier context, in connection with the Baden decree against the Jews of April 1933. The law was directed primarily against Jewish civil servants and politically undesirable civil servants, who could not be relied upon (amongst other things) to act in the national interest. Needless to say, the rector of every German university was required to enforce this law in collaboration with his regional ministry of education. The law, which was accompanied by a whole set of executory provisions, was exceedingly complicated, and offered some room for interpretation with regard to the time allowed for compliance and the provisions governing exemptions. It was the enforcement of this law that signalled the beginning of the great exodus of German intellectuals, severely depleting all the scientific disciplines, not least the natural sciences and medicine.

The full rigour of the law was felt particularly by Jewish junior lecturers and scientific assistants, who in most cases were not old enough to qualify for exemption (anyone who had been a civil servant prior to 1 August 1914, for example, could remain in office, or anyone who had been a front-line soldier). They were stripped of their teaching qualifications and dismissed in droves – even Heidegger's assistant, Dr Werner Brock, as we have already seen. Likewise a certain young scientist and junior lecturer in the Faculty of Medicine, Dr Hans Krebs, later Sir Hans Krebs, who won the Nobel Prize for Medicine in 1953 for his discovery of the citric acid cycle. Most of the groundwork for his prize-winning research was done in Freiburg. (See his dispassionate and dignified account '*Wie ich aus Deutschland vertrieben wurde*' ['How I was driven out of Germany'] in the *Medizinhistorisches Journal*, Vol.15, 1980, pp. 357–377.)

How difficult and contradictory the situation was for Rector Heidegger appears from a letter of his to the ministry of education, dated 12 July 1933, in which he appealed for exemption on behalf of two internationally renowned professors who were now under threat: Eduard Fraenkel, professor of classical philology – we have already encountered him in an earlier context – and Georg von Hevesy, professor of physical chemistry (and winner of the Nobel Prize in 1943). Heidegger pointed out that he was writing 'in the full awareness of the need to enforce to the letter the law for the re-establishment of a permanent civil service'; but the cause of upholding and strengthening the international standing of German universities and German science would be ill served by alienating foreign opinion. Yet such must surely be the consequence – particularly 'in leading intellectual and political circles abroad' – of dismissing these two Jewish professors, whose extraordinary scientific standing was beyond doubt.

In the person of Georg von Hevesy Heidegger held several trump cards. The internationally renowned natural scientist, many times honoured, had received substantial funds from the Rockefeller Foundation which he had used to purchase apparatus and equipment for the new Institute for Physical Chemistry. These funds would be lost and the chemical industry as a whole would be seriously damaged, since von Hevesy was virtually irreplaceable in the field of applied research and in the training of young chemists. Furthermore, von Hevesy's family belonged to Hungary's ruling political elite. One brother, Paul von Hevesy, had served as Hungary's ambassador to Madrid. As Heidegger put it, the outright dismissal of these two professors 'would deal a severe blow to the standing of German science in general and of our borderland university in particular – a blow from which neither would recover for a very long time.' Furthermore, both men were Jews of the better sort, men of exemplary character; he could vouch for the irreproachable conduct of both men, 'in so far as it is humanly possible to predict these things.'

The ministry dismissed Fraenkel but kept von Hevesy, doubtless because of his usefulness and for reasons of political expediency. (As it happened, von Hevesy left of his own accord a year later, resigning from the Baden civil service on 1 October 1934 to take up a post in Copenhagen.) Here again Heidegger simply had to accept the ministry's decision: his room for manoeuvre was extremely limited.

Should he perhaps have protested, or even resigned in protest? But then: are these appropriate questions to ask about a rector who was fundamentally convinced 'of the need to enforce to the letter the law for the re-establishment of a permanent civil service'? In *Facts and Thoughts* Heidegger characterized his personal position thus: 'The only good thing, even if only in a negative sense, was that in the so-called 'purge', which often threatened to overstep the limit and exceed its mandate, I was able to prevent acts of injustice and minimize the damage to the University and my colleagues.' But then he goes on: 'These preventive efforts on my part were not apparent to an outsider, nor indeed was it necessary that my colleagues should know anything about them. Certain of my distinguished colleagues in the faculties of law, medicine and the natural sciences would be astonished to learn what the authorities had in store for them.' Heidegger is alluding here to article 4 of the law. He hints at something dark, threatening and hidden, shrouded in mystery and secrecy. Yet there is no evidence of such intentions in the official record. It could be objected that official records are not everything; but it is my belief that there was nothing sinister 'in store' for these men. At least in the initial phase of the Third Reich the authorities acted pretty much in accordance with the law – however inhuman the laws themselves may have been.

As far as the University of Freiburg is concerned there is only one recorded instance – during Heidegger's period of office as rector

– of a full professor being dismissed from his post in accordance with article 4 of the above-mentioned law. This was the moral theologian and Catholic scholar Franz Keller, whose pacifist views prior to 1933 cast grave doubts on his reliability as a loyal supporter of the national state. The ministry in Karlsruhe had acted on its own initiative in the case of Keller – formally, at least – and had instigated the dismissal proceedings against him. We know for a fact that denunciation by anonymous informers was rife during those weeks of political turmoil. On 2 June 1933 Pro-rector Sauer noted in his diary that he had learned from the rector's office that a complaint had been lodged against the dogmatist Engelbert Krebs; furthermore, that serious charges were pending against the economist Adolf Lampe – a name we shall come across again – on the grounds that during the campaign for the Hindenburg election of 1932 he had described the economic programmme of the National Socialists as 'nonsense'. Sauer also learned that 'very serious charges indeed' had been brought against the moral theologian Keller on account of his anti-nationalist and pacifist sentiments.

The case of Hermann Staudinger or 'Operation Sternheim'

A classic example of collaboration within the National Socialist cadre of university teachers

There are doubtless moments in the life of every scholar when he really does not know what to do for the best when he encounters certain situations. Take the case of the historian who unearths a piece of source material so astounding that at first he thinks – has no choice but to think – that it is not genuine. When I was looking through archive material in the course of my research on Heidegger I was surprised and taken aback to come across a clear instance of political denunciation by Rector Heidegger in the files of the Baden ministry of education. For days I could not decide whether I should leave this group of documents undisturbed, or whether I should publish it with a commentary and analysis. Let some other historian who comes after me agonize over such things! Let him be branded a bloodhound if he likes! The trade of the academic toiler – as practised by the historian in the eyes of the philosopher (or rather, of certain individual philosophers) – is not always easy. The discovery was complicated by the fact that there was no obvious explanation for Heidegger's motives; even now it seems that we cannot produce a totally satisfactory explanation – short of entering the realms of depth psychology.

On 11 October 1933 the following memorandum was placed in the files at the ministry of education in Karlsruhe: 'On the occasion of his visit to Freiburg on 29 September 1933 the secretary for higher education was briefed by the rector of the University, Professor Heidegger, whereupon he instructed the rector to inquire into the existence of circumstances pertaining to article 4 of the law for the re-establishment of a permanent civil service in the case of Professor Staudinger.' Behind the stilted legal jargon lies a case that offers us an explosive insight into Heidegger's thinking at this time.

The documentary record has survived intact and gives a remarkably detailed picture of events. It clearly shows that on 29 September 1933, when Professor Fehrle, Baden's secretary for higher education, was staying in Freiburg (primarily in connection with Heidegger's forthcoming appointment as *Führer*-rector, with effect from 1 October), Heidegger, in his capacity as rector, had informed the secretary of some politically damaging allegations against the chemist Hermann Staudinger, already a world-famous figure in his field. These allegations related to events that had taken place at the time of the First World War and in the immediate post-war years. Fehrle took immediate action on 30 September 1933 'in order to meet the deadline' – proceedings on political grounds had to be instituted by this date – and preferred charges against Staudinger at police headquarters in Freiburg. 'The proceedings are now officially in motion', states the record. Acting on this information, the Gestapo office in Karlsruhe took up the case and reported back to the ministry within a matter of days, on 4 October 1933: 'In order to avoid confusion I take the liberty of pointing out that the case of Professor Staudinger in Freiburg will be referred to under the code name "Sternheim".' The rector of the University, the report continues, had not been able to furnish the Gestapo with any concrete evidence; he had simply notified the authorities on the basis of certain rumours that were in circulation. But were there really any such rumours?

At all events, 'Operation Sternheim' was executed smoothly and silently, as befitted such an enterprise. The Gestapo struck gold, not only at the district government offices in Karlsruhe – Staudinger had worked at the Technical College in Karlsruhe until 1912, first as an assistant, later as a junior lecturer and extraordinary professor – but more especially via the Foreign Office in Berlin. Three fat bundles of files were collected over the next few months, based on reports from the German Consulate General in Zurich and the German Embassy in Bern.

The Gestapo office in Karlsruhe wanted prompt action to resolve the matter, sending a reminder to the Foreign Office, which in turn put pressure on Germany's ambassador in Bern. By Christmas 1933 the incumbent ambassador, Baron von Weizsäcker, who later became secretary of state under Reich Foreign Minister von Ribbentrop, had sent the documents requested by special courier.[148] What was behind this flurry of activity?

Having worked as a full professor at the Federal Technical College in Zurich since 1912, Staudinger had applied for Swiss citizenship in 1917, while at the same time retaining his status as a German national. To all outward appearances this was purely a formality. However, the German Consulate General in Zurich had recommended that Staudinger's application be refused in the light of the following circumstances. In 1915, having been rejected in 1904 as permanently unsuitable for military service, Staudinger was sent to be assessed by the Consulate General in Zurich for possible use in some other military capacity. He was less than delighted by the prospect, particularly as he had already adopted an essentially pacifist position. An examination by the army medical board resulted in his provisional exemption from service. At this point the German military intelligence section at the embassy in Bern released intelligence information indicating that Hermann Staudinger was advising hostile foreign governments on the manufacture of chemicals that were vital to the war effort, notably in the area of dye production. In order to forestall a call-up order – Staudinger was liable for military service – he applied for Swiss citizenship while at the same time remaining a German national. His application was refused on the grounds that his patriotic loyalties could not be fully guaranteed. He had, it was claimed, given evidence of anti-military sentiments, and refused to support his country in any way during the present war. It was well known in Zurich 'that he is determined not to answer the call to arms if it should come, nor to comply if directed to work for the fatherland in some other capacity'. Such were the arguments that led to the rejection of his application for citizenship in 1918.

Staudinger remained under a cloud of suspicion even after the war, particularly with regard to the charge that he had never attempted to conceal his total opposition to the nationalist movement in Germany, and had repeatedly stated that he would never support his country by taking up arms or performing any other kind of labour service. Such was the gist of the report sent to Berlin by the German Consulate General in May 1919, when Staudinger made a renewed application for dual citizenship – this time with success. While a number of suspicions had proved unfounded (the report went on), the fact remained that 'the stance adopted by Professor Staudinger in time of war was likely, particularly in view of his position as a university professor, to do grave damage to Germany's cause in the eyes of the international community'. The recent political turmoils in Germany, it was claimed, only served to bear out this assessment. Such was the state of play in 1919. From their study of the diplomatic files the Gestapo also learned that in 1917 Staudinger had intervened on behalf of the pacifist Professor Nicolai, who had refused to take the oath of allegiance during the war, and had collaborated with him on a literary project. (Strictly speaking we would need to digress here in order to fill in the somewhat complicated background to Staudinger's political stance. Suffice it to say that he and his wife were associated with the religious-pacifist and religious-socialist circles around the Protestant churchman Leonhard Ragaz, who was subsequently stripped of his clerical appointment.[149])

This together with other complaints gave the Gestapo sufficient material, and on 25 January 1934 a letter was sent to the ministry from Gestapo headquarters in Karlsruhe: 'I enclose the files sent by the Foreign Office, together with the files relating to the applications for citizenship. I assume that the contents of the diplomatic files will furnish sufficient grounds for the institution of proceedings.'

On 6 February 1934 the ministry of education sent the files to Rector Heidegger and asked him to submit his comments as a matter of urgency: 'In view of the fact that any proceedings under article 4 of the law must be initiated not later than 31 March 1934, I would urge you to give your prompt attention to the matter.' The rector of Freiburg was nothing if not prompt, sending in his report on 10 February and returning the files at the same time. It was a devastating report, and a revealing one, written on the official letterhead of the

Freiburg rector's office – albeit without a reference number and riddled with typing errors, as if typed by an unskilled hand. After careful study of the extensive documentary evidence, Heidegger summarized the complaints against Hermann Staudinger under four headings and commented as follows:

> On the question of whether article 4 of the law for the re-establishment of a permanent civil service is applicable to Professor Staudinger, a study of the files relating to his case reveals the following points: (1) All the reports prepared by the German Consulate General in Zurich during the war, and most notably the memorandum of 15.10.1917 drafted by embassy secretary von Simon, speak of St. handing over information about Germany's chemical manufacturing processes to (hostile) foreign powers. (2) In January 1917, i.e. in the hour of the fatherland's greatest need, St. applied for Swiss citizenship when there was no pressing need, professional or otherwise, for him to do so. The application was blocked by the German Consulate General. (3) On 9.1.1919, i.e. immediately after Germany's collapse, St. renewed his application for Swiss citizenship, 'in view', as he put it, 'of the recent changes in Germany's domestic and foreign political situation'. He was granted Swiss citizenship on 23.1.20, without the written approval of the German authorities. St. claimed to have received written approval from the Baden district government office on 15.1.1919, but no longer had the letter in his possession. The letter in question will be found in the first set of files (I). It contains nothing that could be remotely construed as an approval. The reports from the Consulate General in Zurich dated 12 January 1918 and 1 May 1919 contain material that is damaging in the extreme. According to these reports, Staudinger had 'never attempted to conceal his total opposition to the nationalist movement in Germany, and had repeatedly stated that he would never support his country by taking up arms or performing any other kind of service'. Significantly, the future Marxist ambassador Adolf Müller describes Staudinger as an 'idealist'. Equally

> damaging is the fact that in 1917 Staudinger wrote a letter of petition from Zurich on behalf of the pacifist Dr Nicolai, who had refused to take the oath of allegiance.

Heidegger's final verdict was an outright condemnation, which speaks for itself:

> These facts are sufficient in themselves to require action under article 4 of the law for the re-establishment of a permanent civil service. And since these facts became widely known inside Germany following the debate about Staudinger's appointment to the post at Freiburg in 1925/26, and have not been forgotten since, the good name of the University of Freiburg also requires that action be taken, especially as Staudinger now claims to stand 100 per cent behind the present national reawakening. I would have thought this was a case for outright dismissal rather than early retirement. Heil Hitler!

Properly speaking, of course, we would need to consider each accusation point by point. But suffice it to say that the charge that Staudinger had betrayed manufacturing secrets was not upheld even by the Nazis themselves. Yet for Heidegger it was an established fact. According to him, the good name of the University of Freiburg demanded the dismissal of this world-renowned chemist, a member of numerous scientific societies in Germany and abroad and a future Nobel Prize winner. Heidegger was alluding to the controversy preceding Staudinger's appointment in 1925, when his pacifist attitude during the First World War was publicly aired, together with his post-war quarrel with Fritz Haber, the discoverer of the synthesis of ammonia, over Germany's role in the use of poison gas; in short doubts had been cast on Staudinger's patriotism. An internal inquiry was conducted at the time, in which Staudinger explained his attitude to the Germany of his day. He produced documents in support of his statements, and was able to reassure even the more stridently nationalistic professors in the Faculty of Mathematics and Natural Sciences at Freiburg.[150]

As far as can be established, Staudinger's one-time political sympathies were in no sense an issue in Freiburg University circles.

The professor pursued his scientific labours and was a successful head of the university's Chemical Laboratory. But there were wheels within wheels, as we are about to see. Meanwhile, acting on the recommendation of Freiburg's rector, the minister of education submitted the following request on 22 February 1934: 'The Ministry of State hereby submits to the *Reichsstatthalter* a recommendation that Professor Hermann Staudinger be dismissed from the civil service of Baden with effect from the day he is informed of this decision.' When questioned, it was alleged, Staudinger had failed to rebut the accusations of un-German behaviour during the war. 'On the basis of these facts Professor Staudinger can no longer be considered a suitable person to be entrusted with the education of Germany's academic youth; I consider the criteria for dismissal from the University of Freiburg in accordance with article 4 of the law . . . to have been met', concluded the minister, enclosing the transcript of the interrogation with his letter.

This request had been preceded by a 'hearing' at the ministry of education, where the professor of chemistry was summoned by telephone on 17 February 1934. The interrogation became an exercise in the humiliation of this physically huge man, who was built like a guards officer: the totalitarian state was beginning to lay aside the mask and show itself in its true colours. Hermann Staudinger, who never found out that his whole 'case' was ultimately the result of Heidegger's action, was confronted with the charges and forced on to the defensive unprepared. He was fighting with his back to the wall, unable to refute the charges by a straightforward denial.

For example, he had published an article in the *Friedenswarte* in 1917 that concluded with this observation:

> A future war could bring with it unimagined destruction and annihilation. That being the case, the question of a really lasting peace is a task confronting the whole of mankind. It is an issue that must be resolved today, today more than ever before, if the civilized nations are not to be threatened with total destruction. A peace that is merely a kind of temporary armistice would be the very worst fate that could befall Europe.

This and other statements were now cited against him. Defending himself now, in 1934, Staudinger explained his views thus: he had not been a pacifist in the narrowly religious sense of the Quakers or conscientious objectors, but a pacifist 'on the basis of my understanding of the role of technology in modern warfare'. For the rest he pointed out that he had long since dissociated himself from his earlier political views. Since he had taken up his duties at Freiburg, nobody could accuse him of 'unpatriotic sentiments'. On the contrary: he had 'greeted the outbreak of the national revolution with great joy'. Furthermore, the new National Socialist state offered an 'extraordinarily broad scope for my activities'. In industry, too, people no longer held his earlier views against him.

Picking up on this point, Staudinger sought to tip the balance in his favour by emphasizing the contribution he could make as a scientist to the new Germany's bid for economic self-sufficiency. Within a few days, on 25 February 1934, he had published a long article in the Düsseldorf *Völkische Zeitung* under the title 'The importance of the chemical industry for the German nation', sending a personally inscribed offprint to the minister of education at the beginning of March 1934 as a gesture of goodwill. The National Socialist mayor of Freiburg, Dr Kerber, also intervened on Staudinger's behalf, having evidently learned somehow of his impending dismissal. And even Rector Heidegger sent a handwritten letter on 5 March (again written on the letterhead of the rector's office, and again without a reference number: consequently there is no record of the correspondence in the Freiburg University files). Heidegger referred to the settlement that had been reached in the case of the moral theologian Keller, who had been forced to take early retirement rather than being dismissed. 'On mature reflection it seems to me advisable to adopt a similar course of action in the case of St., given his international standing within his professional field . . . I need hardly point out that this in no way alters *the facts of the case*. It is simply a question of avoiding fresh difficulties in our foreign relations if at all possible.' Such was the view taken by Rector Heidegger, who here recycles the same argument he had already used in favour of Professors von Hevesy and Fraenkel in July 1933. His only concession: a reduction in Staudinger's 'sentence'.

The ministry of education contrived a Solomonic solution that bordered on the grotesque. The recommendation for dismissal was withdrawn, Staudinger was summoned once again by telephone and subjected to a humiliating ordeal. He was forced to submit a formal notice of resignation from the Baden civil service, which would then be placed on file for the next six months. Since the offences of which he was accused had taken place some considerable time previously, his resignation would only be accepted 'if his present conduct gives grounds for renewed concern'. In the event that did not happen because the accused chemist, aware that he was being watched, was careful not to step out of line; above all he took care to advertise his indispensable role in Germany's future policy of self-sufficiency. In October 1934, as agreed, Staudinger was allowed to withdraw his resignation. The case was closed, and Staudinger had got off lightly. Others were not so fortunate, as we shall see below.

What does 'Operation Sternheim' tell us about the subject of our present inquiry? What were Heidegger's real, underlying motives for denouncing a respected member of his university in this unbelievable fashion, and for recommending his colleague's dismissal after a study of the Gestapo's files? Was it Heidegger's deep-seated concern for the national interest that led him to extrapolate from the past to Staudinger's political unreliability for the present and future? I confess that I have no answer to this question. But one thing is clear: this episode, which from start to finish took up nearly six months of Heidegger's time, casts a devastating light on his mental make-up and his basic state of mind. In the very weeks when Heidegger claims to have broken with National Socialism he was playing a dirty game behind the scenes, helping the Gestapo with their inquiries and associating himself with the interrogation and humiliation of a blameless scientist who was very nearly ruined as a result. The only concession the rector was prepared to accept was a lighter sentence: early retirement instead of outright dismissal without pay. And even this act of grace was only a concession to public opinion abroad, where the 'treatment' originally envisaged for Staudinger might well have provoked an outcry. And of course, 'this in no way alters *the facts of the case*'. These are murky waters indeed: but some things appear crystal clear . . .

But we need to dig even deeper, because the Staudinger affair also shows that Heidegger worked purposefully to achieve his long-term aims, leaving nothing to chance. The background to this case is particularly interesting. Following the first publication of my findings, it was pointed out in a number of quarters that it has not been absolutely proved that the rector of Freiburg University actually instigated the action himself; perhaps he had acted at the direction of the ministry in Karlsruhe. Even though the dossier to which I was given access in 1984 establishes beyond a doubt that the initiative was Heidegger's, many commentators remained sceptical. But now the full facts have emerged. An examination of the political records of the German Foreign Office has revealed that as early as July 1933 Rector Heidegger employed the services of a certain Dr Alfons Bühl, a junior lecturer in physics, as his confidential agent to make enquiries in Zurich. A member of staff at the German Consulate General in Zurich placed the following memorandum on file on 28 July 1933:

> Dr Bühl informs me that he has been commissioned by the rector of Freiburg University to collect documentary evidence relating to Professor Staudinger, currently a professor at Freiburg University, where he is the subject of various rumours concerning conscientious objection, etc. After consultation with the Consul General I have suggested to Herr Bühl that he might wish to advise the rector in Freiburg of the fact that the Baden district government office in Karlsruhe has material on file relating to Herr Staudinger from the year 1919.

This episode gives us an insight into the *modus operandi* of the National Socialist group on the staff of Freiburg University. Dr Bühl, born in 1900, had obtained his doctorate at Heidelberg in 1925 under the Nobel Prize winner Philipp Lenard, the founder of 'German-Aryan physics', and completed his habilitation thesis under Gustav Mie in Freiburg in 1929.[151] From the autumn of 1931 to October 1933 he was an assistant in the department of physics at the Federal Technical College in Zurich – the same institution where Hermann Staudinger had earlier been employed. Bühl had retained his principal domicile in Freiburg. He was a member of the inner circle of National Socialist university teachers who had hoisted Heidegger into office just a few months previously. Now, as someone who knew the Technical College in Zurich from the inside, he was the ideal man to investigate Hermann Staudinger's past.

From a nationalist point of view, certainly, Dr Bühl was thoroughly sound. Having participated in the closing months of the First World War, he went to Berlin when it ended and enlisted as a volunteer. From December 1918 to May 1920 he took part in the fight against the *Spartakusbund* as a member of the 'Iron Squadron'. Meanwhile he pursued his studies in physics. From 1921 onwards he was working under Lenard in Heidelberg, and in 1922 took an active part in the Ruhr campaign of passive resistance. His academic performance was not particularly scintillating, but his ideological credentials were impeccable. Dr Bühl, who worked as Rector Heidegger's agent, was one of the few natural scientists who sought to walk in the 'native' or 'authentic' [*arteigen*] ways of 'German' thought, despite continuing interference from the many who remained enslaved to 'Jewish' thinking. He was a pupil of Philipp Lenard's and a willing agitator on his behalf. So it comes as no surprise to learn that in 1934 Lenard used his influence from Heidelberg to get the militant Bühl appointed as temporary professor of physics at the Technical College in Karlsruhe, following the dismissal of the previous incumbent on political grounds. He was later appointed a full professor, contrary to the wishes of the College.

In Karlsruhe, too, a campaign was mounted to dismiss a natural scientist under the terms of article 4 of the infamous law. And it succeeded: in September 1933 dismissal proceedings were instituted on the basis of information received against the distinguished physicist Wolfang Gaede, the founding father of high-vacuum physics. According to the Nazi newspaper *Der Führer*, he had said that Germany would be finished without the Jews: 'The time has come for this professor to make himself scarce.' The proceedings ended in 1934. We have a direct parallel here with the case of Hermann Staudinger, and in the person of Dr Bühl we have the curious link between them, which suggests another interesting possibility: perhaps the incumbent professor at Karlsruhe was denounced in order to make way for a younger man, a 'native' or 'authentically German' physicist. Were the proceedings against Hermann Staudinger motivated by similar considerations, perhaps?

Just twelve years later, in an open hearing conducted by the academic authorities, Heidegger likewise found his career under threat. He was called to account for his actions. Had the case of Staudinger, or 'Operation Sternheim', been known at the time, Heidegger would not have stood the slightest chance of rehabilitation. For this personal and human failure, in the name of political-ideological self-importance, would have stigmatized him for all time.

However, in one of the many versions of his defence, sent to his colleague Constantin von Dietze, the chairman of the denazification commission, on 15 December 1945, Heidegger has some comments that close the file on the denunciation of Staudinger – albeit in a cryptic manner that only makes sense to somebody in the know. From 1935 onwards (he writes) he had repeatedly warned – and argued in a lecture on 'The metaphysical foundations of the modern world view' given in the summer of 1938 – that the sciences were becoming increasingly subservient to technology. The lecture had been the subject of a malicious report in the pages of the Nazi party journal, *Der Alemanne*:

> The articles in the features section of the paper were so arranged that the account of 'the interesting lecture evening' was followed by a brief statement to the effect that the Chemistry Society was currently holding its deliberations in Freiburg and that the University was contributing to these efforts on behalf of the Four-Year Plan. The lecture given by 'Professor Heidegger, who owes his celebrity solely to the fact that nobody can understand him, and who teaches the doctrine of Nothing' (implying that I subscribe to nihilism) was made out to be of no importance compared with the 'vital' work of professional scientists.

Heidegger's recollections of the text were not entirely accurate, but he was certainly right about the general tone of the piece. It was a fact that *Der Alemanne*, once Heidegger's mouthpiece, had chosen to target the philosopher and his particular brand of philosophy, having put its full weight behind the Four-Year Plan in keeping with the prevailing *Zeitgeist*. And there it was: Staudinger the chemist was more in demand and more highly esteemed than Heidegger the philosopher, who had to suffer such humiliation – albeit only at this minor local level. The page compiler had carefully contrived to place the following item immediately below the dismissive report on Heidegger's lecture:

> 'Chemistry and the Four-Year Plan'
>
> On 15 June at 12.15, during University Week, the newly completed premises for the Chemical Institute at Albertstrasse 21 will be officially handed over to the Department. The gathering will be attended by the rector of Freiburg University, the president of the Chamber of Industry and Commerce and a number of persons with a professional interest in chemical research. Afterwards Professor H. Staudinger will read a paper entitled 'Chemistry and the Four-Year Plan'.

The hidden irony of this predicament is something that only Heidegger could have appreciated. And because he understood it only too well, he was so agitated that he took this turn of events as a profound affront. It is tempting, and perhaps not altogether inappropriate, to recall Heidegger's own dictum here: 'Nothing is chance in the unseen history of poetic discourse. All is destiny.'

In the light of these *aperçus*, which cast a bright, not to say glaring light on Heidegger's 'true' attitude to the Third Reich, National Socialism and the *Führer*-state, it is not easy to take with any great seriousness what the philosopher has to say in *Facts and Thoughts* about distancing himself increasingly from National Socialism, and safeguarding science and learning from the dangers of politicization. The same applies to his motives and methods when he came to relinquish the rectorship. There is no avoiding the issue: the apologia of Rector Heidegger has to be read in conjunction with documents from the files and archives, in which Heidegger's mentality is laid bare – and with awful clarity. For the philosopher's apologia paints a picture from memory that has too many subjective elements, and fails to stand up, either on details or essentials, to objective scrutiny based on the historical method.

How ambivalent Heidegger's position in these matters was can be seen from his response when another member of the Faculty of

Natural Sciences and Mathematics was denounced by a colleague, and the ministry was considering instituting proceedings under article 4 of the aforementioned law. The case concerned a senior lecturer and permanent member of staff, Professor Johann Georg Königsberger, a highly specialized geophysicist, who was reported to the Baden ministry in December 1933 by Professor Wilhelm Hammer, who drew attention in the most damning terms to his 'Marxist past'. Rector Heidegger replied on 16 January 1934 (around the same time, therefore, that the Staudinger case was pending) – this time officially, on the record, so that the correspondence is preserved in the University archives. He proposed that no action be taken against Königsberger, since the latter had been living a life of seclusion for many years, was no longer politically active, and devoted himself exclusively to his scientific work. The accused had declared on the relevant questionnaire that he had been a member of the SPD up until the beginning of 1932. Another point to be considered, wrote Heidegger, was 'that nearly all the apparatus and equipment in the Institute of Mathematics and Physics is the personal property of Professor Königsberger, and if he is forced to take early retirement the University is likely to lose the use of it.'[152]

The fact that Heidegger takes a different line here makes it all the more probable that his uncompromising insistence in the case of Staudinger sprang in part from purely personal motives.

The 'academic summer camp' project

The drive to harness the resources of the nation's university teachers and students to the cause of national revolution, culminating in the Berlin training seminar for the Party elite, was intended as the positive counterpart ('The awakening of the German spirit') to the campaign of book-burning launched under the motto 'Against the un-German spirit'. It was to be 'a logical extension of the first campaign', as the head of the central office for the political education of German students, Georg Plötner, put it in a letter of 29 May 1933 to Heidegger's faithful follower at Freiburg, the junior history lecturer Rudolf Stadelmann, with whom he was discussing the concept of the 'political university'. After consultation with Heidegger, Stadelmann had declared his readiness to support the leadership of the German Student Union in the revolutionary struggle. In a letter to Berlin of 25 May 1933 he had observed: 'The political university is not so called because "politics" constitute its central preoccupation, but because it is composed of politically-minded persons.' In his reply, Plötner expressed the hope that Heidegger would make available an extract from his rectorship address or some other equally powerful piece in support of 'The awakening of the German spirit'. And Heidegger would of course be coming to the first conference of student leaders in Berlin.[153]

The purpose of this first training seminar organized by the Office of Science of the German Student Union on 10 and 11 June 1933 was to pave the way for a genuine 'university community, created through a relationship of trust between the university teachers and the students grounded in a new science'. It was intended to facilitate the forging of 'a real bond between the university and the working life of the nation, growing out of close collaboration between the students in their respective disciplines and members of the working population'.

Heidegger's rectorship address, in which he enjoined his threefold conception of service upon Germany's student – labour service, military service, and the service of knowledge – formed as it were the intellectual backdrop to the Berlin conference, at which Heidegger spoke about research and teaching, while Baeumler addressed the delegates on 'The representative bodies of the new university'. The fusion of the new-style student body with the new generation of university lecturers was to be accomplished in the setting of the 'academic summer camp' (*Wissenschaftslager*), a scheme that was entirely Heidegger's brainchild. He declared himself willing to set up a kind of model study camp at Todtnauberg, where a new core of politicized students and lecturers would be trained and indoctrinated. The purpose of the 'academic summer camp' was twofold: to establish a rapport of intellectual trust between university teachers and students through shared academic work, understood as comradeship in the common political struggle; and to introduce students to the working class.

The definition of science as it was discussed in Berlin is the one given in Heidegger's rectorship address: 'Science is the questioning standing-one's-ground [*das fragende Standhalten*] amidst the entities of Being that are forever concealing themselves within the whole. This active waiting and enduring [*dieses handelnde Ausharren*] is well aware of its impotence in the face of destiny.' The original character of science as defined here has been buried beneath the subsequent accretions of the Christian-theological interpretation of the world and the mathematical-technical thinking of modern times. But: 'The beginning continues to *be*', and has irrupted now into the future. 'It stands there as the distant dispensation over us, waiting to claim its greatness again.' Heidegger's anti-Christian attitude – 'the invasion of German intellectual life by Christianity' – was given more weight in Berlin: this was an exercise in training, after all. At the present time we have no way of knowing the actual details of what Rector Heidegger said in Berlin, but it seems to have been so controversial that the manuscript he was asked to submit was for circulation only to a select few, and was not intended for publication in any form.

According to the report that appeared in the August 1933 number of the Nazi journal *Der deutsche Student*, the student of the new university should be firmly anchored

> by his attachment to a science that is born of our spirit; a science that has awoken once again to the living reality of nature and history from the spellbound slumber of an unreal, sterile preoccupation with ideologies of all shades, to which it succumbed following the invasion of German intellectual life by Christianity – and which has now freed itself again from the prison of positivistic fact-mongering.

We cannot say for certain how many of these camps were organized and led by Heidegger during the summer and autumn of 1933. This was a time of great upheaval and change, when even the ruling elite of the National Socialist student union was being replaced and superseded, and much that went on behind the scenes was not apparent to a political rector living in the provinces. Since these 'academic camps' existed in a kind of grey area, without the official status of, say, the paramilitary training camps run by the student SA organization, and were somewhat elitist in character, it is extremely difficult to do them full justice in a historical study of this kind.

Two published memoirs that give an inside view of one of these academic camps may serve as a useful point of departure. On the one hand we have the account of a participant, the theology student Heinrich Buhr, and on the other, Heidegger's own remarks in the 1983 apologia.

The future Protestant pastor Buhr described his own experiences at the camp in Todtnauberg:

> I think it was in the autumn of 1933, in Todtnauberg (I was a young student of Protestant theology at the time) that I heard Martin Heidegger speak for the first time, when he addressed student representatives from the universities of Heidelberg, Freiburg and Tübingen. I was the only theologian in the group – and one who was fully committed to theology. Martin Heidegger made a speech against Christianity (that much I could just about understand at the time) – against Christian

> theology, against this whole interpretation of existence and reality. If one wanted to attack Christianity, he said, it was not enough to confine oneself to the second article of this doctrine (that Jesus was the true Christ). One must start by rejecting the first article, that the world was created and sustained by a God, that what exists is merely an artefact, something that has been made by a divine craftsman. This was the origin of that false devaluation of the world, that contempt for the world and denial of the world – and the source of that false feeling of comfort and security, founded on subjective ideas about the world that are untrue compared with the great noble awareness of the *insecurity* of 'existence'. That was roughly how I understood him at the time and how I have remembered it. It was nothing new to me, having read Ernst Jünger (I am thinking here of *The Adventurous Heart*).[154]

Buhr's recollection was not at fault, as reference to other passages in this book will soon confirm.

Heidegger gave his first detailed account of the Todtnauberg camp in his posthumously published apologia; these passages do not appear in the records of the 1945 denazification proceedings. He presented this camp – which was a failure – as part of his efforts to swim against the tide of Party doctrine and to eradicate 'the influence of Party officials'. In order to understand the complicated situation, it is necessary to quote at length from Heidegger's account:

> A rather curious preliminary to the winter semester of 1933/34 was the 'Todtnauberg camp', which was intended to prepare teachers and students for the work of the semester proper, to explain my views on the nature of science and scientific work and to put those views up for discussion and debate.
>
> Those selected to attend the camp were *not* selected because they were Party members or National Socialist activists. When the plan for the camp became known in Karlsruhe, Heidelberg also expressed a strong desire to send a few people down; Heidelberg also made arrangements with Kiel to that end.

> In a lecture on science and the universities I sought to clarify the central argument of the rectorship address and emphasize the importance of the university's role in the light of the aforementioned dangers. This led to fruitful discussions within the individual groups about knowledge and science, knowledge and faith, faith and ideology. On the morning of the second day the district student leader Scheel and Dr Stein suddenly arrived unannounced by car and had a deep discussion with the Heidelberg contingent at the camp, whose true function now began to emerge. Dr Stein asked if he might give an address himself. He spoke about race and racial principles. The address was listened to by the camp participants, but no subsequent discussion took place. The Heidelberg group had been given instructions to break up the camp. However, the real target was not the camp but the University of Freiburg, which did not intend to have its Faculties run by Party loyalists. This led to unpleasant and sometimes painful scenes, which I simply had to endure if I did not want the entire winter semester to be ruined even before it began. Perhaps it would have been better to resign there and then. But at that time I had not anticipated what emerged soon afterwards. I am talking about the increased hostility both from the minister and the Heidelberg group who influenced his views, and from the ranks of my own colleagues (Heidegger 1983, p. 35ff.).

The concept and internal structure of the first academic summer camp for faculty and students in Todtnauberg, arranged for the week of 4–10 October 1933, emerges fairly clearly from the private papers of Rudolf Stadelmann, then a junior lecturer in history at the University of Freiburg.[155] Heidegger assumed responsibility for the lecturers and assistants – though not in his capacity as rector – and 'for this first experiment' he had selected a small hand-picked team from the many who expressed an interest in the project. This group of instructors would carry out the 'work of the camp' in conjunction with the students, who had likewise been chosen with care, following the established ritual of Nazi camp discipline: marching off in a body from the

University, 'the company will proceed to the destination on foot' – no mean achievement, since the camp was a good distance from Freiburg – 'SA or SS service uniform will be worn; the uniform of the *Stahlhelm* (with armband) may also be worn'. The daily duty roster began with reveille at 06.00 hours and ended with the tattoo at 22.00 hours. The company would also make the return journey on foot, of course. In a letter of 22 September 1933, addressed to university teaching staff eligible to attend the camp, Heidegger outlined the goals as follows:

> The real work of the camp will be to reflect on ways and means of fighting for the attainment of the university of the future for the German mind and spirit. That requires:
>
> 1. The fostering of an awareness of the present state of our university system (students, teaching staff, regional structures and central government).
> 2. The lively inculcation of the aims of a National Socialist revolution in our university system.
> 3. Preparation for the next stages in the task immediately confronting us (internal organization of the 'house of comradeship' [*Kameradschaftshaus*]; organization, limits and provisional status of the student and teacher bodies [*Fachschaften*]; function of the faculties and their role in preparing students for a higher vocation).
>
> The work of the camp must not be dictated by an empty formal schedule. It must grow out of genuine leadership and allegiance, and must impose its own structure on that basis. A few formal lectures, delivered before the assembled camp, will serve to create the appropriate ambience and attitude. The vital group discussion sessions will furnish a basis and inspiration for general discussion later.
> The success of the camp depends on how much new courage we can muster, on our clarity and alertness for what is to come, on disencumbering ourselves as far as possible from the past, on the strength and resolve of our will to loyalty, sacrifice and service. From these powers springs true allegiance. And this alone can sustain and strengthen genuine German fellowship.

The leading figures – under the actual *Lagerführer* or 'camp leader', Heidegger – were the junior lecturers Stadelmann (representing the Freiburg contingent and also the Heidegger line), Johann Stein (head of the Heidelberg contingent)[156] and Otto Risse (leading the group from Kiel). They also stood for very different ideas about the direction the new-look German university ought to take. Otto Risse, a lecturer in medicine, was in a rather special position. Having qualified as a university lecturer at Freiburg in 1930, he was now head of Freiburg's Institute of Radiology; but he was regarded as a 'Kiel man' because from 1925 onwards he had worked for a while at Kiel University's Institute of Physiology and belonged to the group of Nazi activists at the University.[157]

Heidegger's uncompromising aim of 'a National Socialist revolution in our university system' evidently encountered opposition, which was directed more at the means than the end: witness the so-called 'Kiel action programme', which Heidegger viewed as nothing more than radical agitation. At any rate, it turned out that Heidegger's claim to leadership was not accepted without question. We need to recall that only days beforehand, on 1 October 1933, Heidegger had been officially designated as the '*Führer*-rector' of Freiburg University; although he had just declined the second offer of a position from Berlin University, he was to remain in close contact with 'government agencies' in Berlin with a view to exercising certain political leadership functions. In short, Heidegger's position was one of relative strength.

After just a couple of days tensions at the camp at Todtnauberg erupted into acrimonious dispute between the factions, and Heidegger seriously thought about calling an early halt to the whole venture. At the urging of Dr Stein and Dr Risse he decided not to take this drastic step; but he did expel a group of participants.

At all events, the picture that emerges from the original sources clearly contradicts the account given above by Heidegger. From the correspondence of October 1933 between Heidegger and Stadelmann it appears that the former used his follower Stadelmann as a kind of sacrificial lamb. Stadelmann was due to give a general introductory lecture on the new science, but was forced to abandon it on Heidegger's order and was even expected to leave the camp, stealing away secretly at dawn, before morning reveille, with no explanation given. Clearly this was a tactical move on Heidegger's part, aimed at defusing the atmosphere in the camp by the sacrifice of his trusted aide – in a disciplinary action that was very much Heidegger's style.

Stadelmann bowed to this 'order', but felt profoundly ill-used in his loyal allegiance by his '*Führer*' Heidegger. The exchange of letters referred to arose out of this incident, and from them it is possible to reconstruct the history and conception of the Todtnauberg camp.

The 'allegiance' repeatedly invoked by Heidegger was for him the key element in the structure of the new National Socialist university, as exemplified in the interplay between leadership and what he called, in the rectorship address, the genuine 'allegiance of those who are of new courage'. This 'allegiance' took on a profoundly personal quality for Stadelmann. Like the penologist Erik Wolf – and yet again in a different way from him – Stadelmann had placed himself at the disposal of the 'leader' Heidegger, the man who bore 'a distant dispensation' within him. Was it a total surrender? Presumably not, if the young historian, with his romanticized view of history, took proper account of Heidegger's dictum in the rectorship address: 'Every leader must allow his followers their own strength. But all allegiance contains within it an element of resistance. This fundamental contradiction in the relationship between leaders and followers must not be obscured, let alone obliterated.'

Immediately after his return from Todtnauberg Heidegger rounded sharply on Stadelmann in a letter: 'I had assumed you would leave the camp the following morning, and was consequently surprised to see you still in conversation with Risse.' The atmosphere at the camp had been damaging for everyone, he wrote. 'It became a test, both for those who stayed and for those who went.' As the 'leader' (*Führer*) Heidegger expected his loyal henchman to display the appropriate 'responsibility': 'We must now learn to "think different things together" in our minds: for example, that I advised you to leave the following morning, on the very day you had a special task to perform, and that notwithstanding this I assured you of my continuing confidence in you.' He knew very well, he wrote, that he was asking a great deal. 'But we must not avoid such situations; on the contrary, if they did not constantly recur we would have to seek them out and create them.' This yearning for tests and trials of strength, for subordination and confrontation, among grown men, scientists and academics, who willingly submitted to the discipline of camp life: to us, today, it all seems very strange and difficult to comprehend. Heidegger even uses the key word 'think' to convey the paradoxical nature of this experiential world [*Lebenswelt*]. There is undoubtedly a strong element of posturing here; at the same time, such thinking lies at the heart of Heidegger's philosophy of science and education. Of course, he himself pronounced sentence on the quality of his concept of 'the university of the future for the German mind and spirit', with which 'the real work of the camp' was concerned. But what does all this grand talk really amount to – this poking around in a fog of verbiage that bears little relation to any recognizable reality? The goal was simply defined as 'the lively inculcation of the aims of a National Socialist revolution in our university system'. What else besides? The work of the camp must not be dictated by an empty formal schedule. 'It must grow out of genuine leadership and allegiance . . .' Loyalty, sacrifice, service, allegiance, fellowship, 'German fellowship': was this anything more than hollow rhetoric, full of sound and fury, signifying nothing?

Heidegger ends his letter to his follower Stadelmann with the exhortation: 'Learn to become hard!'. Here it is again, that curious yearning for hardness and rigour – a fundamental characteristic of Heidegger during these years.

Called to account in this fashion his loyal henchman stood his ground, having learned from Heidegger that 'all allegiance contains within it an element of resistance'. In his reply of 16 October 1933 (Heidegger was then at another 'study camp' in Bebenhausen, not far from Tübingen, engaged in imparting his ideas to his Württemberg colleagues and students) Stadelmann elaborated on the fundamental dilemma of revolution and allegiance. He was not concerned about being in the right, or feeling sorry for himself (he wrote): there were more important things at issue. 'I don't suppose anyone actually passed the "test" of the camp' – not even Heidegger, therefore. 'But everyone came away with the great awareness that the revolution is not yet at an end. And that the goal of the university revolution is the SA student.'

He would supplant the earlier forms of student life. The goal was clear and in sight. 'And all those who desire it belong together. If they have a leader who is leading them towards that goal, they form a following – his following. And he can demand and expect a great deal from them.' Heidegger's position as 'leader' (*Führer*), he went on, was beyond question. The follower had to trust in his leader, to render dutiful service in the interests of saving the 'camp', even if it meant acting against his own better knowledge and conscience. And as leader Heidegger had renewed the bond of loyal service with a handshake, thereby reaffirming the basis of allegiance. Stadelmann had believed in this allegiance; but Heidegger had overstepped the mark and demanded only discipline. The disappointed Stadelmann ended his embittered letter thus:

> I shall never seek to evade this discipline. And in Todtnauberg I realized more clearly than ever before that my place is in the camp of revolution, not among the ranks of the opposition or the carping spectators. I shall maintain discipline – but I had hoped for more, I had believed in the possibility of true allegiance. That is why I was so saddened and shaken by the way things turned out.

It was out of this quasi-masonic amalgam of *Wandervogel* movement, Stefan George literary coterie and National Socialist revolutionary thinking that the transformation of the German university system was supposed to emerge. It is astonishing to think that in the hour of new awakening this insipid brew was dished up in lieu of the more nourishing fare needed if radical change was seriously being contemplated; as food for the journey it was woefully inadequate. The warrior heroes entered the lists only to indulge in shadow-boxing. They had no spectators, performing solely for their own benefit. The great deeds were being accomplished somewhere else – and soon turned to dark doings. Here they talked big – and did nothing. Stadelmann was not to be outdone by Heidegger in the endless formulation of revolutionary definitions: reflecting on the 'historical character of the German revolution', he declared 'that we are now in the midst of a third revolution in things German, which will finally establish definitive forms of national and ethnic existence for Europe.'[158] The 'revolution' of Martin Luther and the German uprising of 1810 were for him the quintessential German revolutions, both of which transformed the balance of power in Europe. And so it is today, as the Germans rediscover their national identity: 'It is a tremendous moment when the dormant soul of the nation becomes aware of its own power, and under a heroic leadership takes the step from passive association as a people to a dynamic national community capable of action.'

Stadelmann ends his article with the following words, taking himself by the hand, as it were: 'Only in revolution does the German realize his true character, for it is only in action that a man shows what he is made of' – precisely catching the rhythm and flow of Heidegger's thought and language.

But at that time, in October 1933, Heidegger's concern was to repair the damage and restore the bond of trust with Stadelmann, to explain why he had picked on him, as it were, to be excluded from the camp community, sacrificing him to the demands of the Kiel and Heidelberg factions. 'I know that I must now set about winning back your allegiance, which remains as important to me as it ever was': with this confession Heidegger ends his letter of 23 October 1933.

The close bond of allegiance between the two men was re-established, and Heidegger entrusted Stadelmann with the opening lecture in the University's public lecture series entitled 'The role of intellectual life in the National Socialist state'. Stadelmann's theme, 'The historical self-awareness of the nation', remained indelibly imprinted on Heidegger's mind, so much so that in July 1945 he was still – or perhaps then more than ever before – deeply preoccupied with this question, as we saw in the very first sentence of this book. 'The Germanic concept of allegiance as the starting point for a new national order': a characteristic quotation from this essay. Ideas of substance had been perverted into an ideology.[159]

The turning-point of the rectorship

In this chapter we shall consider Heidegger's rectorship from another perspective, namely as a test of the new pattern of university leadership. For although he still aspired to play a role at the national level, in fact the rector's activities from the autumn of 1933 were confined to his own university, which was seen as a test case for 'self-affirmation'.

A brief digression is needed here in order to fill in the broader picture. In *Facts and Thoughts* Heidegger puts forward one line of explanation for his resignation. As we saw above, he claimed there was a plot to have the Todtnauberg 'academic camp' of October 1933 broken up by a group from Heidelberg, sent by the district student leader Dr Gustav Scheel, who later became the Reich leader of students and university lecturers. In reality, Heidegger tells us, this campaign was directed against his own university of Freiburg, where Heidegger refused to have his deans appointed along party political lines. Heidegger relates the Todtnauberg episode to subsequent events during the winter semester of 1933/34, which was to bring him such bitter disappointments: specifically, the conspiracy between his Freiburg colleagues, the minister of education and the Heidelberg faction that controlled him – by which Heidegger meant *Gauleiter* Scheel and the rector of Frankfurt University, Ernst Krieck. 'It had become abundantly clear that certain circles within the University, who vigorously opposed anything that smacked of National Socialism, did not scruple to conspire with the ministry and the faction that controlled it in order to force me out of office.'[160] Whatever we choose to make of this, the truth of the matter remains unclear, and we have to accept the Heidelberg sabotage theory advanced by Heidegger. True, there is no mention of this in the official letter of explanation that he addressed at the beginning of November 1945 to the denazification commission and the rector's office at Freiburg; and he certainly says nothing that would even hint at the conspiracy theory. One could understand him suppressing this part of the story for tactical reasons: but why keep quiet about the business at the camp and the machinations of Dr Gustav Scheel, who in 1945 – as *Gauleiter* of Salzburg as well as 'Reich leader of students and university teachers' – was now a totally spent force? From 1945 onwards Heidegger's apologia went through various stages of revision, fine tuning, modification, shifts in emphasis and downright falsification, until it finally emerged in the version universally disseminated as *Facts and Thoughts*. This needs to be borne in mind throughout.

One claim that runs throughout, albeit with minor variations, concerns the two deans appointed by Heidegger on 1 October 1933, namely Erik Wolf (Faculty of Law and Political Science) and Wilhelm von Möllendorff (Faculty of Medicine). Heidegger's account of the affair in 1945 is succinct:

> The growing disapproval of my work as rector by the ministry was soon manifested in the impertinent request that I replace the deans of the Faculties of Law and Medicine (Professor Wolf, Professor von Möllendorff) with other figures, on the grounds that they were politically unacceptable. I refused to accede to this request and tendered my resignation.

In *Facts and Thoughts* this episode is elaborated in colourful detail. The conspiracy theory is documented by 'a smirk spreading across the face of student leader Scheel', who, according to Heidegger, was present at the meeting called in February 1934 at the ministry in Karlsruhe. What we *do* know has already been recounted at some length above: for it was during these same days of February that another professor from Freiburg, namely Hermann Staudinger, was summoned to the ministry for interrogation, as the record shows in precise detail (we even have the original shorthand protocol of the proceedings, which was placed in an envelope and attached to the other documents). Strange to say, there is no record of the episode that Heidegger relates: but an alternative version of events – no doubt the true one – has survived, as we shall see in due course.

When Heidegger claimed that certain circles within the University conspired with the minister against him, he was referring primarily to the Faculty of Law and Political Science, which numbered among its members such conservative professors as Grossmann-Doerth, Walter Eucken (both highly decorated veterans of the First World War), Baron von Bieberstein and Baron von Schwerin. Heidegger does not name names, but these are the people he had in mind; it was not Heidegger's style to be specific in these matters. Yet none of the above-named ever collaborated with the authorities in Karlsruhe, let alone engaged in a conspiracy to topple Heidegger. Heidegger did not have a very high opinion of his Faculty of Law. 'Things here are pretty grim, I'm afraid', he wrote in a letter of thanks to Carl Schmitt of 22 August 1933, acknowledging receipt of the latter's essay *Der Begriff des Politischen* [*The Meaning of Politics*] (then in its third edition – and suitably adjusted to the prevailing political climate).[161] Thus introduced, the two scholars struck up a dialogue. Emphasizing Schmitt's affinity with Heraclitus, Heidegger sought to enlist his aid: 'Let me just say today that I very much hope you will play a key role when the time comes to restructure the teaching of law from the inside, both in terms of its academic content and its pedagogic principles.'[162] In Freiburg, by contrast, Heidegger saw nothing but obstinacy and recalcitrance.

Meanwhile the thirty-one-year-old penologist Erik Wolf, a confidant of Heidegger's, whom he had appointed dean on 1 October 1933, soon became a target for criticism, not so much on political grounds, but because of the manner in which he exercised his office. Heidegger knew perfectly well that Wolf, who had served on the University Senate during the summer semester, was by no means universally popular in his Faculty. At the end of the 1933 summer semester, during the last meeting of the Senate, Eucken and Wolf had a bruising public quarrel, which prompted a frontal attack on the economist Eucken in the Freiburg student newspaper: there was no place for professors like him in the new era, wrote the paper's correspondent. On 8 August 1933, just a few days after the Senate meeting, Eucken called on the pro-rector, Josef Sauer, and complained that nearly everyone in the Faculty was outraged by Wolf's conduct. Wolf had fallen prey to fanaticism and failed to grasp the legal situation; and now he was so lost in admiration for Heidegger that he no longer had any time for normal human feelings. Eucken also reported that when Wolf was asked about the fate of his colleague and friend Gerhart Husserl his only response was a shrug of regret. And when Gerhart Husserl ran into Erik Wolf not long before, Wolf had apparently said: 'It's very regrettable, of course, that you are in such an unpleasant predicament. But it is a martyrdom sent by God, which you must bear with dignity, and bear alone: nobody else can help you.'

We now know a great deal about Wolf's private attitudes and feelings during those months of political turmoil from a long letter he wrote (but never sent) to Karl Barth;[163] and we know that he repressed everything that went on in those days, taking his cue in this from 'the master', Heidegger: illegal acts, harassment of individuals, dangerous tendencies. The fate of Gerhart Husserl, with whom he had been friends for years, is not even discussed.

This was the Erik Wolf who lost his way and became deeply involved during this period of the Third Reich, publishing two studies on legal theory: 'True law in the National Socialist state'[164] and 'The ideal of law in the National Socialist state'[165] – not yet the Erik Wolf of the Confessional Church. It was only later that he extricated himself at agonizing cost. But let us look at a few lines from the letter to Karl Barth:

> You were one of the first who came to me from Basel in 1945. We spoke, worked and lived together for much of the time. I felt impelled to offer you some kind of explanation for things that surprised you when you first heard about them: my involvement in Heidegger's rectorship and my two essays attempting to define the philosophy of law under National Socialism – and the various consequences that ensued. It is not a question of justifying myself. When I saw where the fallacy lay, I fought against it.

When Heidegger pointed out in 1945 and thereafter that in appointing Erik Wolf as dean in 1933 he was effectively appointing an opponent of National Socialism, he was superimposing the later Wolf on the earlier one. The Erik Wolf of 1933 followed the Party line – and indeed went even further.

The primary cause of the tension and strife within the Faculty of Law, and of the eventual isolation of Wolf, was the reform of the law syllabus and timetable, which has to be seen in the context of SA duties, paramilitary training camps and all the other extracurricular commitments that students had to meet – or would be required to meet as from the summer semester of 1934. Wolf clashed so furiously with his Faculty colleagues that on 7 December 1933 he wrote to the rector offering to resign. It is worth quoting from this letter because it throws an interesting light on the prevailing atmosphere and intellectual background. He is taking this step (he writes) in the very painful awareness 'that I am thereby creating momentary difficulties for Your Magnificence in the implementation of those aims and aspirations that Your Magnificence cherishes for this University'. From the enclosed letter to a Faculty colleague the rector would understand and accept his reasons, especially since he was also having problems with his health. He was suffering mental torments.

> I must leave it to the judgement of Your Magnificence, which discerns the deeper reasons that others cannot see, to determine whether the failure of my efforts to execute the functions of my office with success is due to the inadequacy of my abilities, my own personal shortcomings, my lack of diplomacy, or to the fact that the tasks entrusted to me met with a degree of opposition that could not be overcome, given the nature of the persons and the issues involved (Freiburg University archives).

What we have here, in essence, is an internal Faculty dispute that has been dramatized and blown up out of all proportion.

As far as Rector Heidegger was concerned, there was no question of Wolf resigning his deanship. 'The whole point of the new constitution and the struggle we are presently engaged in is that you enjoy *my* confidence first and foremost, and not so much that of the Faculty. It is precisely because you do enjoy my confidence that I am unable to release you from your very important office.' This was the real test case: leader and follower united by the bond of Germanic allegiance in the face of 'the struggle we are presently engaged in' – meaning the struggle against an outmoded Faculty structure. A little later, on 20 December, the rector notified all the Faculties and the entire academic staff that 'the fundamental issue of how far the Faculty will show by its future deeds that it is willing to co-operate in a positive spirit *has yet to be settled*.' And then, returning to the attack as if threatened by the imminent possibility of defeat, Heidegger formulated his credo thus: 'Since my first days in office the determining purpose and ultimate goal – which can only be attained by gradual degrees – has been the radical transformation of scientific education *in line with the dynamics and dictates of the National Socialist state*.' A merely formal overhaul of the university system was not enough. What was needed was 'an *inward restructuring*' of the lecture programme. He was grateful for 'any assistance, however small' that would advance 'the cause of the universities as a whole.

> It is unclear how much of our transitional work will survive. What *is* clear, however, is that no work or achievement, however successful, can ever become an occasion for thrusting individual accomplishment and zeal into the limelight. What is also clear is that only the unbending will to tackle the tasks that lie ahead can give meaning and substance to our present endeavours. The individual, whatever his place, counts for nothing. The destiny of our nation within the state counts for everything (Freiburg University archives).

As far as I know, this is the only such official declaration issued by the rector – a kind of general directive for the inner restructuring of the University 'in line with the dynamics and dictates of the National Socialist state.' This was no grand 'distant dispensation', but one born of helplessness; bogged down in vague generalities, it was merely a vehicle for militant slogans, and therefore worthless. All that was left of the 'self-affirmation of the German university' was the new constitution: and even this was only a formal amendment in Heidegger's eyes.

By December 1933 Rector Heidegger found himself caught in a vicious circle within his own university. Over the Christmas break, he tells us, he took the decision to resign the rectorship at the end of the winter semester. Small wonder in the light of such declarations, which already signalled Heidegger's farewell to a university reform based on authenticity. Indeed, his attitude is quite logical, given that Heidegger had aspired to a leading role – perhaps *the* leading role – within the new university system of the Third Reich. That hope had proved chimeric.

Only a few weeks had passed since he had informed the academic staff at Freiburg, with supreme optimism, that he would endeavour, through Baden's new university constitution, 'to pave the way for the co-ordinated development of the new national constitution for all German universities', keeping 'in very close touch' with the government authorities in Berlin.

The 'leadership principle' (*Führerprinzip*) was to be the driving force behind such endeavours. But in November the Party had imposed its own choice of leadership on the rectors' conference and its own policy on the reorganization of the university system, elevating mediocrities – but old Party stalwarts to a man – to the top positions. The 'Reich Association of German Universities' was formed, under the direction of a professor of psychiatry from Würzburg, who in turn nominated the leader of the German Conference of Rectors: he chose the rector of Jena University. Heidegger's name was not mentioned any more. The National Socialists did not want him on board, and had no further use for him. His bid for the leadership of the German university system in the new Reich had failed, and with it his hopes of mobilizing the will to carry out 'the historic spiritual and intellectual mission of the German nation as a nation that knows itself within its state'. Suddenly he was confined – and this time for good – to his own university, whose allegiance it was his business to 'dis-close' [*entbergen*]. But that was precisely where he failed. And how pallid, anaemic and vacuous they seem, the stage directions issued by Heidegger in December 1933.

'The failure of the rectorship – a thorn in my flesh', as Heidegger describes it to Jaspers in 1935. But was it his fault? In 1983 we read that his colleagues at Freiburg had abandoned him in his aspirations. The rectorship address had been 'spoken into the wind': 'During the whole of my time as rector none of my colleagues ever discussed my address. They were content to stick to the old, well-worn paths of faculty politics.' Heidegger complained that his address had been studiously ignored. Writing to the rector's office in November 1945, for example, he notes: 'The rectorship address, which was not printed in larger numbers than my 1929 inaugural address, was still not sold out in 1934.' Rightly so, perhaps? Was it not essentially an ephemeral piece, already out of date when Heidegger entered on the second half of his rectorship year? Be that as it may, the fact remains that as far as Heidegger himself was concerned, there was never any question of losing his way, going astray or erring. What he had spoken was the truth. The blame lay, simply and solely, with the failure of his listeners to understand.

But what of the question that is always asked: *was* Heidegger a National Socialist, in fact? In the introduction to the reprint of the rectorship address we are told that the words 'National Socialism', '*Führer*', 'Reich Chancellor' and 'Hitler' are not mentioned once in the text. This is specious nonsense, of course, since Heidegger uses these words often enough in his other speeches, proclamations and writings; we have seen examples enough already. The last speech Heidegger made outside Freiburg was on 30 November 1933 in Tübingen, under the title: 'The university in the National Socialist state'. The local newspaper, the *Tübinger Chronik*, carried a detailed report, which allows us to reconstruct the shape, tenor, line of argument and many actual passages from the speech: the following quotations from the newspaper report may be taken as highly authentic. The Tübingen speech was Heidegger's farewell to the mood of pioneering optimism that had prevailed in May 1933. 'One of the most forceful champions of National Socialism among German academics' – as Heidegger was described – was throwing in the towel. People talked about political students and political faculties, he said, but that was simply old wine in new bottles, an attempt to pay lip service to certain results of the revolution while the old system nodded along in its somnolent, introspective way. But the revolution was at an end, and had given way to evolution – as the *Führer* had said. And according to the newspaper report, Heidegger took flight again at this point:

> But the revolution in Germany's universities is not only *not* at an end, it has not even begun. And if indeed we have reached the stage of evolution, as the *Führer* has said, then it can only take place through struggle and in the midst of struggle. The revolution in Germany's universities has nothing to do with changing externals. The National Socialist revolution is and will become the total re-education of the people, the students and the coming generation of young university teachers.

Heidegger then unfolded the whole tableau of the rectorship address once again, reverting constantly to the theme of struggle. He ended his speech in Tübingen with these words:

> We of today are in the process of fighting to bring about the new reality. We are merely a transition, a willing sacrifice. As the warriors in this struggle we must be a hard race, that cares for nothing of its own, that rests firmly on the foundation of the people and the nation. The struggle is not about individuals and colleagues, nor about empty tokens and general measures. All genuine struggle bears some permanent mark of the image of the combatants and their work. Struggle alone reveals the true laws whereby things are brought into being. The struggle we seek is one in which we stand shoulder to shoulder, man to man.

Such was the gist of his speech according to the report in the *Tübinger Chronik* for 1 December 1933. But did he also identify with National Socialism as such – the conventional brand of National Socialism, so to speak? In *Facts and Thoughts* Heidegger himself tells us that following the rectorship address Baden's minister of education accused him of pursuing a kind of 'private National Socialism', which circumvented the aims of the Nazi Party's official programme and was not based on the concept of race. At all events, by virtue of his function as a thinker and custodian Heidegger certainly elevated the *Führer* Adolf Hitler – surely the ultimate arbiter in matters pertaining to the Party programme – to a kind of superhuman status with his declaration to the students (already quoted more than once): 'The *Führer* himself and he alone *is* the German reality, present and future, and its law.' Anyone with even a rough idea of the character of Heidegger's thought knows that the emphasis on 'is', this copula of logic, signifies in Heidegger's case more than just a conjugated form of the verb 'to be'; rather it intimates the presence of Being itself: 'the German reality, present and future'. It may be helpful for a discussion of this central issue to quote Jaspers' assessment in his reference of 22 December 1945. What he has to say is certainly relevant to our theme:

> I can accept to some extent the personal excuse that Heidegger was unpolitical by nature, and that the special brand of National Socialism he concocted for himself had precious little to do with the real thing. But in response to that I would first of all remind you of what Max Weber said in 1919: children who stick their fingers into the wheel of world history are going to get them broken. Secondly I would add this qualification: Heidegger undoubtedly failed to understand the true dynamics and aims of the National Socialist leadership. The very fact that he thought he could have a will of his own proves it. But his manner of speaking and his actions do have a certain affinity with the manifestations of National Socialism – which makes his error understandable. He and Baeumler and Carl Schmitt are three very different academics who each strove for the intellectual leadership of the National Socialist movement. To no avail. They brought their very real intellectual abilities to the task, only to end up blackening the reputation of German philosophy. So I agree with you that there is a touch of the tragedy of evil about it all.

But does this really take us very much further? The answer to that question will have to wait until we consider these issues again in conjunction with the turn in Heidegger's fortunes in 1945.

For the moment, however, our concerns are still focused on the end of 1933 and the beginning of 1934, when Heidegger claims to have reached his decision to resign from the rectorship and thereby end his political engagement. This did not prevent him from writing a devastating report in the Eduard Baumgarten affair at the end of December 1933, viewed from a wholly National

Socialist perspective and addressed to the Göttingen League of National Socialist University Lecturers. From what we have already seen of this document – 'Here at any rate Baumgarten was anything but a National Socialist' – Heidegger was evidently well versed in the criteria whereby the sheep were to be separated from the goats. No wonder he chose to give vent to his anti-Catholic prejudices around this time. On 22 December 1933, for example, he reported back to Karlsruhe in the matter of appointing a successor to the chair of ecclesiastical history in the Catholic Faculty of Theology, where certain difficulties had arisen. Heidegger took this opportunity to pass general comment on the nature of the Catholic Church:

> As with all future nominations the first question that must be asked is this: assuming the candidates are equally well qualified in terms of their academic record and character, which of them offers the better guarantee that the National Socialist educational ideals will be implemented? Since Catholic dogma places the Church above the state, all Catholic education – in so far as it truly is what it claims to be – must necessarily *subordinate the will of the state and the people* to the will of the Church. Consequently the Church forbids its priests to be members of the Party. This is why it is ultimately pointless to try and rate the candidates in terms of their political merits (State Archives, Freiburg, A5).

Heidegger raises a whole range of issues here. But one thing he has no doubts about is the incompatibility of Catholic doctrine and indoctrination with the fundamental tenets of National Socialism. This line of argument is also fuelled by his aversion to Christian (which for him always means 'Catholic') philosophy, which of course was part and parcel of his own background.

Elsewhere, too, Heidegger made no secret of his anti-Catholic sentiments. In the spring of 1934, for example, following the *Gleichschaltung* of the Catholic student fraternities, one such organization in Freiburg succeeded in getting its suspension at the hands of the Reich student leader revoked – a victory for Catholicism in Heidegger's eyes. His letter of 6 February 1934 to the Reich student leader Dr Stäbel, a boorish and fanatical character who belonged to one of the duelling fraternities, was quoted earlier:

> 'Such a public victory for Catholicism in this of all places [i.e., Freiburg] must on no account be allowed to stand. It represents a damaging blow to our whole enterprise, *the worst that could possibly be imagined* at the present time. I have an intimate knowledge of the circumstances and personalities here that goes back many years . . . People *still* haven't realized how Catholics operate – and one day that will cost us dear' (Schneeberger 1962, No.176).

Here Heidegger may have been thinking of his compatriot, patron and long-standing well-wisher, Dr Conrad Gröber, now Archbishop of Freiburg, who had been in residence since the summer of 1932. Heidegger let the connection lapse, until he looked him up again in December 1945, soliciting his support because his case was in a bad way. He may also have been thinking of certain members of the Faculty of Theology, some of whom he had known for well over a decade. At all events, in his letter to this mediocrity, this 'Reich student leader', he did not hesitate to divulge personal details, prophesying gloom and doom if the machinations of the Catholics were not foiled. Their 'whole enterprise' – meaning the great transformation of the university's teaching role into the embodiment of the nation's own will to knowledge and learning, which the rector had enjoined upon the students at the beginning of the winter semester – would be undermined to an unimaginable degree. It is difficult to write about this with a straight face.

Heidegger has given us an uncompromising account of the closing phase of the rectorship in *Facts and Thoughts*, an account that cannot be verified (or of course refuted) by reference to documentary evidence. No one would now accept Heidegger's dating (he speaks of resigning in February 1934) when it has been established beyond a doubt that he remained in office until 23 April. It might be objected that this is hardly a matter of any great significance. But a detail like this has to be taken together with everything else that Heidegger tells us about the circumstances surrounding the end of his rectorship. In the 1945 account the episode is described in much more matter-of-fact terms.

Nonetheless, as we have already pointed out, Heidegger maintained throughout that the ministry in Karlsruhe insisted on the dismissal of the two deans, Wolf and von Möllendorff – for political reasons. And it was because he could not consent to this (he claims) that he resigned the rectorship.

The facts as they emerge from the documentary sources may be briefly summarized as follows. No objections whatsoever were raised against von Möllendorff by the ministry. There *was* a question mark against Wolf, but not for political reasons. The opposition to Wolf within the Faculty of Law had continued, based only in part on personal animosities, with the result that Wolf's conduct as dean turned into an affair that would not go away – so much so that the minister of education informed Rector Heidegger on 12 April 1934 that 'very considerable misgivings, which I am inclined to think are not entirely without foundation' had been voiced in various quarters about Wolf's conduct as dean of the Faculty of Law and Political Sciences. The minister's letter continues: 'I wish to bring this to your immediate attention, and ask you to consider whether it might not be expedient to appoint a new dean for the start of the summer semester.' Considerable weight should be given to this very sober communication from the minister of 12 April 1934, when placed alongside the extremely dogmatic account given by Heidegger in *Facts and Thoughts*. According to Heidegger, both deans had been the target of a witch-hunt kept up by their respective faculties during the winter semester. He had attributed this to internal rivalries and disagreements, 'until I was summoned to Karlsruhe in the late winter of 1933/34, towards the end of the semester, to be informed by secretary for education Fehrle in the presence of district student leader Scheel that the minister wished me to dismiss these deans from their posts.' He had immediately replied that he was not prepared to accede to this unwarranted request. Fehrle had insisted; so Heidegger announced he was resigning his office and requested a meeting with the minister. 'As I said this a smirk spread across the face of district student leader Scheel. They had got what they wanted.' Then comes the fateful remark about a conspiracy between certain circles within the University and the National Socialists, aimed at forcing Heidegger out of office. So according to Heidegger these dramatic events reached their climax and conclusion in February 1934 with his resignation – at the very time, as we have already seen, when he had taken up such a clear position in the Staudinger affair.

The letter of 12 April 1934 that has just been quoted cannot possibly be squared with Heidegger's version of events, and not simply (to reiterate the point) because of the clash in dates. The remaining documents on file at Karlsruhe also fail to corroborate Heidegger's account at any point. Quite the reverse. From the ministerial letter of 12 April 1934 it is clear that this was the first word from Karlsruhe in the matter of Wolf and the deanship – and it was not even in the form of an ultimatum. But the main point is that the ministry obviously assumed that Heidegger would be remaining in office, and would see to it, in his capacity as '*Führer*-rector', that a new dean was appointed as requested at the beginning of the summer semester. Wolf made many mistakes as dean, which need to be borne in mind as the background to this affair. Added to this was the fierce dispute that erupted between Heidegger and Wolf on the one hand, and members of the Faculty of Law and Political Sciences on the other, over the man appointed to deputize for a vacant chair of economics in the summer semester of 1934.

The chair in question had been temporarily filled during the winter semester by the supernumerary Professor Adolf Lampe, a First World War veteran and a staunch nationalist, who was nevertheless firmly opposed to National Socialism – for which he had already come under fire during the summer semester of 1933. Subsequently appointed to a full professorship in Freiburg, Lampe was arrested in 1944 along with Gerhard Ritter and Constantin von Dietze in connection with the 20 July plot, since he was a prominent member of the so-called 'Freiburg circle', which maintained links with the resistance groups. Mention is made of this here because Lampe became one of Heidegger's fiercest opponents after 1945 and was very much instrumental in ensuring that Heidegger lost his teaching post. So at the time in question Lampe was widely viewed as 'unsound' in National Socialist circles. Heidegger and Wolf were firmly united in their opposition to any continuation of Lampe's interim appointment, on the grounds that he was too liberal and too lacklustre in his nationalism. Since the summer semester Lampe had taken a heavy battering from the National Socialist student body, which attacked the line he took in his teaching and research. As rector, therefore, Heidegger recognized the political importance of appointing the right candidate to the chair – and Wolf as dean loyally followed his lead. When Wolf blocked the renewal of Lampe's appointment at the end of March 1934, the latter went to see the minister and lodged an official complaint against his dean. This led directly to the ministerial letter of 12 April 1934, since now the matter could no longer be ignored. Heidegger's immediate reaction was formulated in his reply of 14 April 1934, which does not expressly take up the minister's unwaranted demand:

> Dear Minister:
> After careful consideration of the present state of our universities I have come to the conclusion that I must now return to my normal teaching duties in the classroom and lecture hall, free from the distractions of office. The new constitution is now in place, the institutional changes required thereunder have been implemented, and the new work has been set in motion. I therefore take leave to request that you appoint a new rector for the University of Freiburg with effect from the start of the summer semester 1934.

Initially Heidegger kept this letter of resignation secret. But when he heard about the official complaint lodged against Wolf, he wrote a sharply worded letter to the minister on 23 April 1934 – his last official communication as rector: 'I regard it as totally unacceptable that lecturers who are candidates for a vacant position should be granted an interview by the ministry while the appointment is still pending, *and* without the knowledge of their rector. After what has occurred I can accept no further responsibility for the appointment of a successor to the chair of economics.' Later that same day, 23 April 1934, Heidegger informed a meeting of the university's 'leaders' (rector, chancellor and the five Faculty deans) that he had tendered his resignation to the minister. However, we now know his real motives and the true circumstances. Heidegger used this essentially procedural issue as an excuse to throw in the towel.

His exit from public university life was less than dramatic.

Heidegger had not achieved his aims. The rectorship had turned out to be a failure – and not only the rectorship. The inflated rhetoric used by Heidegger as rector was without substance in reality: there was no student body 'on the march', as he put it: 'And they are searching for the leaders through whom they will advance their own appointed claim to a well-founded, knowing truth, and place it in the clear light of the interpretative word and work' (from the rectorship address of 27 May 1933). Instead there were just ugly confrontations between the rector and the Freiburg SA office and student functionaries, who behaved in a distinctly high-handed manner and set no store by allegiance.

'Every leader must allow his followers their own strength. But all allegiance contains within it an element of resistance. This fundamental contradiction in the relationship between leaders and followers must not be obscured, let alone obliterated. And struggle alone can keep that contradiction alive . . .' (from the rectorship address). But the scenes of 'struggle' played out during the winter semester of 1933/34 were merely squalid. There was plenty of wrangling and bickering, but no constructive work in that combative confrontation between leadership and followers envisaged by Heidegger – 'struggle' in the sense intended by Heraclitus in Fragment No.53. The claim was hopelessly at odds with the reality. But instead of admitting this, Heidegger looked for scapegoats whom he could saddle with the blame for this failure, driving them forth into the wilderness – the others, those who failed to comprehend the 'inexorability of that intellectual and spiritual mission which forces the destiny of the German nation into the mould of its history.' What else was to be expected from an academic staff for whose benefit Heidegger had had the four verses of the *Horst-Wessel Lied* printed on the back of the official programme for his ceremonial induction as rector on 27 May 1933, in order to get them into the right spirit, the spirit of National Socialism? In the 'address of the incoming [*antretend*] rector', as the programme notes put it – the German verb 'antreten' is also used of soldiers 'falling in' – the closing appeal is couched in the grand words we have already encountered. The 'incoming' rector, who occupied centre stage on 27 May 1933, bears no comparison with the rector who 'stood down' on 23 April 1934. Heidegger had come unstuck, most recently in his efforts to get a representative of the new spirit – the very opposite of the liberal theorist Walter Eucken – appointed to the chair of economics. In this too he had failed. Heidegger's resignation came as a surprise to the University of Freiburg as well. He did not attend the induction of his successor, the penologist Eduard Kern, and arranged for the report on his rectorship to be read out in his absence by the secretary for higher education. After all, he argued, the rector *was* appointed directly by the minister – an arrangement that Heidegger himself had introduced with his new university constitution for Baden.

Setting great store by the accuracy of his recollections, Heidegger claimed in his apologia that his successor Kern was described in the press as 'the first National Socialist rector of Freiburg University, who, as a veteran of the First World War, held out the guarantee of a martial and militant spirit and its propagation throughout the university'. In other words, his own rectorship had been classed as *non*-National Socialist 'by the Party and the ministry, the teaching staff and students' – once again, the sweepingly anonymous attribution. But this is simply not true. On 30 April 1934, following a frantic search for a new rector by the ministry, the relevant daily newspapers, including the local Freiburg Nazi rag *Der Alemanne*, published the carefully worded ministerial statement exactly as it was issued, without editorial comment (and I have compared the original statement word for word with the newspaper notices). The text reads:

> The Minister of Culture, Education and Justice, Dr Wacker, has accepted the resignation tendered by the present rector of the Albert-Ludwigs University in Freiburg, Professor Martin Heidegger, and has expressed to him his thanks and special appreciation for all the hard work he has done during his leadership of the University. The minister has appointed Dr Eduard Kern, Professor of criminal law and trial law, as the new rector of Freiburg University.

This is followed by a detailed resumé of the new rector's career, both academic and military, cataloguing all his military postings and decorations. No more and no less appeared in the Nazi-controlled press, following instructions from the ministry to the NS press office in Karlsruhe. It could hardly be otherwise, given that the resignation of the distinguished philosopher from the office of rector had to be played down as far as possible; he was, after all, the *Führer*-rector, the flagship, so to speak, of the armada. And would a newspaper like *Der Alemanne* have had any special reason to question Heidegger's National Socialist convictions, when only a few weeks before it had published Heidegger's famous radio talk 'Why we chose to remain in the provinces'? On the contrary: in its report on the induction of the new rector *Der Alemanne* made a point of noting that the secretary for higher education from Karlsruhe had thanked the outgoing rector for 'inculcating the spirit of National Socialism in our colleges and universities', and for his 'contribution to the reconstruction of our university system'. There can be no doubt that Heidegger was seen, particularly after his appointment as 'leader-rector', as the key figure in 'bringing the universities into the fold of National Socialist ideology', as the minister of education had put it in a letter to Heidegger of 2 October 1933.

In his diary entry on the induction of the new rector on 29 May 1934, Josef Sauer noted: 'The whole ceremony, including the dinner in the Hotel Kopf afterwards, felt like a suicide's funeral for Heidegger; nobody so much as mentioned his name.' And indeed, Heidegger's withdrawal from public life was more than just the end of a passing phase. Just how much Heidegger the thinker contributed to the stabilization of the Third Reich when he was rector, and how much influence he had on the student generation under that regime, cannot be precisely gauged. But it weighs heavy in the balance, it seems to me.

Heidegger's rectorship had proved a failure, bearing in mind that it was conceived as an enterprise that went beyond the administrative needs of Freiburg University to encompass the intellectual leadership of the New Germany exercised through the new-style university, through the new 'science' of philosophy as Heidegger understood it – a science resting on firm but 'trembling' ground, founded as it was on the ontological thought of the pre-Socratics. Such was Heidegger's real aim. But he was now forced to recognize that his idealistic beginnings had ended in miserable compromise, that the reconstitution of the university system on the 'leadership principle' had got bogged down in organizational trivia, that the crisp outlines of the grand design had become blurred. And so he gave up. His rectorship had collapsed and failed from within. How indeed could it possibly have succeeded, given the inherent banality of the movement? It marched in the spirit of Horst Wessel. It had nothing but scorn for the likes of Heraclitus and Parmenides. At the camps, Heidegger was mocked and derided as an intellectual dreamer. The 'academic camp' at Todtnauberg was a personal disaster for him. The journey to the place where 'the edifice of state' was to be realized was not even begun. Nowhere was there any sign that Germany's scientists, inspired and united, filled with new courage, were intent on creating the future university of the German nation within the state.

So Heidegger was thrown back on his own resources, excluded from the work of creating the German edifice of state, confined to the search for the meaning of history. He found the answer in Friedrich Hölderlin, to whom his thinking, of one company with Hölderlin's poetry, was henceforth devoted. Thinkers who hearken unto the word of the poet, even when it is 'as yet unheard, . . . stored up in the Occidental language of the Germans', know that history is 'that rare and timely moment', the *kairos*, the crucial event: 'History only occurs whenever the essence of Truth is determined in the beginning.'[166] 1933 was that crucial event, the moment of truth – but the Germans failed to recognize it. They turned their backs on the interpreter of the event. So the essence of Truth remained veiled, fleeing from the 'unique and unrepeatable [*einmalig*] time-space' to seek refuge in 'the immemorial incipience of the beginning'. The sacred that Friedrich Hölderlin had once – timelessly – invoked was poised to descend upon Germany in 1933 in the tempest and tumult of the transformation of existence, poised to 'establish by its coming a new beginning and a new history'. But the sacred degenerated into the saving grace of Adolf Hitler. The sacred hid itself from the light of day in the mantle of darkness. Through this kind of self-identification with Hölderlin Heidegger was able to outlive the years of German barbarism, always honouring what he took to be its true spirit.

It was the same with the 'political mission' he had spoken of in connection with the Berlin appointment in the autumn of 1933, which still beckoned after he had turned down the job offer: it began with vague generalities, and came to nothing in the end. The mediocrities of the old guard saw to that, men of the stamp of Ernst Krieck or Erich Jaensch.[167] Naturally the vulgar abuse to which Heidegger was subjected by Krieck as editor of the journal *Volk im Werden* (*Birth of a Nation*) from the spring of 1934 onwards was bound to hit him hard; there was a time, after all, in those heady early days, when he and his fellow-countryman from Baden had worked together on a policy for the universities in the spirit of National Socialism – until things turned sour between them in the winter of 1933/1934. But who was Krieck, anyway? And who paid any attention to *Volk im Werden*? We know that Krieck, a 'jumped-up primary school master', soon disappeared from the political scene when he became a professor at Heidelberg; as one of the 'old guard' he was treated with a certain dutiful deference, but it was as much as the Party could do to pay its respects on the appropriate anniversaries. So Krieck can hardly be said to represent the Party line. As for Alfred Baeumler[168], who had got to know Heidegger in the late 1920s via their common interest in Nietzsche, and had remained on harmonious terms with him for a long time following Hitler's seizure of power, writing him a glowing reference for the Berlin professorship, he may or may not have become Heidegger's declared enemy in Alfred Rosenberg's cultural *apparat*. It remains an open question.

One thing is surely clear: an anti-Heidegger group existed within the Party by the spring of 1934 at the latest, led by his former Marburg colleague Erich Jaensch and Ernst Krieck. They intervened with Rosenberg to ensure that Heidegger was not offered any kind of leading position in Prussia or in the Reich, because they did not wish him to be seen as 'the philosopher of National Socialism' – in the knowledge (and this applied particularly to Jaensch) of Heidegger's political views during the Marburg years and his private life at that time. The philosophical mediocrity represented by Jaensch was in direct ratio to his crude National Socialist ideology. On 26 February 1934 the *Reichsleiter* of the National Socialist League of German Medical Practitioners, Dr Walter Gross, later to become the head of the Party's Bureau of Racial Doctrine, wrote to the

NSDAP's Office of Foreign Affairs: he was constantly hearing reports from different quarters, he said, about Heidegger's activities in Freiburg, and noted that he had managed to acquire a broadly based reputation 'as the philosopher of National Socialism'. Since he had no way of judging Heidegger himself, he had recently solicited the views of Jaensch in Marburg, and in reply had received 'a totally negative memorandum', which Jaensch had prepared in response to a similar inquiry from Krieck. Apparently Heidegger was under serious consideration for the leadership of the Prussian Academy of University Lecturers. Gross proposed that the matter be taken up with Alfred Rosenberg, who should take action in 'this manifestly dangerous business'.[169] And *Reichsleiter* Rosenberg duly intervened. Jaensch's memorandum, formulated as a quasi-official Party report, formed the basis of a whole dossier against Heidegger.

The report from the Marburg philosopher Jaensch that Krieck mentions was placed in the files of the Prussian Ministry of Science, Art and Popular Education (*Volksbildung*),[170] together with the supporting documentation so assiduously assembled by Jaensch. It is a piece of scurrilous pamphleteering, unbelievably crude in its arguments and in its cheap disparagement of Heidegger's person and philosophy. He asserts that the appointment of Heidegger as head of the Academy would be nothing short of a catastrophe. It would fly in the face of reason 'to appoint to a position of crucial importance for German culture a man who has propped up the old system on all the crucial issues of ideology, and who in turn has been propped up by that system.' Heidegger's work and writings manifested 'all the signs of intellectual decay typical of that blighted epoch' to a greater degree than just about any other figure in German academic life. Heidegger was a man, he went on, whose past can never be buried, and who is incapable of even the outward semblance of *Gleichschaltung*, because his true face is set down in black and white in his own work, ready to be taken out at any time and displayed to the public. It would be contrary to all reason 'if what is possibly the most important post in the intellectual life of the nation in the weeks and months ahead were to be filled by one of the biggest scatterbrains and most eccentric cranks we have in our university system: a man about whom men who are perfectly rational, intelligent and loyal to the new state argue among themselves as to which side of the dividing line between sanity and mental illness he is on.' At Marburg Heidegger's thought had had 'a pernicious influence' on the educational process, since it attracted imitators and spread among young academics like an infectious mental illness. And *this* was the man now being groomed for even greater power! 'Is this quintessential decadent and archetypal representative of the age of decay now to be the focus of a movement that seeks to bring health and healing? We cannot believe that such a preposterous notion could ever become reality.'

For as long as Heidegger was a member of the Faculty at Marburg he had been the leader of a Jewish clique (Jaensch continues), and shortly before his departure he had pushed through the habilitation of the half-Jew Löwith, leaving him in place at Marburg. During his time at Marburg he had consistently championed everything 'that was directed *against* the nationalist groups.' He had been 'lauded' by those who perpetuated the old system, 'by Jews, half-Jews and the representatives of a neo-scholastic and distinctively Catholic philosophy'. He had entered into alliance with 'dialectical theology', which, according to Jaensch, had been tellingly dubbed the 'theology of crisis' because of its connections with the preceding era. 'Those who knew Heidegger here were deeply puzzled, therefore, when they heard that he had joined our movement.' A possible explanation lay in the fact that Heidegger always wanted to be revolutionary, always wanted to be at the head of things wherever 'revolution, subversion and the rejection of the *status quo*' are being fomented. This was why, during the past year, he had been able to 'sound off *without the slightest compunction*' to the students, railing against the University, the academic staff and the state of things in general, driven on by a fierce personal ambition that was reinforced, as everyone at Marburg knew, by the 'almost boundless self-importance of Frau Heidegger'.

Jaensch's diatribe continues in this vein for many pages. And evidently he felt the need to summarize his views about Heidegger, which he does in the following propositions:

1. Heidegger's thought is characterized by the same obsession with hairsplitting distinctions as Talmudic thought. This is why it holds such an extraordinary fascination for Jews, persons of Jewish ancestry and others with a similar mental make-up. If Heidegger acquires a decisive influence over the formation and selection of young academics, this will mean with absolute certainty that the selection criteria in our universities and intellectual life will favour those of Jewish stock who remain in our midst. These people, even if the non-Aryan blood entered their family a long time ago, will invariably take up this hairsplitting nonsense with alacrity, developing it and applying it in all kinds of different fields – and of course their academic careers will prosper accordingly, while our fine young Germans cannot compete because their minds are too healthy and they have too much common sense.

2. But the products of Heidegger's thinking – or rather of 'Heideggerian' thinking, since the disease of imitation has already taken hold – are not just the usual kind of hairsplitting sophistries, so familiar to us from the language of the past, but a special sort of sophistry, so extreme as to border on the pathological; one is constantly asking oneself how much of this is just eccentric and misguided in a normal sense, and how much already qualifies as the wanderings of a schizophrenic mind. Since this type of thinking is naturally exploited and propagated by canny scribblers and publishers with an eye to the main chance – a trend that began when Heidegger was offered the positions in Berlin and Munich – we shall end up with our universities in the grip of an epidemic intellectual sickness, a kind of mass psychosis.

To summarize briefly: (1) preference will undoubtedly be given to those of Jewish origins, together with their like-minded friends and followers, and (2) the disease in the higher echelons of our academic life, left behind by the previous era, will not be healed but made worse.

While we cannot go into the Academy project in any detail here (the reader is referred to the more detailed account in Farias, 1987, p.213 ff.), suffice it to say that in contrast to institutions like the Prussian Academy of Sciences or the Académie française the aims of *this* Academy were to be the 'training of young university teachers to become scientists and educators in the spirit of National Socialism' and the 'political motivation of the rising generation of young academics.' In the coming years the principal task of this establishment was to be 'the grooming of a young academic elite'; candidates for university teaching posts would be selected on the basis of their 'physique, character, fighting spirit and ideology, as well as their intellectual and scientific standing and their pedagogic abilities – in accordance, therefore, with the principles of a National Socialist leadership cadre.' If the Academy evolved along the lines envisaged, it could 'become the instrument of this purpose and retain a decisive influence on the whole of our scientific and research establishment even if the German university does become a truly National Socialist institution'. During the period of transition it could exercise a decisive influence on the character of the German university, with the potential 'to become, by its gradual development into a sort of *political* university, an integral part, in the first instance, of certain selected universities, and thereafter of *all* universities – thus placing it in a position to transform our universities from within'.

In his detailed report of 28 August 1934 Heidegger identified himself to a large extent with this conception, taking up a number of his earlier arguments:

> 1. The college for university lecturers must be structured in accordance with its *aim*. That aim is the *education* of those university teachers who are willing and able to bring into being the German university of the future. 2. The *education* of candidates for the teaching profession must seek (a) to awaken and reinforce a genuinely *educative* approach (role of university teacher not just to pass on results of own and others' research); (b) to rethink the science of the past in terms of the concerns and dynamics of National Socialism; (c) to foster a committed awareness of the university of

> the future as an *organic educative community* grounded in a consistent ideology.

Much more important to Heidegger than all the planning and organization were 'the personality, intent and ability of the leaders and teachers'. These were to achieve their purpose primarily through 'what and who they *are*, not what they "talk about"'.

The internal structure he envisaged was a blend of early Greek academy, Roman school of rhetoric, medieval university and monastic elements. He warned against mere training, a system of courses that 'one simply had to get through in order to reach habilitation'. Above all:

> The 'proposals' underestimate the demands and difficulties of 'academic planning'. If the American influence in contemporary academic and scientific life, already far too powerful, is to be effectively countered and eradicated in future, it is essential to allow the restructured academic disciplines to grow in accordance with their own inner exigencies. That never has happened and never will happen, except through the 'determining influence of individual personalities'. This does not mean the one-sided dominance of individual schools and tendencies, but is simply the dictate of 'struggle', which is 'the father of all things' in the intellectual sphere as well – especially there, in fact.

For all that Heidegger's reputation continued to be cultivated at the Prussian ministry of education in Berlin, those National Socialists involved in the formulation of academic policy were now set on to dog his trail. The ministry's plan to put Heidegger in charge of the proposed German academy for university teachers came to nought, as indeed did the entire project: there never was an academy. Meanwhile the National Socialist stalwarts now had the Freiburg philosopher firmly in their sights, and he was under observation from various quarters. Krieck, Jaensch and their cronies busied themselves constantly behind the scenes. A revealing light is thrown on these intrigues by a letter from the '*Führer*' of the lecturers at Frankfurt University to the German Lecturers' Association in Berlin

– written on 30 June 1934, of all days – which contained the following postscript:

> We have not yet entirely succeeded in ridding the Faculty of Philosophy at Frankfurt from the influence of the Tillich clique. There appeared to be links between this clique and the circle around Heidegger. 'You know who' has arranged to have a typical Heideggerian, Krüger from Marburg, appointed as a temporary stand-in for the vacant chair of philosophy this semester. We are keeping the matter under review and will be glad of your support as and when necessary.[171]

Indeed, Heidegger not only arranged the temporary replacement for Paul Tillich (the leading theorist of 'religious socialism'), who had been dismissed from his chair on 6 February 1933: he also pointed the moderates in the Faculty of Philosophy at Frankfurt in the direction of Hans Lipps at Göttingen, to whom the position was duly offered.[172]

PART FOUR

Heidegger's career after the rectorship

Return from Syracuse

In his apologia *Facts and Thoughts* Heidegger gives a concise account of the period that followed his rectorship. The full realization of the consequences of his resignation came after the Röhm *putsch* of 30 June 1934. In the wake of these events, he tells us, he remained under a cloud of suspicion, and was even the target of abuse. According to his account, Krieck and Baeumler were the prime movers of this campaign against him, which culminated in surveillance at the hands of the secret service (SD), checks on him because of his close connections with 'Catholic' pupils (particularly those who belonged to religious orders), restrictions on publishing and on his freedom to travel abroad and attend international conferences, and discrimination against former pupils of his who were candidates for chairs. 'Although they cut me dead in my own country, they tried to use my name for cultural propaganda purposes abroad to get me to undertake lecture tours. I was asked to go on lecture tours to Spain, Portugal, Italy, Hungary and Rumania, but I declined every time; nor did I ever take part in the Faculty's lecture programme for the troops in France' (1983, p. 41f.).

As early as July 1934 Kern, the new rector of Freiburg, applied to Karlsruhe for permission for Heidegger to travel to Italy. He explained that his colleague had been invited to lecture at the Istituto Italiano di Studi Germanici, founded by the Italian government. Heidegger received the congratulations of the government in Karlsruhe, together with the news that his proposed itinerary had been submitted to Berlin for approval. Strict currency regulations were in force, and for political reasons all foreign travel was centrally controlled. The Italian trip had to be postponed for the sole reason that Heidegger became ill. The Institute in Rome renewed its invitation in the spring of 1936, whereupon Heidegger and his family were allowed to travel freely to Italy and Rome – the first time in his life that he had been there. The ministry in Berlin even approved his applications to lecture in Milan and Pisa, though in the event Heidegger did not keep the engagement – probably because he ran out of time.

In Rome he met up with his former pupil Karl Löwith, as we saw earlier. In short, Heidegger's movements were not at all restricted, despite the campaign against him within the Party that was outlined above. Other trips were approved in due course: to Switzerland (Zurich) in 1935/36, to Vienna in 1936. Then came the pressing invitations to Italy, Spain and Portugal during the war (1942), which Heidegger declared himself willing to accept in principle, with the proviso that he could 'not undertake to give such lectures before the end of the winter semester, i.e. in the spring or early summer of 1943, because of my other work commitments on behalf of the *Wehrmacht*'. These work commitments presumably referred to the Faculty lecture programme in Freiburg. In the autumn of 1943 he was preparing for lecture tours in Spain (Madrid, Valencia, Granada) and Portugal; his proposed topics were 'Plato's theory of truth (the analogy of the cave)', 'On Aristotle's *Metaphysics*', and 'Hölderlin and the essential nature of poetry'. That these lecture tours did not take place was due to the progress of the war, which in 1944 was entering its final phase.

Perhaps the invitation issued by the Zurich student union in the autumn of 1935, which was promptly approved by the ministry in Berlin, deserves to be looked at a little more closely, because Heidegger's lecture 'On the origin of works of art' had triggered an illuminating controversy in the *Neue Zürcher Zeitung*. Heidegger gave his lecture on 17 January 1936 before a large, attentive and distinguished audience, as Hans Barth, the editor of the *Neue Zürcher Zeitung*, noted in his report, which he prefaced with a brief 'political' introduction that has an important bearing on the reserved response to Heidegger's political commitment. In effect, this early controversy during Heidegger's lifetime anticipates all the elements of the debate that is going on today. In his introduction Hans Barth wrote:

> A lot of water has flowed under the bridges of the Rhine since Heidegger revealed the recent results of his philosophical endeavours to a wider public. With the best will in the world it is impossible to regard his rectorship address of 1933 on 'The self-affirmation of the German university' as a significant statement of his thought. It is too sketchy for that. For that reason many will have awaited with eager anticipation the lecture he gave before an audience of Zurich students on 17 January. We must obviously account it an honour that Heidegger has chosen to speak out in a democratic state, given that he was regarded, for a while at least, as one of the philosophical spokesmen of the new Germany. Many people still recall, however, that Heidegger dedicated *Being and Time* in 'respect and friendship' to the Jewish philosopher Husserl, and that he offered his interpretation of Kant as an enduring homage to the memory of the half-Jew Max Scheler. That was in 1927 and 1929 respectively. Most people are not heroes, and that applies to philosophers too, although there are exceptions. So we cannot really expect a person to swim against the tide; only a certain sense of responsibility towards one's own past can raise the standing of philosophy, which of course is not just knowledge, but was at one time *wisdom*.

This review elicited a vigorous response from the literary scholar Emil Staiger, then a young man of 28, who accused Barth of having prefaced his piece with 'a political warrant of arrest that guarantees him the applause of the public', and of having picked holes in Heidegger's language only to travesty the argument of his lecture in a couple of sentences.

> 'But Heidegger does not rank alongside Oswald Spengler or Tillich – to name but two philosophers from diametrically opposed camps. Martin Heidegger ranks alongside Hegel, Kant, Aristotle and Heraclitus. And having recognized that fact, one will still find it a cause for regret that Heidegger ever got involved in current affairs – as indeed it is always tragic when different spheres become confused; but one will not, on that account, be shaken in one's admiration, any more

> than one is shaken in one's veneration for the *Phenomenology of Mind* [Hegel] by the image of the Prussian reactionary.'

The assessment of the rectorship address offered by the editor Hans Barth is of particular interest here: 'too sketchy' – a judgement that has much to commend it, as has already been shown elsewhere. It was time for people to stop dwelling exclusively on this one speech, in so far as they understood what was said in it. What mattered was that somebody had spoken thus, and with that claim to authority (*NZZ*, 20/23 January 1936).

It is clear enough that Heidegger was not for the moment in any danger, or in any way disadvantaged, simply because he had stepped back into the second rank and was being closely watched. Nor was he disadvantaged in the sense that his publications suffered; *Being and Time* was published quite freely by Niemeyer of Halle, in a series of editions, the last appearing in 1941 with the omission of the dedication to Husserl (see above, p. 178). It appears that Heidegger was first put under regular surveillance by the German secret service in the early summer of 1936, after his return from the Italian lecture tour. Whether there is any direct connection between this move and his stay abroad remains an unanswered question. But before we discuss the surveillance issue, let us take a second look at the trip to Rome.

Intimately acquainted though he was with the legacy of antiquity, Heidegger was visiting Rome for the first time in his life, taking this cultural experience on board as a man of forty-five. His ten days in Rome were well filled: a lecture on 'Hölderlin and the essential nature of poetry' on 2 April, another on 'Europe and German philosophy' on 8 April in the Kaiser Wilhelm Institute. As a Jew, Karl Löwith was not invited to the latter venue, as we have already seen; for the malign spirit of racism had long been stalking the corridors of an institution that owed its origin to a Jewish endowment. In between his lecturing engagements Heidegger drank deep of Italian culture and the Italian landscape: Frascati, Tusculum, Piazza Navona, the 'Moses' of Michelangelo in the half-light of San Pietro in Vincoli. He had actually spent the entire ten days in Rome in a kind of daze, wrote Heidegger on 16 May to Jaspers (only his second letter to Jaspers after 1933, and the last he would write to him during the Third Reich);

he had felt 'in a way annoyed and almost angry', and put it down to the flood of visual impressions he had absorbed, which were only now beginning to surface in his memory in some semblance of order – more immediate and vivid now than when he was actually in their presence. There is not a word about Karl Löwith in the letter; and yet the meeting with his old pupil from the Marburg days – the close friend and familiar who used to look after his children, the faithful companion whom he got to know through Wilhelm Szilasi in the early 1920s – was clearly an important event. Perhaps this was the reason for his mood of annoyance bordering on anger? We have already referred in this context to the brief chapter in Karl Löwith's memoir 'My life in Germany before and after 1933'. Löwith kept up an intense and insistent barrage of questions during these few days in Rome. There was an obvious point of reference, namely the controversy in the *Neue Zürcher Zeitung*, which was then only a few weeks old. Löwith disagreed with both Hans Barth and Emil Staiger, because he took the view that Heidegger's support for National Socialism was implicit in the very nature of his philosophy.

Löwith found it impossible to reconcile the spiritual world of Hölderlin invoked in Rome in 1936 with Heidegger's political credo. But not for nothing did Heidegger recite the seventh stanza of 'Bread and Wine': 'What to do and say in the meantime'. He spoke as one who *knew*, who stood guard over what was concealed and misunderstood – stood guard and stood fast: for Heidegger the two concepts merged into one, as he would later expound in the 'Letter on Humanism'. 'Standing fast' in an age of impoverishment, 'in the "no more" of the gods who have taken flight and the "not yet" of what is still to come'. In the letter to Jaspers of 16 May 1936 it all has a more prosaic ring:

> In fact we ought to consider it a wonderful state of affairs that 'philosophy' is without esteem, because now our job is to fight for philosophy in a quiet, unobtrusive way – for example, by giving a lecture on an essay by Schelling,[173] which may seem odd in itself. But sometimes it becomes clear what has happened and what we need: namely the serious knowledge *that* we need a lecture like that.

What a capitulation for a philosopher who had turned his back on the cultural critique of the Weimar years, the years of impotence, who went forth with confidence to embrace the radical transformation of the existence of things German and launched into a political adventure in order to win back for philosophy that 'trembling ground' from the 'groundlessness' to which it had been consigned. Now all he could do was to fight 'in a quiet, unobtrusive way' for a 'philosophy without esteem'. The Heidegger who felt dejected and annoyed in the summer of 1936 certainly had grounds for personal irritation: in Rome he had been taken to task by his own pupil Löwith for being able to sit down at the same table 'with an individual like Julius Streicher' at the so-called 'Academy for German Law'. Heidegger's only answer was to dismiss the rantings of the *Gauleiter* of Franconia as political pornography; for the rest he made excuses and insisted once again on dissociating the *Führer* Adolf Hitler from such repugnant characters.

On 14 May 1936 Dr Gerigk, the head of the 'Bureau for the Cultivation of the Arts' and of the 'Cultural Archive' in Rosenberg's Department of Foreign Affairs, asked the National Socialist League of University Lecturers in Munich 'what view is taken of the character of Professor Martin Heidegger, of Freiburg'. (The episode is recorded in the Federal Archives in Koblenz, under reference number NS 15/209.) The reason for his inquiry is not known. But it is curious that the League of University Lecturers should have been consulted, when Rosenberg's organization already contained a resident Heidegger expert in the person of Alfred Baeumler, head of the Department of Science. Dr Gerigk continued: 'His philosophy is closely tied to scholasticism, so that it is something of a mystery why Heidegger is able to exercise such an influence at times even on National Socialists.' Leaving aside this curious characterization of Heidegger's philosophy as being 'closely tied to scholasticism': the reply from Munich must have been so alarming that the Bureau for the Cultivation of the Arts sent the dossier to the Central Office of Reich Security ('Science Section') on 29 May 1936, with the request 'to contact Party comrade Dr Gerigk to discuss this assessment further'. The surviving records from Rosenberg's Department and other Party agencies are full of gaps; most of the files – notably those of the National Socialist League of University Lecturers – were destroyed. But the broad sequence of events seems clear enough: the answer from Munich was so explosive and negative that it was deemed advisable to have Heidegger put under surveillance by the appropriate section of the secret service. It has already been shown that a dossier on (and against) Heidegger was being compiled from the spring of 1934 onwards, based on material supplied by Krieck and Jaensch. And no doubt the file had grown thicker in the meantime.

Exactly what form the surveillance took or how intensively it was conducted we do not know. But in *Facts and Thoughts* Heidegger states that a certain Dr Hancke from Berlin worked in his department in the summer semester of 1937, displaying 'great talent and interest'. He soon put aside the mask: 'He said he could no longer conceal from me the fact that he was working on orders from Dr Scheel, who was then head of the south-west regional section of the secret service.' Heidegger accordingly sees a continuous link between the 'enforced' end of his rectorship (Scheel's role) and the surveillance operation of 1937. What is more likely is that Heidegger came under regular observation in 1936 – and for quite different reasons. The view within the secret service (according to Dr Hancke) was that Heidegger was collaborating with the Jesuits: hence the remark about Heidegger's philosophy being 'closely tied to scholasticism'. But what a hopelessly misguided view![174]

What is man?

Heinrich Ochsner, a friend of Heidegger's from his student days and a witness at the Heideggers' church wedding, had followed the political career of the great philosopher since the spring of 1933, looking on from a distance as a critical and ultimately baffled observer ('The fiasco of Heidegger's rectorship'). In a letter of 25 November 1933 he noted the powerful impression made upon him by Theodor Haecker's little book *What is man?*, which had just appeared:[175] 'One of the best and most profound books to have appeared in Germany in the last few years. There's a sentence in it that illuminates our present situation wonderfully, when he says that two people today can no longer talk meaningfully about the same thing, even though they use the same words.'[176]

It was indeed a courageous book, written against the spirit of the age, a convulsive gesture of defiance, accusing and offering hope, founded as it was on the unshakable rock of Genesis 1, 26. Haecker, an acute and clear-sighted observer of his age, wrote in a mood of underlying pessimism: 'Where are we going at present? What will the world be like, what will the New Order look like, when so much that is still intact on this planet is finally destroyed? We don't know. Nobody knows except God.'[177] The God in question was of course the God of the Old and New Testament, the God of Genesis: not 'a god who can still save us', and certainly not one of the Greek gods. Nor did this God of Christian revelation have anything to do with the *Führer*, to whom Heidegger was even then paying homage, a leader who vouchsafed to his people 'the immediate opportunity to make the supreme free choice: to decide whether it – the entire nation – wills its own existence or wills it *not*'.[178] This nation, which had risen up 'in like-minded allegiance . . . is recovering the *truth* of its will to existence, for truth is the apparency of that which makes a nation confident, clear and strong in its actions and knowledge' – to quote the insights that Rector Heidegger attributed to the German people on 11 November 1933.[179] So he could have no time for a Christian defeatism of the kind propounded by Haecker, who in the midst of the great revolution, when history is in the making, history as *truth*, could only lament: 'In such times, my friends, we must consider betimes what we should take with us out of the horrors of the present ravages. So then: as Aeneas took the *penates* first, so will we take the Cross, for we can always make the sign of the Cross before we are struck down. And after that: well, whatever one loves the most. But let us not forget to slip our Virgil into our coat pocket' (1933, p.17) – in other words, a humanism imbued with the spirit of Christianity.

Haecker's book, which took as its motto Genesis 1, 26, 'Let us make man in our image, after our likeness . . .', was a declaration of war on contemporary philosophy and ideology, and at the same time a comfort and support for thousands of people who were searching for answers.[180] Heidegger read Haecker attentively, and he understood the central issues of the book. He soon produced a fitting response in his course of lectures 'An Introduction to Metaphysics' in the summer semester of 1935.[181] Much of this course, which was published for the first time in 1953, is also a hostile critique of Christian philosophy, and more specifically of Haecker – a point that has been completely overlooked in all the published research. It is true that Heidegger mentions no names, that he speaks in code, comprehensible only to the initiated. Philosophers do not believe that the answer to the central question 'What is man?' is 'inscribed in the heavens'. The task of philosophical man, rather, is to follow in the footsteps of Heraclitus and Parmenides, in whose fragments the truth of Being is revealed: 'But then the essence and the modality of being-human can only be defined with reference to the essence of Being.' Revelation in the sense of Heraclitus and Parmenides is sharply contrasted with the revelation of the self-revealing God in the Word, the *logos*, of the New Testament: 'The message proclaimed from the Cross is

Christ Himself; He is the *logos* of redemption, of eternal life ... this is a world away from Heraclitus.' How very true! But what *kind* of a world might that be? The philosopher Heidegger is likewise a world away from the Haecker who contemplates the Christian life and lays himself open to the revelation of the Christian God. Hence the anathema hurled at him from the lecturer's platform in Freiburg:

> It is true that we now have books entitled 'What is man?'. But the question amounts to no more than a string of letters on the cover of the book. The question is not actually *asked*. Not because someone has been so busy writing books that he has simply forgotten to ask the question, but because he already has the answer to the question – and an answer that says we should not be asking the question in the first place. It is up to the individual whether or not he believes in the dogmas of the Catholic Church, and that is not the issue here. But for someone to put the question 'What is man?' on the cover of his books when he does not *ask* the question, being both unwilling and unable to ask it, is a practice that has forfeited any claim to be taken seriously. And when the *Frankfurter Zeitung* (for example) commends such a book, which asks the question on its cover but nowhere else, as 'an extraordinary, magnificent and courageous book', then even a blind man can see what a pretty pass we have come to.

There was only *one* book written at that time which had as its title the question 'What is man?', and the sneering references to 'someone' can only have been aimed at Theodor Haecker, whose little book enjoyed considerable success and went into its third edition in 1935. In the *Frankfurter Zeitung*, which had not yet been brought under Party control, the book received an excellent review, which so upset Heidegger that he pounced on the various points made by the reviewer and attacked them with unbelievable venom. Here was a different standpoint not his own, of course; hence the fact that he could only pour bitter scorn upon it, given that the asking of *this* question lay at the very heart of Heidegger's thinking. And so he continues:

> Why do I mention such aberrations in connection with the interpretation of Parmenides' dictum? In itself this kind of writing is without substance or significance. What is *not* without significance, however, is the present long-standing state of affairs in which the passionate urge to pose the question has been paralysed. The effect of this state of affairs has been to throw all accepted standards and viewpoints into confusion, so that most people no longer know where the real decisions have to be made and what the options are, if the strength of historical will is ever to be combined with the acuity and originality of historical *knowledge*.

The 'aberrations' with which Heidegger charges the clear-sighted Theodor Haecker, because he bases his interpretation of human existence not on a fragment of Parmenides but on the story of Genesis, can only be understood as the anathema uttered by a prophet. The 'passionate urge to pose the question' – and perhaps the question is only a rhetorical one – can also be seen as a rhetorical device. And perhaps the 'passion' itself turns out to be just empty posturing, as soon as one looks behind the words. With what right can Haecker be accused of 'throwing all accepted standards and viewpoints into confusion' by someone who has himself thrown restraint to the winds by making the *Führer* the standard by which all things and all reality are measured – including the reality yet to come! Heidegger was speaking just after the first anniversary of the Röhm *putsch*, as people were remembering the thousands who were murdered on that June night a year before. The concentration camps had become a tangible presence. Anyone who openly professed his Christian beliefs knew what he stood for and where he stood. The propaganda machine was now running in top gear, preparing the ground for the enactment of the 'Reich Citizens' Law' at the forthcoming Nuremberg Party rally. But during this summer semester of 1935 Martin Heidegger once again laid exclusive claim to a knowledge of 'the inward truth and greatness' of National Socialism and made a dogma of the 'will to know'.

Heidegger's attack on Haecker's book also needs to be seen in relation to other contemporary events – specifically, the vicious attacks to which Haecker was subjected by the National Socialist students at Freiburg University in May 1935.[182] Haecker was courageous enough to accept an invitation from Catholic students to speak on the subject of 'The Christian and history' on 13 May in Lecture Room I (Heidegger's lecture room).[183] The lecture itself was disrupted; but afterwards a group of 'incensed' students gathered outside the theological seminary (the Collegium Borromaeum) to demonstrate against this monstrous provocation. Up went the cry: 'Down with Rome!', 'Line the Catholics up against the wall!', 'Beat the Catholics black and blue!', 'Down with the Catholic dogs!', 'Hang the Jews!' And the Freiburg student newspaper carried a report on Haecker's lecture on 15 May which stated that he had not attacked the ideas of National Socialism directly because 'That is not how it is done. One is much too clever for that. One simply ignores them altogether. It is entirely understandable that the National Socialist students should oppose these methods used by political Catholicism. Their National Socialism required it of them.' The same issue proclaimed: 'We stand together in the fight against political Catholicism, Jesuitism, Jews and freemasons.' Of course I am not suggesting there was any direct connection between these incidents and Heidegger's lecture. But this was the ambience in the small university town of Freiburg, where everybody knew everybody else. By the end of 1935, at all events, Haecker had been officially banned from lecturing.

That same summer of 1935 Heidegger's lecture course 'An Introduction to Metaphysics', which amounted to a basic course in the doctrine of Being, an object lesson for beginners and more advanced students alike, was also attended by two Jesuits, Karl Rahner and Johannes Lotz, who had been sent to Freiburg in 1934 by the head of the Innsbruck order to study for a doctorate in philosophy. They had not been sent to Heidegger as such, but rather to his colleague Martin Honecker, who held the chair of Christian philosophy.[184] The story put about later – that the two Jesuits wanted to take their doctorate under Heidegger – is entirely without foundation, since at that time he would never have accepted a Jesuit as a doctoral candidate. Nevertheless, after 1945 the philosopher claimed these Jesuits as his 'pupils', playing the clerical card, as it were, in order to distance himself clearly from National Socialism. In *Facts and Thoughts* Heidegger writes: 'For a number of semesters the Jesuit fathers Professor Lotz, Rahner and Huidobro attended my senior seminar, and were frequent visitors to my house. One need only read their writings to recognize the influence of my thought – which indeed they do not deny' (1983, p. 41f.).

As a historian I can only presume to comment on Heidegger's lecture course 'An Introduction to Metaphysics' from a historical point of view. These lectures, primarily directed against the 'falsifying tendencies' of Christian philosophy, were obviously directed also against Heidegger's close colleague, Martin Honecker, who occupied the chair that proclaimed its commitment to that philosophy. It was only to be expected, given Heidegger's past career and background, that the phenomenon of Christian philosophy would be a constant sore point with him, and that he would attack the chairs of Christian philosophy that were guaranteed by concordats under international law. And that was precisely the situation he encountered at the University of Freiburg.

Since Heidegger's appointment in 1928, Honecker had had to endure this powerful and domineering thinker and teacher as his neighbour. The tension, already pretty intolerable, grew worse after 1933 as a result of Heidegger's political involvement, and thereafter he let it be known that the days of these ideologically affiliated chairs of philosophy were strictly numbered. The best way to ensure their demise was to work on the rising generation of young scholars. Consequently some of Honecker's pupils, particularly if they intended to embark on a university career, got caught up in this highly charged political field – most notably two aspiring candidates and near-contemporaries, Max Müller and Gustav Siewerth. Before we address ourselves to their 'fate', in so far as it was influenced by Heidegger, let us pause briefly to look at the careers of the two Jesuits Rahner and Lotz; for the sequel gave rise to a good deal of muddle and confusion that still determines our view of those events today, not least because of Rahner's failure to obtain his doctorate at Freiburg.

Honecker assigned topics from Thomist philosophy to the two Jesuits: for Rahner it was an epistemological problem from the

Summa theologica[185] of Thomas Aquinas, for Lotz it was a metaphysical explanation of the scholastic proposition 'Ens et bonum convertuntur'.[186] Rahner's draft thesis was rejected by Honecker, not, as is often claimed, because it was 'too much influenced by Heidegger', but because it failed to meet the academic standard demanded by Honecker. Lotz's dissertation, on the other hand, was accepted without question, praised by Honecker in a lengthy and detailed assessment that noted 'the serious intellectual effort displayed here'. Lotz, wrote Honecker, had submitted an excellent study, which 'deliberately kept within the framework of scholastic, or rather Thomist thought . . . without however subscribing to any of the rather feeble discussions of Thomism that abound in so-called neo-Thomism'. What Heidegger thought of such work – he served *ex officio* as second examiner for his colleague's doctoral candidates – may be judged from the following sentences, which constitute the sum total of his report:

> Within the prescribed framework, which decides the outcome in advance, the dissertation is an excellent piece of work as a product of a scholasticism that tailors itself to the times. The real 'systematic' roots of the problem of the 'interchangeability' of *ens* and *bonum* could only be brought to light by going back behind the existing body of scholastic doctrine and seeing the problem as originally one of Greek – and more specifically Platonic-Aristotelian – thought.

This is a typical ploy of Heidegger's, whereby he routinely discredited any philosophical inquiry that drew its inspiration from the Christian tradition. While the writer's achievement is acknowledged as 'an excellent piece of work', it is fundamentally flawed by the fact that he is partisan and in no sense unbiassed, trapped inside a system of thought in which the real question, the authentic question, cannot be asked. He is obliged to work within a prescribed intellectual framework, within which the answer to an essentially *un*authentic question is already given. Of course, this examiner's report was of a piece with the lecture course Heidegger gave in the summer semester of 1935, when he was himself wrestling with the faith of his birth. In these lectures he freed himself once again from 'extra-philosophical allegiance', meaning allegiance to Christian doctrine. Someone who accepts the divine revelation as truth cannot ask the Heideggerian question, since he always knows the answer in advance and accepts it as such: namely that God, as the creator of all that exists, is Himself uncreated from all eternity. The Christian believer, weighed down by this fundamental pre-conception, is not able to go down the same intellectual path – not unless he first abandons his faith. 'He can only act "as if".' In Heidegger's eyes all believing Christians are suspect anyway, since they have succumbed to indifference: they neither believe nor question – because it is easier not to. Their faith is 'an arrangement they have come to with themselves to cling on to the doctrine in future as a kind of legacy that has been handed down to them'. Heidegger's view is unequivocal: 'A "Christian" philosophy is neither fish nor fowl, a fundamental misconception.'

So it was high time to let young scholars in this discipline know what the score was and close off the un-philosophical path once and for all. The subject was to be killed off by starving it of resources. At his own university Heidegger proved himself an implacable enemy of Christian philosophy, taking belated revenge for the injury done to him earlier. That revenge came easily enough in the age of totalitarian National Socialism, which was only too glad to encourage anti-clericalism in the universities. When it was a matter of withholding teaching appointments – and hence security of livelihood – from newly qualified lecturers, the appropriate agencies within the Party were extremely obliging.

The full force of this harsh academic policy instigated by the National Socialists at Freiburg University was felt by two of Honecker's most gifted and hopeful candidates for habilitation, namely Gustav Siewerth (1903–1963)[187] and Max Müller (born 1906), who nurtured the untimely ambition of wishing to qualify as lecturers in philosophy of a specifically Christian kind. At the same time neither of these young scholars was to be outdone in his enthusiastic admiration for Heidegger, and in both their subsequent philosophical *oeuvres* the debt to him is almost total. Despite their treatment at Heidegger's hands, he remained their true mentor, whom they venerated and venerate still. Max Müller has given us a masterly account of the complexities of this relationship, and the pure paradox of it all.[188] He tells how Honecker had to fight hard to push through their habilitation in 1937 in the face of fierce political and ideological opposition. In the case of Siewerth's habilitation the difficulties started with the examiner's report – even though the candidate in question was one of Heidegger's most fanatical admirers, who in May 1930, when Heidegger had declined the first offer of a professorship in Berlin, had marched in the students' torchlight parade and found words of boundless jubilation for his hero. The report written by Heidegger as second examiner on Siewerth's application for habilitation was frosty and to the point: it cannot be said of this piece of work (he writes) 'that it pursues science without presuppositions; it is just that its presuppositions are of a somewhat curious kind. That it will continue to be academically represented at the University is guaranteed by the concordat.' Siewerth had attempted a speculative discussion of the principles of the Thomist system within the rigid framework of Catholic dogma, drawing on concepts and approaches developed by modern thinkers. At this point Heidegger waxes categorical and sits upon the judgement seat to deliver his verdict:

> But since any science, and philosophy in particular, is determined by the way it perceives and understands Being itself and the essential nature of truth and the place of man; and since these ideological presuppositions serve to shape not merely the content but also the approach and methodology of a science, and of philosophy in particular, the terms of reference for evaluating the present case are perfectly clear.

Heidegger, the prophet of the truth of Being, now hands down his verdict, even though he does not really consider himself competent to judge, and therefore expresses himself only in qualified terms: 'For a considered opinion as to whether this habilitation thesis meets the required academic standard within its own frame of religious reference the Faculty must consult my colleague, Professor Honecker, who alone is competent to judge.' However, Heidegger's judgement is in no way inhibited by his self-confessed lack of competence:

> My judgement must be stated in the following terms: if such arguments and analyses are considered valid and worthwhile

> as a means of defending and developing the Catholic faith, then the candidate's work must be accounted a significant academic achievement. But at bottom that is not a meaningful judgement, because the heart and substance of it, namely the underlying conditions on which any academic assessment of the work is predicated, are not for me to determine.[189]

This is quintessential Heidegger: at odds with himself, contradictory, underhand, simmering with resentment, equivocal in his judgements, refusing to accept responsibility.

Seven years previously, on 29 May 1930, when Siewerth was a student preparing for his final examinations, he had delivered himself of the following paean to Heidegger during a torchlight parade of students:

> But we would ask you to believe that the divine [*gottentstammt*] spark of 'enthusiasm', of all spirits the most productively creative, plays its unseen part in the dancing rows of lights you see before you, that this collective movement is the outward expression of what has been growing to fruition in the stillness of your labours: namely a sense of awe-struck reverence in the presence of the treasures and depths of the 'logos', which holds divine sway in the infant genius of language and in the eternal creation of the Greeks, in the ecstatic spirituality of the Middle Ages and the great intellectual battles of German Idealism.

Heidegger's grateful acknowledgement of the students' vote of confidence culminated in these words: 'You have owned your allegiance not to some philosophical point of view or system, but to the imperative that bids us hold on in the midst of existence. We must see things for what they are: in the present age there is no objective and universally valid knowledge or power we can hold on to; the only way we can hold on today is to hold our heads high.' The imperative of the hour was to 'hold on in the midst of existence'. But that implied the need to 'fight'. 'It is the business of the young to maintain the fight.'[190]

Such was the exalted rhetoric in 1930, when it was a matter of trying to 'hold one's head high' in a power-less and impotent age. By

1937 the issue of power had long since been resolved. But in 1933 the doctoral candidate Siewerth had failed to heed the *Führer*'s call, and had not entered the ranks of his followers to perform loyal service. He stuck to his own point of view. Perhaps Siewerth, who did the bare minimum in political terms, namely a mandatory stint in the Labour Corps, but was not a member of any Party organization, had actually believed what Heidegger said in 1930: that the philosophical system as such did not matter. Siewerth was no timeserver, and he bore the consequences that ensued from the 'political' reference submitted by Heidegger to the leadership of the League of University Lecturers: his application for a lectureship was denied on political-ideological grounds. Max Müller suffered exactly the same fate, as emerges very clearly in conversation with him. Once again, the examiner's report from Heidegger bristles with ambivalence. Although the writer had stated, in his preface, that he was 'not a Thomist', Heidegger disagreed:

> The writer is very much a 'Thomist', in so far as he begins by affirming the decisive theological questions that lie behind 'philosophy', and not only fails to call them into question, but actually rearticulates them, albeit dressed up in contemporary patterns of thought. So although there is a lot of talk about 'problematic questions' in this piece of work, this relates only to the particular field of dogmatic theology, which in itself is not at all 'problematic', and in which the crucial philosophical questions are not asked because they cannot be asked.[191]

Heidegger regarded the applicant Müller as 'eminently suitable for a Catholic professorship'. Nonetheless Heidegger effectively blocked the Faculty's requested lectureship for Müller on ideological-political grounds, again by sending a report in 1938/39 to the head of the Freiburg office of the League of University Lecturers, in which he emphasized Müller's negative attitude to the National Socialist state. Furthermore, he took the opportunity to reaffirm his basic position once more: a Christian, if he was honest, could not be a philosopher, and a genuine philosopher could not be a Christian. The universities must reject Christianity in favour of a radical new order. As long as Christianity continued to claim a contemporary relevance, it must be actively opposed. For that reason, he argued, no committed Christian of any denomination should in future be admitted to a university lectureship.[192]

An opportunity soon came to abolish one of the ideologically affiliated chairs of philosophy that he so hated, when his colleague at Freiburg, Martin Honecker, died prematurely in October 1941. Together with other elements in the Faculty who shared his distaste for a 'subjective' philosophy tied to a particular ideology or creed, Heidegger acted in accordance with the 'objective spirit' of National Socialist ideology to ensure that the vacant chair, guaranteed under international law, was now renamed. The new incumbent, appointed on 1 September 1942, was the psychologist Robert Heiss[193], in whom Heidegger found a loyal supporter – especially after 1945, when he did his best to sustain Heidegger's position.

Now that the late Honecker's chair had been stripped of its original function, his former pupil and departmental assistant Dr Heinz Bollinger was allowed by Heiss to remain at his minor post. In March 1943 Bollinger was arrested in Freiburg because of his association with the 'White Rose' resistance group, and on 19 April 1943 he was given a long jail sentence for complicity and failure to report the conspiracy to the authorities. His friend Willi Graf was executed, following in the footsteps of Professor Huber at the University of Munich, Sophie Scholl and her brother Hans Scholl.[194] Huber, Graf and Bollinger were all products of the Catholic youth movement, the two last-named being members of the 'New Germany League'.

To this day the identity of the informant who first tipped off the Gestapo about Heinz Bollinger remains unknown. There is a good deal of evidence to suggest that the information came from someone inside the department now chaired by Heiss at the Faculty of Philosophy. In their interrogation, which lasted several weeks, the Gestapo were mainly concerned with investigating the circle around Bollinger in Freiburg, in order to establish the extent of the 'conspiratorial activities'.

Those who were arrested and investigated were confined to a small circle of Bollinger's acquaintances. Robert Heiss, with whom Bollinger had had 'political' discussions on occasion, was spared, although Bollinger had told the Gestapo about his conversations with Heiss. There was no question of extending the investigation to Heidegger, because he and Bollinger had never had a conversation or indeed any contact of any kind. Heidegger had simply served earlier as the second examiner for Bollinger's dissertation (on Max Scheler), which had been supervised by Honecker. Heidegger's report followed the now-familiar pattern, reiterating the usual deep-seated reservations about a Christian, and therefore non-philosophical approach. Heidegger's philosophical institute and Heidegger himself were never investigated in connection with the 'White Rose' affair. Consequently, there is no basis for the 'exonerating' remarks in Heidegger's *Facts and Thoughts* – like the claim that the Gestapo conducted an investigation into the background of Dr Bollinger 'in connection with the student protest staged by the Scholls in Munich, for which they expected to find supporters and sympathizers in Freiburg and in my lecture courses' (1983, p.42). Heidegger's own domain remained completely untouched by these events, being quite distinct and separate from the department formerly chaired by Honecker – and not just in terms of its intellectual orientation.

Meanwhile Professor Heiss was furious that Bollinger, the departmental assistant bequeathed to him by Honecker, should have stooped to this shameful treachery and brought the chair – his chair – into disrepute. In the circumstances Honecker's widow Irmgard wrote a letter to Heiss on 29 April 1943 that she considered sufficiently important to keep a copy for her own files.[195] 'I cannot get the Bollinger affair out of my mind. I have thought back over our acquaintance, and I am bound to say that he seemed a quiet, steady and reliable sort of person. And I am quite sure that my husband appointed him as departmental assistant because of these qualities.' Other people in Freiburg also recalled Bollinger as a modest, retiring man. Frau Honecker went on: 'That is how I remember him too, and I can only think that he got drawn into this affair through a certain naïvety on his part. But it would distress me greatly if the name "Bollinger" were to cause you any embarrassment or difficulty in your department, and I hope that you will be able to tell me more about the whole business at some convenient time.' So the widow of Heiss's predecessor felt it necessary to justify herself because a potential 'enemy of the people', a member of Catholic circles, had been appointed by her husband as departmental assistant. Such was the grim reality of everyday life during the dark years of Nazi terror, when human beings and human life counted for nothing. We might recall the words that Heidegger wrote to his loyal follower, Erik Wolf, in December 1933: 'The individual, whatever his place, counts for nothing. The destiny of our nation within the state counts for everything.'

Philosophy derided, or: what is humanism?

At the Central Office for Science in Rosenberg's Department, the staff of the 'Bureau for the Observation and Evaluation of Science' (such were the complexities of the bureaucratic machinery) were also busy collecting information on Heidegger. Several files have been preserved, which illustrate the methods of intelligence-gathering. These range from observations on Heidegger's influence on the university curriculum and notes on Heidegger's membership of editorial committees to the imposition of specific censorship restrictions, which Heidegger mentions in his apologia. Specifically, the censorship order applied to the essay 'Plato's theory of truth', which appeared in the second volume of the *Jahrbuch für geistige Überlieferung* (1942) edited by Ernesto Grassi: a directive banning any mention of the essay in reviews was issued by the Reich Ministry of Propaganda.[196]

This act of censorship was only one aspect of a much larger issue, which had to do with the regulation and prescription of language in the whole area of the humanities. Grassi, at that time honorary professor at the University of Berlin, had been an enthusiastic student and disciple of Heidegger's since 1928, and was a close friend of the Hungarian Jew Wilhelm Szilasi, of whom we have already heard a great deal, and whom we shall meet again later. His main job in the Reich capital was to act as a kind of cultural go-between or promoter of Italian culture, and to arrange interdisciplinary meetings between German and Italian scholars working in the humanities. Sponsored by Italy's Fascist government, the enterprise naturally had a foreign policy dimension as well, and Grassi acted with the full support and backing of his government. After the Second World War Grassi made a distinguished career for himself: after an interlude at the University of Zurich – where people had considerable reservations about him on account of his political past – Grassi became a professor at the University of Munich and director of the University's Institute of Philosophy and the History of Humanistic Thought.

Grassi's cultural programme was viewed with suspicion by the Central Office for Science in Rosenberg's Department, which followed his activities with an eagle eye. At the outbreak of war Grassi succeeded in founding the *Jahrbuch für geistige Überlieferung* in close collaboration with Walter F. Otto, who was teaching at the University of Königsberg, and the classicist Karl Reinhardt, who was teaching in Berlin. The first volume appeared in 1940, and addressed itself to specific aspects of the classical heritage, seeking to 'illuminate the thought and world-view of the classical world', to 'explain the nature and character of humanism and the Renaissance and their interpretation of the classical world' and finally to 'examine nineteenth- and twentieth-century attitudes to the classical world'. In November 1941 the *Nationalsozialistische Monatshefte*, a journal closely identified with Rosenberg's Department in which Grassi himself also published, printed a lengthy review of this first volume by the Protestant theologian and student of religious thought Wilhelm Brachmann, a leading member of the 'academy' planned by Rosenberg. The principal purpose of Brachmann's essay, entitled 'The ancient world and the present. Notes on the problem of contemporary humanism in Germany and Italy', was to establish a standard terminology and set of definitions for the term 'humanism'.[197] Brachmann subsumed the work of Grassi, Otto, Reinhardt and others under the concept 'contemporary humanism' in its Italian and German forms, though still taking the term in its conventional sense, as opposed to the concept of 'political humanism', which now had to be explained and promoted. Citing the work of the notorious racial theorist Hans F.K. Günther, he wrote that the 'natural affinity of the Greeks and Romans with the German people' needed to be emphasized: 'Classical antiquity as the supreme example in world history of what Indo-Germanic culture is capable of achieving – that is the central concept of political humanism.' (p.926) What was at stake was 'the real presence of an Indo-Germanic culture imbued with a sense of its own destiny'. Finally Brachmann advanced the following sweeping claim:

> From this it follows that the term 'humanism' will have to give way to the term 'Indo-Germanic intellectual history', which expresses more clearly than any other term the specific concerns of German political humanism. It stands guard for the genetically-determined intellectual inheritance of Indo-Germanic culture in general, and thus also – and indeed primarily – for the legacy of classical antiquity. At the same time, what we call 'contemporary humanism' is certainly to be welcomed as a pioneering guide to the profound riches of this legacy. (p.932)

This definition was elevated to a quasi-official status by Rosenberg's Department. And with this definition in mind, manuscripts were subjected to a particularly close critical scrutiny – including Heidegger's piece on 'Plato's theory of truth', an interpretation of Plato's cave analogy (*Republic* VII) which he had presented for the first time as a lecture to the Benedictine monks at Beuron in the autumn of 1931.[198] In this extremely difficult text, which presupposes a sound knowledge of the Greek language and Greek thought, the censors came across a passage that provoked their wrath. Heidegger had dared to kick against the pricks by not following the line laid down by Brachmann, choosing instead to modify terms and definitions at will in line with his own thought:

> The beginning of metaphysics in Plato's thought also marks the beginning of 'humanism'. This term is to be seen as essential to this discussion and therefore understood in its broadest possible meaning. In this sense 'humanism' refers to the process encompassed by the beginning, development and end of metaphysics, whereby man moves in all kinds of different ways, but always *knowingly*, towards a centre of being in the world, without thereby becoming the highest entity of being in the world. The term 'man' can refer here to mankind or humanity in general, the individual or a community, a nation or a group of nations. Within the fixed metaphysical framework of being in the world the aim is always the same: to lead 'man' as defined here, the *animal rationale*, to the liberation of his capacities,

> the certainty of his destiny and the safeguarding of his 'life'. This may take the form of shaping his 'moral' attitudes, redeeming his immortal soul, developing his creative powers, training his intellect, cultivating his personality, awakening his public spirit, disciplining the body – or a suitable combination of some or all of these 'humanisms'. Each of these endeavours is a metaphysically determined attempt to encompass man in greater or smaller orbits. As metaphysics comes to perfection, so too 'humanism' (or anthropology, to use the Greek term) strives to take the most forward and extreme – which is to say, absolute – 'positions'.[199]

Dr Erxleben, an administrator in the Central Office for Science who knew Heidegger from the old days, having met him through the German Student Union in the summer of 1933, did everything he could to block Heidegger's article, seeking to have it withdrawn completely; he was not content simply to omit the offending passage on humanism. In a letter of 17 June 1942 to Dr Lutz at the Reich Ministry for Popular Enlightenment and Propaganda (Goebbels' ministry), he writes that he has tried in vain to contact him by telephone over the last few days, and is now enclosing his office's written report on Heidegger's article:

> I think it would be wise if Professor Grassi were to drop Heidegger's article altogether. The position adopted by Heidegger on the central problem of humanism is calculated to lend support to the claims now being made by the Italians even with regard to German scholarship. By saying that it does not essentially matter whether 'humanism' is understood in a Christian-theological sense or a political sense, he is undermining the position recently set forth by Party comrade Dr Brachmann in the *NS-Monatshefte*, in which he states quite clearly that there is no longer such a thing as 'contemporary humanism' for us in Germany, and goes on to explain the concept of 'political humanism' as distinct from contemporary humanism. This is a position that we wholeheartedly endorse.

> Heidegger's tendency to support Grassi's efforts aimed at a revival of contemporary humanism within the German intellectual world as well can only serve to confuse the issues in the current debate. It seems to us that the removal of an isolated sentence from Heidegger's text will not suffice to eradicate this tendency. So while we fully acknowledge the scholarly significance of Heidegger's essay, we are unable to approve its publication in Grassi's journal.[200]

The basic bone of contention, then, was Heidegger's relativizing concept of humanism, which could not be reconciled with the 'political humanism' favoured by Rosenberg's staff. Indeed, nothing could better illustrate the total incompatibility of primitive Nazi ideology with Heidegger's thought than this central comparison between Brachmann and Heidegger. We can only speculate as to whether Heidegger wrote his piece in the knowledge of Brachmann's review, or indeed whether he ever learned – from Grassi – the true facts about the workings of the National Socialist propaganda machine. For in this case Rosenberg's officials failed to get their way. *Il Duce* intervened in person, via his ambassador in Berlin, Alfieri, and secured an assurance from Goebbels that the *Jahrbuch* would be published without cuts – and therefore, as Erxleben noted with irritation in a memorandum of 3 July 1942, '*with* Heidegger's essay'. On the other hand (and this was a compromise of sorts) Goebbels' ministry did agree, on the recommendation of the Rosenberg group, to 'issue a directive on terminology and usage for the press, whereby reviewers were forbidden to mention Heidegger's essay'. This was the censorship order to which Heidegger later refers – and that was the story behind it. Erxleben's memorandum went on to record that Lutz in the Ministry of Propaganda had told him of the plan 'to bring out an Italian translation of Heidegger's complete works. The two ministries agreed to keep in touch on this matter.'

Compared with this frontal assault on Heidegger's philosophy at the highest Party level, the attacks on him by the local Party organization in Freiburg look pretty muted and provincial. But since it was to these that Heidegger pointed in 1945 as particularly damning evidence of the regime's hostility towards him, they will bear closer examination. It all revolves around an incident that we have already touched upon in connection with the Staudinger affair.[201] The point at issue was a lecture entitled 'The metaphysical foundations of the modern world view' that Heidegger gave in Freiburg on 9 June 1938, as part of a lecture series on 'The foundations of the modern world view' organized by the Society of Art, Natural Science and Medicine in Freiburg. The organizers wanted to 'make a valuable contribution to the great work of our times', which *Reichsleiter* Alfred Rosenberg had defined as 'the vanquishing of the Middle Ages' – as the written record shows. Indeed, they were convinced they had thereby 'done a signal service to National Socialism here in Freiburg'. Small wonder, given that the author of *The Myth of the Twentieth Century*, *Reichsleiter* and chief Party ideologist Alfred Rosenberg, had addressed a major rally outside Freiburg Cathedral the autumn before, in October 1937. He spoke in full view of the archbishop's palace, where the unloved archbishop Conrad Gröber resided – Heidegger's compatriot and former patron: and his words were intended to strike fear into the hearts of Freiburg's Catholics. The press reports were correspondingly fulsome, particularly those in the local Nazi daily *Der Alemanne*, and the Freiburg Party leadership was still basking in the afterglow of its triumph at the beginning of 1938.[202] Such were the events that set the scene for the lecture series in the summer semester of 1938. Heidegger's lecture was to be the climax and conclusion of the series. The organizers billed his address as 'an intellectual event of extraordinary significance'.

All the greater, therefore, was the irritation provoked by the offending report in *Der Alemanne* – a scurrilous and mediocre hatchet job on Heidegger's lecture: worse than that, it was a vicious blow below the belt aimed at Heidegger's philosophy, indeed at philosophy in general, delivered by a miserable hack who was out of his depth and wrote from the perspective of the gutter – albeit with the backing of the paper's chief editor, a certain Dr Goebel, who had already allowed his paper to be used as a mouthpiece for an attack on Heidegger.[203] The first few sentences will convey the flavour of the piece:

> Nobody sits an exam in philosophy in Germany, apart from those whose heart's desire is to get a doctorate in philosophy. It is up to the individual to read it up from books at home, or to enrol at a university under one of the thirty-six full professors still teaching the subject. This is an important point that should be borne in mind, because there have been a lot of changes in this field compared with how it used to be. Philosophy as such has fallen out of favour, and people no longer worry their heads unnecessarily about metaphysics and system-building, or get involved in fruitful or fruitless arguments about fancy made-up words. The fights and squabbles between philosophers are not of the slightest concern to us, and who can take it amiss if one professional philosopher chooses to write about another in these terms: 'It is a matter of presentation that one of the most successful anthropologies, Heidegger's *Being and Time*, describes itself as a fundamental ontology, which looks at man in his passive role rather than his active one. The spurious originality of Heideggerian existential analysis is quickly disposed of with a classic observation of Kant's: "The coining of new words, when language does not lack for terms to describe existing ideas, is a childish attempt to stand out from the crowd, if not by having true and original thoughts, then by putting a new patch on an old coat." Very few people are in a position to judge whether Heidegger's fellow philosopher is right, for Heidegger's works are not written in a style that makes them accessible to all – and for that reason his fame has been much enhanced.'

The reporter, a certain Party comrade Graf from Württemberg, proceeded to make merry over the lecture in this manner, though he had to admit that the event was very well attended: 'The general interest in philosophical problems is still very strong in Freiburg, and in a certain form has never gone out of fashion there.' His closing ploy was to interweave views and reactions from younger members of the audience –

> to give some idea of *what* the audience actually took in. It was stimulating and interesting, and afterwards some people were saying how much is going on in science at the moment and how vital it is for the success of the four-year plan, where each individual is required to place all his energies in the service of the nation, and where there is no time to waste on contrived word-games and idle speculation. But it was a pleasant enough diversion for everyone on Thursday, despite the sultriness of the weather.

It was immediately beneath this report that the editor had placed the announcement of a lecture by Hermann Staudinger on 'Chemistry and the Four-Year Plan', as we have already seen. How circumstances had changed! Of course, only Heidegger himself was in a position to appreciate the full irony of the juxtaposition.

What Heidegger probably found particularly galling was the triumph of sheer utilitarianism that he saw all around him, and the fact that the professor of organic chemistry, Hermann Staudinger, whom he had denounced during his rectorship as patriotically suspect and politically unsound, was now the man of the moment in the new phase of rearmament and mobilization of all the nation's economic resources. What a contrast between this and the lofty aspirations with which Heidegger, as rector and philosopher, had once entered the arena of academic politics!

In his various statements, culminating in *Facts and Thoughts*, Heidegger blamed the university authorities for leaving him in the lurch, as it were, and failing to show him any support or solidarity. But this was not the case. The organization primarily responsible was the Freiburg League of National Socialist University Lecturers, of which Heidegger had been a member since its inception. The University's press office – which was also the press office for the Freiburg League of NS Lecturers – acting in conjunction with the League's local leadership, moved swiftly to lodge an 'official protest' about the style and tone of the newspaper article, threatening to report the matter to the national executive of the League of University Lecturers. They extracted an agreement whereby 'attacks of this kind on university teachers and members of the League of Lecturers in particular will in future cease, and the editor of *Der Alemanne* will in future consult the League's press office prior to the publication of any personal comments on the teaching activities of a university lecturer'. Following this understanding the threatened report to the national executive was shelved, and the League was content simply to 'emphasize once again the scandalous circumstance of this attack'. The author of a letter from the scientific society that had planned and organized the lecture series commented as follows: 'We need hardly tell you who in Freiburg will have taken the most delight in seeing a Party comrade and pioneer of National Socialism at our University publicly attacked in the pages of the leading Party journal.'[204] Here Heidegger is unequivocally ranked as a Party man – more than that, as a 'pioneer of National Socialism at our University', who was now being dragged in the dirt, turned into a laughing-stock, his philosophy publicly pilloried, to the delight and glee of the despised Catholics (whom the writer does not mention, but clearly has in mind) – men like Archbishop Gröber and his followers, for instance.

So Heidegger was perceived as a National Socialist, at least by the outside world, and however that term may have been understood. And he remained loyal to the Party, wearing its insignia – as Heidegger's pupil Karl Löwith was shocked and affronted to note in Rome. At no time did he adopt a position of protest, not even after the *Reichskristallnacht* of November 1938. Yet in his apologia he tells us that his own change of heart came after the Röhm *putsch*, and he ascribes a share of the responsibility to anyone who 'accepted a senior position in university administration after that time'; for such a person 'could not fail to know the kind of people he was getting involved with'. But was not the wearing of the Party insignia likewise an expression of identification with the totalitarian regime, which was now showing itself more and more clearly in its true colours? Was it not an open declaration of support for the world-view of the *Führer*? Did it matter whether he held an administrative post in the university – particularly when his own entry into the Party had been carefully stage-managed for maximum effect?

The Freiburg historian Gerhard Ritter, who later served on the post-war denazification commission, having been arrested after the events of 20 July 1944 because of his links with Carl Goerdeler and other resistance groups, was quite categorical in a letter to Jaspers at the beginning of 1946: he knew for a fact, he wrote, that Heidegger 'was secretly fiercely opposed to National Socialism after 30 June 1934, and had completely lost the faith in Hitler that had led him into his fatal error in 1933.'[205] Heidegger himself kept coming back to the same few facts, repeated over and over again in different versions and variations: since resigning the rectorship, he claimed, he had used his lecturer's podium as a platform for a sustained intellectual critique of 'the anti-intellectual foundations of the "National Socialist world-view"'. In his draft letter to the editor of the *Süddeutsche Zeitung* of July 1950, for example, already referred to more than once in this account, he claimed that thousands of his students had been imbued by him with respect for the traditions of Western democratic culture; they would have picked up on his true feelings of opposition. Certainly the Party had been quick to note that opposition, and had responded by making life difficult for him in all kinds of different ways.

The time has not yet come, I think, to pass final judgement on these matters. Only when everything that Heidegger wrote in those crucial years has finally been published will we be in a position to draw valid conclusions. For the present we must content ourselves with what has been published, even if the manner of its editing and publication does continue to raise questions in the minds of serious scholars. One text that will always be cited in any discussion is the course of lectures from the summer semester of 1935 entitled 'An Introduction to Metaphysics', which we have already considered at length. Published for the first time in 1953, the text contained a particularly contentious passage, namely the one that speaks of the 'inward truth and greatness of the movement'. Immediately following its publication Jürgen Habermas, then still a student, triggered the controversy with a lengthy critical review.[206] The lectures have subsequently been reprinted in the complete works, with a postscript from the editor that contrives to obscure, rather than clarify, the truth behind the controversial passage. That we now know what really happened is due to the impressive work done by Otto Pöggeler, who in researching the evolution of Heidegger's thought has been through this whole episode once again with a fine tooth-comb.[207] As we now see quite clearly, what Heidegger actually said in the summer semester of 1935, in the brief disquisition on 'Being and duty' (or in other words, the problem of a philosophical ethic) that he gave towards the end of his lectures, was as follows:

> With the being of values we have reached the highest pitch of confusion and dislocation. Since the term 'value' is now beginning to look rather worn, especially as it is also used in the science of economics, people have now taken to calling values 'totalities': but all this amounts to is a change of label. Of course, these so-called totalities soon show themselves for what they really are, namely half-truths. But in the realm of intrinsic reality, half-truths are invariably more dangerous than the much-dreaded void. In 1928 someone published Part I of a complete bibliography of writings on the concept of 'value'. The bibliography lists 661 titles; that number has probably swelled to a thousand by now.[208] *And if we apply the science of resolving paradoxes to the theory of values, it all becomes even more bizarre – a complete nonsense, in fact.* Yet this is what passes for 'philosophy'! What is nowadays touted as the philosophy of National Socialism, but in fact has nothing whatsoever to do with the inward truth and greatness *of National Socialism*, is given to fishing around in these murky waters of 'values' and 'totalities'.

What Heidegger did *not* say, contrary to the later expurgated version he authorized for publication, was this: 'but in fact has nothing whatsoever to do with the inward truth and greatness of the movement (namely with the encounter between technology on a planetary scale and modern man).' Such distinctions may appear insignificant, but to the historian the original document or version is all. So when *did* Heidegger's alleged change of heart take place? In this section on 'Being and duty', which takes issue with the philosophy of values, Heidegger deals summarily with those philosopher colleagues who claim to offer a philosophy of National Socialism, but are in fact ignorant men who fish around – therefore – in the murky waters of a philosophy of values: for none of this has anything whatsoever to do 'with the inward truth and greatness of National Socialism', a truth and greatness that are vouchsafed only to the philosopher of Being, the initiate, who has dis-closed (*entbergen*) and is forever dis-closing the inward truth – albeit in a nation that is becoming increasingly confused, among people who for the most part 'no longer know where the real decisions have to be made and what the options are.' But this was not the fault of the philosopher of Being: in 1935 'the inward truth and greatness of National Socialism' was a given fact, hallowed ground, untouchable, beyond all questioning, and therefore in no need of interpretation; for the lecture audience was bound to associate these words of the philosopher's with the slogans that had been endlessly hammered into them, that were ubiquitously present in their songs and hymns, at the solstice ceremonies and in all kinds of other contexts. That was in 1935. But in 1953 the wording is changed: instead of 'National Socialism' we read 'the movement' – less specific – and the term itself is glossed: 'namely with the encounter between technology on a planetary scale and modern man.' But this kind of *ex post facto* exercise in hermeneutics has no relevance for the historian. What he will note is the persistence of the stereotype. In his lecture on Hölderlin's *Der Ister*, given in the summer semester of 1942, Heidegger inveighed against those academics (again unnamed and unspecified) who viewed Greek culture and history only in political terms, and presented the Greeks 'as pure National Socialists', thereby 'doing no service to National Socialism and its unique historical status – not that it stands in need of such favours.' When are we finally going to see a study of stereotypes in the writings of Martin Heidegger? As late as 1942 the term 'National Socialism' has been carried through intact, like an erratic boulder – interpreted in terms of its 'unique historical status'. So as far as one can see, there had been no movement in Heidegger's thinking between 1935 and 1942. Yet National Socialism had plunged the world into war and ruin, while the crimes against humanity perpetrated in the name of National Socialism were numbered in their millions. Meanwhile Theodor Haecker sat writing his *Tag- und Nachtbücher*, close to despair and profoundly afflicted by the gift of prophecy that had been visited, Cassandra-like, upon him.

After the war, the void

Wandering hither and thither
Not knowing where to turn
He comes upon the void.
But from Death's dread pursuit
Nothing can ever save him,
Though he have thus far 'scaped
The pangs of lingering affliction.

(Sophocles, *Antigone*, V 357-361)[209]

For the winter semester of 1944/45 Heidegger had announced his intention of giving a course of lectures on 'Philosophy and poetry', plus a seminar on Leibniz ('Leibniz: the twenty-four theses'). With reference to the lectures he noted tersely: 'Course suspended after third lecture on 8 November, because drafted into *Volksstum* by Party leadership.' He went on: 'Following the actions of the National Socialist Party no further teaching activities from November 1944 until 1951, since the occupation forces imposed a teaching ban in 1945'[210] – perversely implying a continuity of purpose between the NSDAP and the French occupation forces. One detects an undertone here that is repeatedly sounded by Heidegger, though elsewhere it is less muted – as for example in the draft letter to the editor of the *Süddentsche Zeitung* in 1950: 'And finally the decision to draft me into the *Volkssturm*, which was unusual for someone of my age and in my position, and from which many far younger colleagues were exempted'. In his various 1945 statements, it is true, Heidegger forbore to say that his conscription into the *Volkssturm* was the final bitter turn of the screw by the Party, an attempt, indeed, to send him to his death amidst the destruction of the German nation, as David once sent forth Uriah the Hittite. In the autumn of 1945 these events were still too much in the recent past. But the *Volkssturm* topos runs right through every account in English and French, reaching its apogee in Petzet's version of events:

> More shocking than all the other intrigues and unbelievable plots that were hatched against the philosopher, who was now *persona non grata*, was the episode he relates from the closing phase of the war. Classed as 'dispensable' from his university teaching duties, he was one of the first to be put on the list for drafting into the *Volkssturm* – no doubt in the hope of getting rid of him for good: an act of perfidy that signally failed of its purpose. In the end, as the Allies were preparing to march into the ruins of Freiburg, he set out to make his way home to Messkirch on a bicycle.[211]

The period of time spanned by these events is a relatively short one – from 8 November to mid-December 1944 – but these were weeks of momentous significance for the Upper Rhine region and the University of Freiburg. The war was turning into a rout. Since the summer of 1944 the western front had been moving ever closer to the ridge of the Vosges Mountains, and in November the first French units crossed the line of the Vosges. Hitler's attention was diverted from his southern flank by the Ardennes offensive, into which he now threw all his forces. On 27 November 1944 Freiburg was devastated in raids by British and American bombers, and normal life in the city came to a standstill. The University effectively shut down. The first and only priority was to dig the dead and wounded out of the ruins.

Heidegger had already been conscripted into the *Volkssturm* before this, on 8 November 1944, in the first wave of call-up orders following Hitler's decree of 18 October 1944, when he ordered all able-bodied males between the ages of sixteen and sixty to be drafted into this last-ditch reserve. Heidegger was fifty-five years old, and consequently fell into the category from which the *Volkssturm* was to be drawn. There was no question of his being exempted on the grounds that he was in 'a reserved occupation', and his general eligibility followed automatically from his fitness to work: 'Doubtful cases will be resolved by a medical practitioner to be appointed by the local Party *Kreisleiter*.' The age-groups eligible for call-up were conscripted via 'the local Party branches of the NSDAP, using existing records and procedures' with no bureaucratic obstacles and no need to set up separate registration machinery. Heidegger came under the jurisdiction of the Freiburg-Zähringen local Party branch. According to the story put about by Heidegger, 'they' wanted to get rid of him once and for all. To establish the truth of this, we would have to examine in detail the conscription procedures that were in place during those hectic days – which we cannot possibly do here. So we will have to reserve judgement on this interpretation. The only evidence to support Heidegger's reading of events is the fact that he was conscripted right at the beginning, while his younger colleagues were spared. In the end it was probably simply a question of local variations in the conscription procedures adopted by the different area Party branches in Freiburg. Certainly Heidegger was not the only university teacher to suffer this fate. But he did have two sons serving at the front; the Heidegger family had sacrificed enough, and their commitment to the war effort was beyond question. But the word had gone out: the nation was rising up, the storm was breaking loose . . .

On 23 November 1944 Heidegger marched off with his *Volkssturm* unit in the direction of Breisach, with the intention of taking up positions in Neu-Breisach on the left bank of the Rhine and holding it as the most forward outpost and last bastion of defence, fighting 'to the last man' if necessary. But the marching column did not get very far, advancing scarcely beyond the massive anti-tank ditches by the so-called Mengener Bucht, which had been dug by gangs of young entrenchment workers in the late autumn of 1944, just as the succulent grapes were ripening to full sweetness on the slopes of the nearby Tuniberg. For Neu-Breisach had already fallen, and on the very day of their departure, 23 November, French armoured units were sweeping into Strasbourg. So at least Heidegger was not called upon to prove himself in battle.

The news of the departure of the *Volkssturm* battalion for the nearby front on 23 November 1944 was the signal for a flurry of activity in what was left of the Faculty of Philosophy. The acting dean, the art historian Kurt Bauch, who had been closely associated with Heidegger since 1933, did everything he could to secure the philosopher's release from the ranks – 'to get *him* out, at least', as Bauch wrote. This was how Eugen Fischer, a member of the leadership group around the head of the Reich League of University Lecturers, Dr Scheel, came to petition on Heidegger's behalf, as we saw earlier. Writing to *Gauleiter* Scheel, who was still carrying on his duties in the idyllic surroundings of Salzburg, he ended with a characteristic oath of allegiance: 'In submitting this petition in these most difficult times, knowing that the enemy is now in German Alsace, less than fifty kilometres from our city, we hereby pledge our faith in the future of German science. And we swear to give our all in the service of the *Führer* and the Reich, in the certain assurance that we shall yet achieve victory.' On the day he wrote these words the 'enemy' was barely thirty kilometres from Freiburg, on the opposite bank of the Rhine.

The *Gauleiter* of Salzburg and head of the Reich League of University Lecturers did not reply to his executive colleague until 12 December 1944, writing from the safety, still, of Mozart's native city, the cornerstone of Hitler's alpine redoubt. He apologized for the delay in replying: the last few days had been impossibly hectic. Fischer's telegram had arrived just as the news came in of Strasbourg's occupation. 'There was nothing I could do for Heidegger at the time because the situation was so confused. I do hope that the Heidegger business has now been cleared up. If not, then please let me know.' It was no longer necessary for Heidegger's arch-enemy (as he never tired of presenting him) to intervene in order to secure the philosopher's discharge from his military unit on the strength of a doctor's certificate – the only possible way out – and to ensure that he was not called to the ranks again, let alone required to take part in any last-ditch stand by the local *Volkssturm* contingent in his home district of Zähringen. For the 'Heidegger business' had indeed been cleared up in the meantime, and Heidegger himself was busy organizing the removal of his manuscripts to safekeeping in Bietingen near Messkirch, in the days following the heavy air raid on Freiburg. From his native town he wrote to the University's rector to request leave of absence, which was readily granted, since nobody wished to place any obstacles in the way of his desire to remain at home 'until teaching activities can be resumed in some new location' – especially as everything in Freiburg was in a state of turmoil, and nobody could give thought to anything else.

Officially lectures had not been suspended and the University remained open. Selected examinations were still conducted, though in some cases they had to be sat in school buildings that had survived the bombing. The machinery of administration also continued to function after a fashion, in so far as the buildings were still standing. But the distress was acute. Bodies still lay buried under the rubble of the heavily-hit, defenceless city of Freiburg; people's faces were still marked by the terrible memory of the devastating air raid of 27 November 1944; and every ounce of strength and energy was needed just to ensure survival in the hard winter ahead. Now the rapid approach of the western front, signalled for weeks past by the sound of distant gunfire, forced the adoption of further and even harsher measures. In the face of the inevitable catastrophe for which this terrible war was plainly headed, and which only a few deluded optimists still hoped to avert, the immediate priority was to move as many institutions as possible back to places of putative safety in the rear – including Freiburg University, which was such an integral part of the city's life.

It was no ordinary university gathering that took place on 31 January 1945 in the hastily repaired university buildings. The meeting of the Senate and plenary session turned into a thoroughly stormy affair. In theory such assemblies had been abolished under the new university constitution initiated by Heidegger; but now the hour of need had come. The rector of the university and the mayor of the city of Freiburg – the latter in his capacity as 'head of emergency measures' – worked out the plans for the transfer of the University to a new location, although the rector insisted that the seat of the University would remain *de jure* in Freiburg, even though some of its faculties would have to be evacuated temporarily from the city. He made it clear that this was only a temporary measure; there was no question of the city losing its university for good, but in view of the fact that teaching had effectively been suspended since the air raid – which had also inflicted heavy damage on the University – and only the most important examinations and administrative functions could be sustained, it made sense to organize the forthcoming summer semester (1945) away from Freiburg.

This was not the first time in the history of Freiburg University that the professors and students had been forced to leave the city at the approach of war or in the face of raging pestilence and epidemics. And always the staff and students had returned to the ancient capital of the Breisgau. This time, however, the situation appeared more hopeless than ever before. Disaster threatened on a national scale, and many things were likely to be sucked down into the maelstrom. The swiftest reaction came from the Faculty of Philosophy, which, as appears from Heidegger's letter to the rector of 16 December 1944, had already discussed plans for moving down to the Hegau region. Now, at the beginning of February 1945, it had shifted to its attention to the valley of the Upper Danube, and specifically to Beuron or the surrounding area. The Faculty of Theology also wanted to relocate to the famous Benedictine monastery there, which would have made obvious sense in view of its excellent library facilities. But Beuron had long since been taken over as a reserve field hospital, and was now overflowing with people who had sought refuge from the approaching front. Reichenau Island on Lake Constance could have furnished a sanctuary for the entire university if only they had started planning for it early enough – as the theologian Sauer, Heidegger's predecessor in the rectorship, noted in his diary. The other faculties were also making plans, but here the law of inertia prevailed, and they took no concrete action. Only the scholars of the Faculty of Philosophy had acted decisively to book their accommodation in the self-contained fortified complex of Castle Wildenstein, having entered into an informal arrangement with the owners, the Fürstenberg family of Donaueschingen. So they set to and packed up their books, and before the end of March a small advance party had made its way east across the Black Forest, followed by a larger group of teachers and students, leaving only an 'outpost' behind in Freiburg. It was a hazardous journey, using every available means of transport, which usually meant bicycles: through the Black Forest jam-packed with refugees and retreating troops, up the Höllental, across the Baar and along the Danube, until first Beuron came into sight, nestling in the broad valley basin, and then, around the next bend of the river, Castle Wildenstein itself, gazing down proudly from the commanding heights. Eventually some ten professors and thirty students, the majority of them women, were installed in the castle.

Above the valley of the infant Danube, which carved a path through the limestone rocks of the Swabian Jura millions of years ago, Castle Wildenstein rises up on a precipitous, free-standing outcrop, a double cone of rock: an ancient fortified place, built of hewn blocks of stone that seem to grow out of the native rock, a fine, well-preserved example dating back to early times, impregnable – for here nature and the art of military engineering had entered into a harmonious union. The castle and fortifications changed hands repeatedly, and for many years formed part of the dominions of the lords of Messkirch, until in the seventeenth century they came into the possession of the house of Fürstenberg, who used the site for a variety of purposes: it served as a place of refuge on numerous occasions, as an arsenal, a place of safekeeping for valuables, but also as a state prison and place of banishment. The fine altarpiece in the castle chapel, painted by the Master of Messkirch, had been removed by the Fürstenbergs and placed in the family art collection at their residence in Donaueschingen.

The view extends far across the valley to the north, into the heartland of the Swabian Jura, to the Grosser Heuberg, while down below, clinging to the rocks above the river on the opposite bank, is St Maurus, built by the Benedictines from nearby Beuron and home to the Beuron school of art. Grey herons leave their colony to sail above the meandering Danube in search of prey. Pairs of swans glide through the air to alight elegantly on the waters of the river. This was the landscape that Hölderlin had in mind when he wrote '*Der Ister*':

The rock needs the pick
And the earth the plough,
Inhospitable if not worked.
But what *it* does, the river,
Nobody knows.

And above Beuron, before the water has scooped out a narrow valley, the river almost stands still:

But it seems almost
To flow backwards.
Methinks it must come from the East.[212]

An idyllic scene, especially at the approach of springtime. The clamour of war had not yet penetrated this far up the valley, even though the omens were not good and bad news was rife. It was in these contemplative surroundings that the worthy members of the Freiburg Faculty of Philosophy now established themselves, taking up quarters in the rambling castle complex and bringing new life to the castle kitchens and alehouse. Here they made their preparations for the 1945 summer semester, far removed from Freiburg. Precious few will have believed in victory still. Most of the professors took lodgings in the nearby village of Leibertingen, where the road forks for Messkirch and Kreenheinstetten, the birthplace of Abraham a Sancta Clara.

And now Heidegger went to join the professors who had come to the valley of the Upper Danube. He was officially on leave, but felt his place was with the group – a sort of pathfinder, who knew the country like the back of his hand: after all, this was home to him. The choice of Castle Wildenstein as a last place of refuge was probably Heidegger's idea. For the philosopher was accompanied by one of his students, the Princess of Sachsen-Meiningen, who was returning to her adoptive home, to the forester's lodge at Hausen im Tal, not far from Castle Wildenstein and within sight of the castle at Werenwag on its imposing eminence. There Heidegger enjoyed the hospitality of Prince Bernhard von Sachsen-Meiningen, the Princess's husband, making frequent journeys to Messkirch to ensure the safety of his manuscripts. The Freiburg University contingent were still settling into their craggy retreat amidst the white limestone cliffs of the Upper

Danube valley when the leading units of French armour, swinging south via Freudenstadt and Horb, came through the valley on 21 April 1945; they took Beuron without a fight and advanced on Sigmaringen, the family seat of the Hohenzollerns, ignoring Castle Wildenstein and scarcely noticing the place where Heidegger was billeted. They met with no further resistance, no more than they had in Freiburg, which had been occupied by a French colonial infantry division on 22 April, having fortunately been declared an open city.

The idyll seemed to endure, even if the scattered remnants of the Freiburg Faculty of Philosophy had to adjust to a new set of circumstances – and specifically to the proclamations issued by the French military government, one of which ordered the closure of all educational establishments until the teaching staff and curriculum had been politically purged under the process known as '*l'épuration*'. Although this proclamation was also posted up at Castle Wildenstein and in the surrounding villages, the programme of lectures went on inside the castle, like a fire that glowed still beneath the ashes of the lost war. The historian Clemens Bauer, the indirect successor to the Catholic Heinrich Finke, Heidegger's early patron, was lecturing two hours a week on 'Periods in medieval history', while the philosopher Gisbert was giving a series of lectures on Kant's *Critique of Pure Reason*. There were also practical courses and exercises – in Old High German, for example. And all of this went hand in hand with the novel experience of working on the land – helping to bring in the hay harvest, for example; after all, by assisting the farmers in their labours during May and June the Faculty in exile was providing for its own sustenance, at a time when the reports from Freiburg, sparse though they were, told of dreadful conditions in the badly damaged city, now overrun by the French occupation forces and French civilians. While they managed to live reasonably well up here, the academic exiles felt rather as if they were confined within a roomy prison, because the French kept a tight rein on their movements through the strictly enforced permit regulations. But through it all this little group of students and teachers regarded themselves as the legitimate succession, so to speak, to the university in Freiburg, of whose fate nothing was known beyond vague rumours.

As the summer wore on, the end of the idyll was already in sight.

But before the Faculty-in-exile dispersed they were invited to a festive evening at the Castle, with fine food (the like of which they would not see again for many a day), after-dinner speeches, and even dancing in the great hall. This event was held on Sunday 24 June. But the real farewell ceremony took place three days later on Wednesday 27 June, down in the valley at the forester's lodge in Hausen, where Prince Bernhard von Sachsen-Meiningen and his wife Margot had invited them to a piano recital and a lecture to be given by their distinguished guest, Martin Heidegger. It was a formal, solemn occasion. The audience, sun-tanned from their recent labours in the hayfields, listened to the last lecture Heidegger would give as a full professor of the Faculty of Philosophy at Freiburg University. He chose as his theme a quotation from Hölderlin, the authenticity of which has been called into question by modern Hölderlin scholars: 'All our thoughts are concentrated on the things of the mind. We have become poor, that we might become rich.' Whatever the verdict on Hölderlin's authorship, it was a richly suggestive topic, full of contemporary associations and overtones.[213] It was also Heidegger's farewell to the Upper Danube, the 'Ister' of Hölderlin's poem, on which he had lectured in the summer semester of 1942, taking in Hölderlin's translation of Sophocles' *Antigone* as well: the Sophoclean view of man. One wonders whether he still had in mind the remarks he had made to his audience back in the summer of 1942 – well before Stalingrad:

> Today it is virtually impossible to read an article or book about the ancient Greeks – for those who do such things – without encountering at every turn the claim that here, among the Greeks, 'everything' is 'politically' determined. In most so-called research findings the Greeks are presented as pure National Socialists. In their excess of zeal the scholars in question do not appear to have realized that by putting forward such 'findings' they are doing no service to National Socialism and its unique historical status – not that it stands in need of such favours.[214]

This lecture, delivered in June 1945, was to be Heidegger's last official engagement for a very long time. For back in Freiburg in recent weeks a very different future had started to take shape, and men were marching to the beat of a different drum. A new age had dawned, both for the city and for the University. But in the festive atmosphere of the princely household it was hard to imagine such things. Like his colleagues, Heidegger was able to return to Freiburg at the end of June – much earlier than some have claimed; Petzet, for example, places his homecoming in the snowbound days of December 1945.

PART FIVE

The post-war years: Heidegger under scrutiny

Coming to terms with the political past

Just three days after the entry of the French troops into Freiburg, on 25 April 1945, those professors who had remained in the city – full professors, that is – convened in plenary session to elect a rector and pro-rector after a twelve-year interregnum, and to appoint new deans and senators – except for the Faculty of Philosophy, which was not yet back in residence. This was intended to demonstrate the University's determination to continue functioning as an autonomous, self-governing body even under the conditions of military occupation by an enemy power. More than anything else, though, the professors wanted to make it known that they were taking up the unbroken tradition of the University of Freiburg as it existed prior to the summer semester of 1933, rejecting the constitutional changes that had supervened: the surrender of the city on 22 April 1945 also marked the end of the '*Führer*-university'. So the purpose of the meeting three days later was to signal a new beginning and initiate the work of internal legal reform. This is duly reflected in the minutes of this plenary session: 'The plenary assembly further takes the view that the rector and pro-rector should seek to enter into dialogue with the new mayor of the city of Freiburg im Breisgau, the chief of police, the Lord Archbishop and above all the city commandant. In this process the University of Freiburg im Breisgau will act in its capacity as an autonomous self-governing body.'

The gap between the pretension and the reality soon became apparent. The French naturally rejected the University's claim to special status, and instead subjected this self-proclaimed 'autonomous body' to the full rigour of the occupation laws and regulations – and this against the background of the university system then in force in France. Despite this, the University of Freiburg did all it could to safeguard its own freedom of action, as did the other university within the French zone of occupation, namely Tübingen. Much depended on the individual personalities involved – who was representing the university, and which people on the French side really called the shots both on the ground, locally, and at military government headquarters in Baden-Baden. Before long a third element entered the picture, as the new administrative machinery for education in Baden began to take shape in a very slow and unfocused way under the legal and professional supervision of the occupation authorities. Initially this was based in Karlsruhe, acting for both the regional occupying powers, the Americans and the French; but from the summer of 1945 onwards a parallel organization was developed in Freiburg as it gradually became clear that the southern part of the province of Baden, in the French zone of occupation, was destined to become a distinct territorial entity, with Freiburg as the seat of government and subsequently of the regional parliament. The figure of Dr Leo Wohleb, hitherto a headmaster in Baden-Baden, entered the political arena in the autumn of 1945 and became one of the chief architects of Freiburg University's history during these difficult post-war years, which also decided the fate of Martin Heidegger. In order to understand what follows, it must be borne in mind that after the fall of Hitler's Germany and during the initial development of democratic life, there were forces at work that were not always driven by the purest of motives, forces that operated in a shadowy, twilight zone. We will endeavour to present the sequence of events in a way that highlights the most important phases, the principal issues and the final decisions.

All teaching activities at the University were suspended for the present on Allied orders. Once the machinery of academic administration was in place, therefore, the professors were able to give their undivided attention to the problems of reconstruction, both external and internal. The series of plenary assemblies and Senate meetings that now followed were very soon dominated by the problem of '*l'épuration*', the term coined by the French occupation forces to denote the purging of the political past – in other words, denazification. As early as 5 May 1945, following a motion tabled by the economist Walter Eucken, the Senate was discussing the issue of

'Party men'. The atmosphere was highly charged, following the new rector's announcement that the French city commandant had taken him severely to task over the role of the German universities during the Third Reich. The rector had pointed out that Freiburg was one of the universities least tainted by the spirit of National Socialism – quite the reverse, in fact: Freiburg was an acknowledged bastion of clericalism. Walter Eucken, whose academic star was only now in the ascendant following the publication of his seminal book *Die Grundlagen der Nationalökonomie*, took a very different line. According to him, the German universities in general shared much of the blame for what had happened in those twelve calamitous years, because of their failure to resist vigorously right at the beginning.

Positions began to harden and the battle lines were drawn. On 8 May, the day of Germany's unconditional surrender, the Senate was embroiled in a lengthy debate on the matter in question, and it was decided to conduct an internal survey of university opinion and draw up a list of criteria for evaluating the political past of university staff. The intention was clear: the University was anxious to retain control of the proceedings as far as possible, and to limit outside interference to a minimum. Three categories were proposed: informants (for the most part agents of the *Sicherheitsdienst*, the internal secret service); functionaries (leading officials within the League of National Socialist University Lecturers and suchlike); and rectors or deans. An investigation into the conduct of rectors during their period of office was postponed for the time being. The most prominent figure on the list was of course Heidegger – who was still not back in Freiburg. Indeed, the absence of the Faculty of Philosophy during those critical first weeks of the university's internal reconstruction undoubtedly had significant negative consequences. The psychologist Robert Heiss and the Romance scholar Hugo Friedrich were both arrested by the French for their membership of a unit set up to monitor radio communications between enemy aircraft, while the historian Gerhard Ritter, now the leading survivor of the resistance movement associated with the bomb plot of 20 July 1944, had not yet returned from imprisonment in Berlin. Most of the Faculty, meanwhile, was under wraps at Castle Wildenstein. And Heidegger was deep in conversation with Friedrich Hölderlin.

It should be emphasized once again that the University was concerned from the beginning to exercise some control over the dismissals and punishments meted out by the military government by setting up its own internal procedures. For in Freiburg as elsewhere there was a good deal of licence and high-handedness in the first few weeks following the collapse, given the uncertainties surrounding a new democratic beginning; conspiracy thrived in a general atmosphere of shady dealing, and blacklists were drawn up for compulsory labour duties and the requisitioning of homes – if these could be designated as former 'Party residences'. The rector and the Senate did everything they could to clarify the rights and wrongs of these cases, especially as the number of persons affected grew steadily larger. This meant, of course, that the University was represented on the committees and councils set up for this purpose by the provisional city administration on orders from the military government.

Martin Heidegger, together with his family, was one of the first to come under scrutiny in this way, even as he sojourned still, all unsuspecting, in the valley of the Upper Danube. The requisitioning of Heidegger's house at Rötebuck 47 – a modest dwelling, built when the philosopher accepted the invitation to come to Freiburg in 1928 – is an episode that merits our attention, not because we wish to pry into private matters, but because it has already been touched upon in an earlier study[215] – albeit on the basis of inadequate source material, which was made available through somewhat curious channels. The main reason for reviewing these events is that the fate of the philosopher's private library, and hence his ability to do any kind of serious research, hung in the balance for the next few years. There is no problem about reconstructing the actual sequence of events; the sorry episode is well enough documented.[216] But what also emerges is the vigour with which the University intervened on behalf of its members, not least Heidegger, in order to ensure that the proceedings against them were conducted in a proper legal manner. Nor was such action confined to the University authorities: other individuals also spoke up on Heidegger's behalf, principal among them his compatriot the Archbishop of Freiburg, Dr Gröber, whom we shall encounter frequently from now on.

Heidegger's house was put on a blacklist in mid-May 1945 by

a provisional city administration that was acting at the direction of the French military government – along with many other homes of university professors and ordinary people alike, who all had one thing in common: they had been branded as Nazis. Inquiries were immediately initiated into the size of the house, its structural condition, furnishings, the number of occupants, and so on, and a provisional requisitioning order was issued. On 10 June Frau Elfride Heidegger lodged a formal protest, requesting that the matter be held in abeyance until Heidegger returned to Freiburg from 'his current posting to Wildenstein on the Danube'. Since the official from the housing office had observed in passing that it might be a political matter, Frau Heidegger added a word of explanation – just to be on the safe side:

> My husband became a Party member after 1933, but was never at any time active within the Party or any of its organizations. In 1933 he held public office as rector of the University of Freiburg (having been duly elected by the general assembly), but resigned from the position in 1934 in protest against the government. Since then he has devoted himself exclusively to his philosophical studies. All discussion of his books was banned years ago by the Party authorities, and for the past three years there has been a ban on the publication of any new editions of his books. For a more detailed account of my husband's political position you will have to ask him, of course.

Here we can discern in broad outline the basic elements of the self-justificatory line that Heidegger was later to follow; we have a sense of language being deployed with studied care. The acting mayor of the city rejected the appeal on the basis of a memorandum that read: 'Heidegger is regarded in the city as a Nazi (his rectorship); his name is world-renowned (?), he should be able to work. He will be able to find accommodation with colleagues. Request denied.' Apart from the general mood of the times conveyed in this formulation, the characterization of Heidegger as 'a Nazi' is particularly significant, indicating that he was now classified as '*nazi typique*' – to use the official French terminology. Heidegger's house was henceforth regarded as a so-called 'Party residence', available for the needs of the French occupying powers – an important preliminary decision, not to say pre-emptive verdict, reached before the official proceedings got under way. The Freiburg city administration now found itself in a real dilemma, since it could not satisfy the exorbitant demands of the French for accommodation in a city that had been so extensively destroyed. This emerges clearly from the reply that the acting mayor sent to Frau Heidegger on 9 July:

> The military government requires the city to make available a large number of homes for the needs of the government and the accommodation of priority categories. The directive issued by the military government states that the homes of former Party members are to be requisitioned first. In view of the fact that Professor Heidegger was a member of the Party, all the criteria for the issue of a requisitioning order have been met.

The arguments put forward, he wrote, would not suffice to avert the requisitioning order. He saw no possibility of removing Heidegger's house from the list. He then went into further particulars, explaining that the scope of the requisitioning order was governed by the specific needs in each case, and could extend to the requisitioning of the entire building. In such cases the housing office would allocate alternative accommodation. They must also be prepared to leave behind essential fittings and items of furniture when they left the house. But there was no question of expropriating the furniture and fittings. The mayor then instructed the housing office: 'The petition submitted by Dr Heidegger is hereby refused, and the matter is closed.'

No sooner had Heidegger returned home from the idyllic surroundings of the Upper Danube valley than he found his whole livelihood under threat, and had to endure the impertinent behaviour of petty officials who intimated, amongst other things, that he would have to leave his library behind, since he would not be able to practise his profession for the time being anyway. With this threat hanging over him, Heidegger wrote a letter himself to the mayor on 16 July,[217] setting forth for the first time the basic pattern of defence that was to be sustained in many subsequent versions, right through to the apologia published after his death in 1983. He claimed that he had been unfairly victimized by the requisitioning order, which threatened both his home *and* his library:

> On what legal grounds I have been made the target of such an unheard-of proceeding I cannot imagine. I wish to protest in the strongest possible terms against this attack on my person and my work. Why should I have been singled out for punishment and defamation before the eyes of the whole city – indeed before the eyes of the world – not only by having my home requisitioned in this manner, but also by being stripped of my employment? I never held office of any kind within the Party, and was never active in the Party or in any of its organizations. If there are those who regard my rectorship as politically compromising, then I must insist on being given an opportunity to defend myself against any charges or accusations, made by whomsoever – which means being told, first and foremost, what *specifically* has been alleged against me and my official activities.

In the spring of 1934 (the letter continues) he had been the only one among all the university rectors of his day who dared to resign, thereby registering his attitude to the Party in no uncertain terms. The Party must have had its own reasons for conducting a mounting campaign of harassment against him and denigrating him to the point of gross abuse in its publications (and Heidegger cites the journal *Volk im Werden – passim –* and *Der Alemanne*). The Party must have had its reasons for depriving him of substantial earnings by banning reprints of his books and prohibiting the mention of his name. Later in the letter, Heidegger complained bitterly about this city of Freiburg in which he had resided since 1906, where he went to school at the Bertholdsgymnasium, spent his entire student years at the local university, and, 'apart from a brief interlude at the University of Marburg', had worked more or less continuously in the employ of the University.

> In 1930 I declined an invitation to the University of Berlin, in order to carry on running one of the most internationally

> renowned departments of philosophy here at my home university, and indeed to enhance the standing of the University in general. For the same reason I declined a second invitation to the University of Berlin in 1932, as well as an invitation to the University of Munich.

His published works since 1927 had appeared in several editions in all the major international languages, and had even been translated into Rumanian, Turkish and Japanese.

> And now action is to be taken against me here in the city of Freiburg on the basis of accusations whose content and source are unknown to me – action of a kind hitherto undertaken only against senior Party officials. I object in the strongest possible terms to being associated in any way with these people, with whom I had absolutely no political or personal dealings – nor during the period of my rectorship, not most certainly not thereafter.

His protests were directed against the denigration of his person and his work. It went without saying that he was quite prepared to bear his share of the general hardship and to confine his needs to the bare minimum; he fully accepted, for example, the preferential treatment accorded to former inmates of concentration camps. Self-denial was not a problem for him, since he came 'from a poor and simple home', had struggled through his years as a student and junior lecturer 'with many sacrifices and privations', and had 'always lived a simple life' at home: '. . . so I do not need a lecture in what it means to think and act in a socially responsible manner.'

The danger that his books would be confiscated was averted for the present, and the family were permitted to carry on living in their own home, subject to tight restrictions. The extremely cramped living conditions were to last for several more years, since two families were sometimes quartered in the none-too-roomy house at the same time. The University of Freiburg protested vigorously against such actions, not least in the light of the proceedings undertaken against Heidegger; they had no legal foundation, and in the prevailing atmosphere of hatred, victimization, embitterment and petty tyranny they led to what were effectively proscription lists. Rector Janssen had expressed the University's deep misgivings in a letter of 13 August to the mayor, which was followed by a protracted exchange of reports and further statements of views between the University and the city authorities. The gist of all these communications was that the use of housing policy as an instrument of political retribution was poisoning public life and hindering the re-establishment of a democratic legal order. The settlement of political accounts, a necessary task in itself, became perverted if it was left to the whim of any and every authority and its political advisers. It was not for the city of Freiburg or any of its agencies to decide which members of the University of Freiburg were to be regarded as politically tainted. 'If justice is to be done, the responsibility for establishing political guilt must, in the view of the University, rest with a proper court of law, even if the legal consequences that follow from the establishment of guilt are of an administrative nature.' This key statement from the University, based on the considered legal opinion of the respected and politically irreproachable Franz Böhm, who made a significant contribution to post-war German reconstruction, must be seen in the context of an intensive search for a just system of law. The official records for the summer and early autumn of 1945 would make a fascinating case study in the philosophy of law.

So only a properly constituted court of law was competent to establish political guilt, insisted the University in August 1945. Yet the *de facto* responsibility rested with the French military government – however the latter chose to define political guilt and mete out punishment. Purging the political past – '*l'épuration*' – was the order of the day. We have already noted how the University of Freiburg reconstituted itself, so to speak, a few days after the French occupation of the city, and elected its self-governing bodies; but in so doing it was standing on very shaky constitutional ground. It was up to the military government whether it acquiesced in the University's view of its own constitutional status. That acquiescence was far from total.

In late July 1945 the liaison officer for the University appointed by the military government, the so-called '*curateur*', approached the three professors who had been released from captivity in Berlin, namely Constantin von Dietze, Gerhard Ritter and Adolf Lampe, with the request that they act as the University's representatives from now on *vis-à-vis* the military government – a request that the Senate was required to approve by passing a vote of confidence in these three colleagues. The real function of this so-called commission was to speak for the University in dealings with the French military government, whose confidence it enjoyed. But this was closely connected with the political purging of the academic staff. There was provision for enlarging the commission, and a professor from the Faculty of Theology and the botanist Friedrich Oehlkers duly joined its ranks. Oehlkers and Karl Jaspers had known each other for a number of years now; their friendship grew out of the circumstance that both men were married to Jewish wives, and had accordingly lived in great fear throughout the Third Reich. The main task of this commission, whose precise legal status was difficult to define, turned out to be the compilation of reports and references for denazification proceedings – a laborious and thankless task indeed, since there were so many cases to deal with: but the most important and distinguished case before them was that of Martin Heidegger, who, having just been threatened with eviction from his house, was instructed to appear before the commission on 23 July. It should be said at once that its members were by and large well disposed towards Heidegger. The proceedings were conducted along the lines of a trial, except that there was no formal indictment as such, since the actual facts of the case were taken as read. The accused was interrogated, witnesses were examined, and their testimony for and against was evaluated on the basis of their recollections. Relatively little evidence was adduced from the files of the rector's office. But as things stood, Heidegger was faced with the imminent prospect of losing his livelihood. News of this had also reached Tübingen; and it was during these days that Rudolf Stadelmann wrote to Heidegger and offered him a lifeline, a chance to withdraw from the Freiburg front – as we saw in the opening chapter of this book.

So it was on 23 July 1945 that the proceedings against Heidegger got under way, the start of a journey that would scale the heights and plumb the depths. If, however, one follows the account given by Heinrich Wiegand Petzet, which treats this episode only cursorily and using Heidegger's own words, then a totally different picture emerges.

Petzet begins by relating his version (already referred to above) of how Heidegger was drafted into the *Volkssturm* and how he set out by bicycle from Freiburg, which had been heavily bombed and was about to be taken by the Allies, in order to make his way to Messkirch. The account then continues in the following vein, claiming that the worst was yet to come, awaiting Heidegger on his return:

'I was then interrogated without warning – this was in December 1945 – before the Faculty, subjected to a twenty-three question inquisition. When I collapsed afterwards, the dean of the Faculty of Medicine, Beringer (who had seen through the whole business and realized what my accusers were up to) called round and promptly drove me off to Badenweiler, to Dr Gebsattel. And what did he do? He took me on a hike up through the forest in the snow. That was all. But he showed me human warmth and friendship. Three weeks later I came back a healthy man again.'[218]

This account is completely untrue. It was a blazing hot summer when Heidegger was interrogated by the denazification commission, which had absolutely nothing to do with the Faculty of Philosophy as such. The proceedings were drawn out – to reiterate the point once more – over a period of months; it is no longer possible to reconstruct the individual phases. And it was not until the spring of 1946 that Heidegger went for treatment to Baron Viktor von Gebsattel, then the head physician of a sanatorium in Badenweiler.

He was certainly in need of help, and he sought it from people to whom he had not been particularly close for a very long time. From Romano Guardini, for example, to whom Heidegger wrote on 6 August 1945, not receiving a reply to his letter until 14 January 1946. Heidegger had encouraged Guardini to take over the chair known as the 'Concordat Chair' (Philosophy II) in the Faculty of Philosophy, the same chair that Heidegger had been instrumental in renaming after Honecker's death in 1941. One of the first acts of restitution after Germany's collapse and defeat was the restoration of the Concordat Chair to its original status and purpose in the summer semester of 1945. So swift had been the change in political climate and circumstances: and so rapidly had Heidegger adapted to the new conditions, now extending the hand of welcome to Guardini, who was favoured by Catholic circles in the University and town alike. 'You must think me very ungrateful not to have replied sooner to your letter, which showed such a friendly interest in the matter of my appointment', writes Guardini, relating the story of his appointment to the chair at Tübingen, where he began lecturing in the winter semester of 1945/46 – on particularly favourable terms. And Guardini added: 'How I should love to talk with you on all manner of things. It is so long since we last saw each other. I still have the clearest memories of my visit to Zähringen, and of your wonderful study.'[219] The visit in question, it should be said, took place back in 1930.

As has already been stated, the commission was well disposed towards Heidegger. True, in Adolf Lampe the denazification commission had one member who was critical of, even perhaps bitterly opposed to, Heidegger – which was understandable, since the Lampe affair in 1934 had been one of the causes of Heidegger's quarrel with the ministry in Karlsruhe, as we saw earlier. In close alliance with Lampe was Walter Eucken, Heidegger's most resolute opponent within the university in 1933. He had been openly opposed to National Socialism from the outset, and was closely associated with Husserl; now he was intent on calling Heidegger to account. The Lampe-Eucken team became the real driving force behind the 'Heidegger case', pushing it through to a final decision in January 1946 (as far as the University's involvement went).

Heidegger himself had realized at his initial interrogation by the commission on 23 July that he needed to address his defence primarily to Lampe; and with this in mind he sought an immediate private meeting with him, which took place on 25 July 1945. Lampe prepared a detailed memorandum on their two-hour conversation, copies of which were sent to the commission and to Heidegger himself.[220] He made it clear to Heidegger at their meeting that his own personal experiences in the spring of 1934 had not prejudiced him in any way; rather, he had come to a negative conclusion about Heidegger based entirely on the facts of the case – namely his conduct during his rectorship – and therefore wished to put that conclusion to the test through an open exchange of views. Lampe summarized the charges against Heidegger: he had severely damaged the reputation and standing of the University, particularly by his proclamations to the students and his circulars to the academic staff, which had jeopardized 'vital university interests'. These matters could not be ignored: that would be entirely unfair to other members of the University who were under investigation. Lampe insisted that Heidegger's international renown as a scholar should not be seen as a mitigating circumstance, but on the contrary as an aggravating one:

> *For one thing* because his voice was heard far beyond the walls of the University, and indeed beyond the frontiers of the Reich, thereby affording vital support for the particularly dangerous tendencies in National Socialism that were then emerging; *and for another* because one naturally expects a scholar of such standing to adopt a supremely responsible attitude in matters of educational policy as in all else.

Heidegger had already been asked at the first interview whether he had read Hitler's *Mein Kampf*, to which he replied that he had only been able to read the book in part 'because I found its contents repugnant'. Lampe regarded this as a seriously damaging admission, because it meant that Heidegger had placed himself in an impossibly contradictory position by glorifying Hitler in his proclamation to the student body at Freiburg on 3 November 1933. Heidegger offered the following arguments in his defence:

> He had looked upon support for National Socialism as the last remaining bastion against the spread of Communism.
>
> He had accepted the rectorship only with extreme reluctance and solely in the interests of the University.
>
> He had remained in office, despite constant difficulties and harassment, solely in order to prevent worse excesses.
>
> Due allowance must be made for the extremely unsettled conditions under which he had had to conduct his rectorship.
>
> On many occasions he had indeed been able to prevent a situation from getting dangerously out of hand: but nobody gave him any credit for that now.

> He had found no support among his colleagues for the real aims he wished to pursue.
>
> He had subsequently been openly critical of the regime in his lectures, particularly his Nietzsche seminars.

Lampe went on to note that it had not been possible to come to a satisfactory understanding. He had reminded Heidegger that during his rectorship he had sustained the 'leadership principle' with such radical consistency that any attempt at constructive collaboration in the Senate was doomed to failure from the outset. Heidegger, he said, must assume full personal responsibility for all that had occurred; he could not take refuge in allegations of intrigue and chicanery or the claim that he was powerless to act in the face of higher authority. As for his subsequent criticisms, which were articulated only indirectly, they could not be regarded as adequate restitution, given the unassailability of his personal position as a scholar of world renown.

Heidegger then pointed out that an unfavourable verdict from the commission would effectively outlaw him. He had touched here upon an important point; the French powers of occupation took drastic action in certain cases, and ordered university professors to be detained under concentration camp conditions. In July, for example, the Professor of Anatomy at Freiburg University had been placed under arrest in a concentration camp on the grounds that he had been a member of the SD. Lampe spoke with disapproval of this incident, and made it plain that if necessary he would resign from the commission; certainly he was not prepared to do anything that 'could remotely expose Heidegger to a similar fate'. Be that as it might, it had become clear to Heidegger that he was not going to come out of these proceedings unscathed: whatever happened, he would not be allowed to carry on as before in his present teaching position. Lampe suggested that Heidegger might curtail the proceedings 'by voluntarily accepting an honorary professorship', after Heidegger had mentioned in passing that the most important thing for him was not the actual teaching, but the opportunity to 'publish the results of his work which had been suppressed by the National Socialists'. But he did not want to take the initiative; the responsibility for that must rest with the commission. If their verdict was less than totally positive, he was not averse to accepting an honorary professorship, 'provided an assurance can be obtained beforehand from the French military government that the "H. case" is regarded by the occupying powers as closed, so that no further interference with his research work, and more particularly with his opportunities for publishing, need be feared'. So notes Lampe in his memorandum.

This summary neatly defines the parameters of the argument and the proceedings. On the strength of this, Heidegger could expect to be let off lightly. He was confident of the outcome. From the letters he wrote to Stadelmann during those summer weeks of 1945, which were cited at length in the opening pages of this book, it does become clear, however, that he was more likely to receive favourable treatment from the French than from the Germans, who claimed to have uncovered 'certain incriminating facts' about his rectorship. So it behoved him to keep in with the French – by turning his attention to French philosophy, for example. According to his former friend Heinrich Ochsner, writing on 5 August 1945, Heidegger was planning to set up a small study group on Pascal under the title 'L'esprit de géometrie et l'esprit de finesse'. He himself had been invited to take part, provided the French authorities allowed it. Nobody knew, of course, when the University would open again. Ochsner expressed the hope that the danger for Heidegger was 'now pretty well past'.[221]

A seminar on Blaise Pascal, on some aspect of the *Pensées*, was a clever tactical move. Although nothing came of the proposed study group under the circumstances, Heidegger could have pointed to Pascal as a philosopher to whom he had paid signal tribute in *Being and Time*. And Heidegger was undoubtedly acquainted with Pascal's *Mémorial*, that subjective *confessio* which begins: '*Feu. Dieu d'Abraham, Dieu d'Isaac, Dieu de Jacob – non des philosophes et des savants*.' Was it also a sign of religious conversion?

For the moment, therefore, it looked as if Heidegger might emerge from the proceedings relatively unscathed. It was a matter of weighing his political involvement via the rectorship in 1933/34 against a kind of 'inward emigration' to which Heidegger could lend some credibility, supported notably in this by Gerhard Ritter, who 'knew for a fact' that Heidegger was secretly fiercely opposed to

National Socialism after the Röhm *putsch* and had completely lost his faith in Hitler. But the problem of the rectorship remained. With the best will in the world the denazification commission could not dismiss this crucial episode in Heidegger's academic career so completely as to exonerate the philosopher altogether. In September 1945 the commission's report was ready for submission to the University Senate. A doctored version, in English translation, appears in the dissertation by Moehling, so toned down that the reader is surprised to learn that such a report could have led to the imposition of *any* sanctions against Heidegger, however mild. Heidegger saw to it that only the favourable passages were released into the English-speaking world; the damaging portions were suppressed. In order to set the historical record straight, therefore, the report is published here in its entirety:

> Report of the Denazification Commission, Sept. 1945
> Members: Prof. v. Dietze (chairman), Ritter, Oehlkers, Allgeier, Lampe.

> Prior to the revolution of 1933 the philosopher *Martin Heidegger* lived in a totally unpolitical intellectual world, but maintained friendly contacts (in part through his sons) with the youth movement of the day and with certain literary spokesmen for Germany's youth – such as Ernst Jünger – who were heralding the end of the bourgeois-capitalist age and the dawning of a new German socialism. He looked to the National Socialist revolution to bring about a spiritual renewal of German life on a national-ethnic basis, and at the same time, in common with large sections of the German intelligentsia, a healing of social differences and the salvation of Western culture from the dangers of Communism. He had no clear grasp of the parliamentary-political processes that led up to the seizure of power by the National Socialists; but he believed in the historical mission of Hitler to bring about the spiritual and intellectual transformation that he himself envisaged.
>
> He did not join the Party until 1 May 1933, and then only on condition that he would never be required to assume any

office within the Party or any of its organizations, since he considered himself ill-suited for dealing with problems of a practical political nature. His entry into the Party was directly associated with his assumption of the rectorship, which he was persuaded to accept by his friends and admirers. Many hoped that this famous philosopher's academic distinction (his works have been translated into many of the world's major languages, including some non-European ones) would enable him to maintain a certain independence *vis-à-vis* the Party in his conduct of the University's affairs, and to protect our University from the unacceptable demands of radical elements. For that reason he was elected to the rectorship by a majority of full professors from all Faculties. And he did indeed prevent the crude campaign of persecution against the Jews mounted in April 1933 from being carried on to the University premises, while in his rectorship address on 'The self-affirmation of the German University', which caused a considerable stir both at home and abroad, he set out his own programme for the reform of the university system. In that speech he eschewed any discussion of 'racial policy' and other Party slogans, and chose instead to develop his own concept of true science and scholarship, which at bottom was far removed from mere subservience to the tactical dictates of the day. But in so far as he placed 'labour service' and 'military service' on an equal footing with the 'service of knowledge' for the student, he himself provided Nazi propaganda with the lever it needed to exploit his speech for party political purposes. While he hoped to bring about a spiritualization, deepening and reorientation of German academic life in line with his own philosophical metaphysics (ideas that he also developed at length before the assembled audience of university lecturers), the Party merely used the fact that a scholar of his intellectual standing had joined its ranks, and now acclaimed its victory in public speeches, as a very welcome instrument of propaganda. He himself made the Party's task easier in that he was rash enough, in his efforts to secure a loyal following among the students, to incite them against university teachers who had been branded as 'reactionary'. In this way he hoped to further his own reform plans and also to win a position of respect within the Party that would enable him to pursue his own course – and possibly even to exercise a beneficent influence on the internal development of the Party. Needless to say, these hopes were very soon dashed. The students became insolent and arrogant, the majority of the professors were deeply offended by his decrees, which were frequently seen as heavy-handed and overbearing, and they very quickly aligned themselves against him. Meanwhile the Party gradually withdrew its support from him as it came to recognize the irreconcilable differences between his plans for the future of the universities and its own. Not even the fact that he played an active part in transforming the university constitution in line with the 'leadership principle' and in introducing the outward forms of Hitlerism (e.g. the Hitler salute, the so-called '*deutscher Gruss*') into academic life, nor the fact that he penalized or sacrificed persons who were opposed to the Nazis, and even contributed directly to National Socialist election propaganda by issuing proclamations in the press: none of this affected the growing estrangement felt on both sides. And so his rectorship came to an early end at the conclusion of the winter semester 1933/34, when he clashed sharply with minister of education Wacker for a combination of political and administrative reasons. Thereafter Heidegger devoted himself exclusively to his philosophical studies, becoming increasingly estranged from the Party, and in the end fiercely opposed to it in private – although this was not at all apparent to an outside observer. His lectures, seminars and addresses were placed under surveillance by the SD, his literary work was proscribed in the Party press at the instigation of the pseudo-philosopher Krieck (Heidelberg), a Party creature whom he despised utterly; a ban was even placed on the printing of some of his books, and any discussion of them or reference to his name in official Party literature was suppressed as far as possible.

Despite this later estrangement there can be no doubt that in the crucial year of 1933 Heidegger consciously placed the full weight of his academic reputation and the distinctive art of his oratory in the service of the National Socialist revolution, and thereby did a great deal to justify this revolution in the eyes of educated Germans, to raise the hopes people had placed in it, and to make it much more difficult for German science and scholarship to maintain its independence amidst the political upheaval. The label 'Nazi' ceased to be applicable to him after 1934, and there is no danger that he would ever again promote the ideas of Nazism. It would be a serious and lamentable loss if our University were to lose this renowned scholar altogether because of his political past; on the other hand, we feel it would be unthinkable to leave his formal position at the University unchanged in the wake of such disastrous political errors. The best solution in our view would be to give him emeritus status, which would allow him to undertake a limited amount of teaching work, while excluding him from active participation in university administration, the conduct of examinations and the examining of candidates for habilitation. One member of our commission is of the opinion, however, that the facts of the case as described above justify more serious action.

Recommendation: Category B.[222]

It was Adolf Lampe who called for more serious action to be taken. The turning-point in the Heidegger affair can be pin-pointed fairly precisely. When this report, with its recommendation that Heidegger be given emeritus status and allowed to retain his teaching rights, was approved, it emerged that the French military government had merely declared the philosopher to be *disponible* – the mildest of measures, which left open the possibility of complete rehabilitation. At the same time it was rumoured in Freiburg that Heidegger had been invited to lecture before groups of French officers and others in Baden-Baden. This information prompted Lampe, Walter Eucken and the pro-rector Franz Böhm to draft a lengthy statement to the rector's office at the beginning of October, with the aim of quashing the commission's report and preventing Heidegger's reinstatement. Their arguments may be briefly summarized: Heidegger's responsibility from the early days of the Third Reich was so enormous that no other member of the University could possibly be called to account if Heidegger was allowed to get off more or less scot-free. Yet the fact was that action had already been taken against two former rectors (meaning Metz and Süss) while several professors in the Faculty of Medicine had been dismissed from their posts and even interned in French camps under extremely harsh conditions. It would be a mockery if Heidegger, the intellectual Pied Piper who had led so many young scholars astray – Böhm mentions Stadelmann, Heimpel and Schadewaldt amongst others – were to be treated in this lenient fashion, because it meant that none of the others involved could be brought to book. Böhm, the future education minister for Greater Hessen, threatened to resign his pro-rectorship if Heidegger was restored to his post, or even if he was granted emeritus status: '... for that, too, is a form of reinstatement'. In his detailed arguments Böhm went straight to the heart of the matter:

> In view of the fact that in a whole series of cases the military government has imposed harsher penalties than were recommended by the University and the examining commission, it makes me very bitter to think that one of the principal intellectual architects of the political betrayal of Germany's universities, a man who at a critical moment in time, from a position of prominence as the rector of a leading German borderland university and a philosopher of international standing, became the vociferous spokesman of an intolerant fanaticism, charted the wrong political course and preached a catalogue of pernicious heresies, heresies which to this day he has never repudiated – that a man such as this should merely have been subjected to the stricture of '*disponibilité*', and clearly feels no need at all to answer for the consequences of his actions' (Letter to the rector's office, 9 October 1945).[223]

Notified by the rector of this change of mood, and of the threat of a Senate resolution calling for stiffer action, Heidegger submitted an immediate request (on 10 October) for the grant of emeritus status, couched in the following terms:

> In implementing its policy of *épuration* the French military government has decided that I should retain my academic post and remain available. The holder of that post heads one of Europe's leading departments of philosophy. Since the events of the last few months have persuaded me that the Faculty attaches no importance to my contribution as an official member of staff, I would be glad if you would make a formal application to the ministry on my behalf for the grant of emeritus status.

The alleged disinterest of the Faculty was in fact just a pretext. The balance of opinion in the Senate had shifted against Heidegger, so that instead of insisting on full reinstatement – the only really acceptable solution for Heidegger, particularly in later years – it seemed advisable for tactical reasons to pursue a middle path and indicate a readiness to compromise.

But what was the story behind these lectures in Baden-Baden? The rumours had exaggerated, but there was an element of truth in them. As the rector noted in the minutes of a meeting with Heidegger on 5 November 1945, the Freiburg philosopher had been invited to Baden-Baden to meet Jean-Paul Sartre in person, 'who is the main proponent of existential philosophy in France. They have told him that if, as a result of these philosophical discussions, an opportunity should arise to give a more general exposition of existential philosophy before a larger audience, Professor Heidegger would be at liberty to do so.' As yet Heidegger had reached no decision in the matter. But more important was the disclosure that followed, namely the visit of a young lieutenant attached to the French military government in Baden-Baden, who was also a correspondent for the *Revue Fontaine*. This was Edgar Morin, then just twenty-four years of age, who had a special interest in philosophy and sociology, and was currently engaged in the writing of *L'an zéro de l'Allemagne* (published in 1946). Morin visited Heidegger at the end of September, bringing with him a letter from Max-Pol Fouchet, the director of the *Revue Fontaine*, which contained a series of proposals. Fouchet invited Heidegger to submit an essay or a fragment of a longer work to be translated into French and published. He was also interested in bringing out one of Heidegger's books in the series of titles published under the journal's own imprint, or alternatively some lectures or compilations of essays by Heidegger. He himself would see to the French translation, and Heidegger and the editorial team would have an opportunity to check the text before it went into print. Correspondence could be conducted through Lieutenant Morin, or alternatively through Général Arnaud, head of the press and information office at French military government headquarters in Baden-Baden – not exactly by the back door, therefore. If, however, Heidegger was unable to make available any of his past work either for the journal or for the book series, they would dearly like to have an essay from him giving his views on the present situation or on the state of philosophy in France – an original contribution, in other words. The journal's circulation figures were such that he was guaranteed a wide audience.

The rector disclosed this letter and the letter of invitation from the Gouvernement Supérieur in Baden-Baden at the Senate meeting of 21 November 1945 – the crucial turning-point in the Heidegger affair; for in the meantime the mild-mannered original report of the denazification commission had been revised, and in its second, much harsher version would form the basis of the Senate's resolution of 19 January 1946.

Again it was Eucken, Lampe and Böhm who gave vehement voice to their indignation that the *Revue Fontaine* should offer to publish Heidegger's observations on the current situation. 'It was not my understanding at the last Senate meeting', wrote Lampe in an extended statement of his views,

> that Herr Heidegger had appreciated or acknowledged the utter impossibility of any such involvement in contemporary affairs. It must therefore be concluded that Herr Heidegger – contrary to what is assumed in the report placed before us by our denazification commission – has *not* undergone that radical change in his political thinking which has hitherto been taken as read. In the absence of such a change we had no business to exonerate Herr Heidegger from the consequences

> of his actions as first rector of our University during the Third Reich to the degree that we did in the commission's final report (From Lampe's personal papers).

In other words, he was demanding a complete revision of the original report – which was what then happened. Lampe summarized his view succinctly: if Heidegger seriously supposed 'that he of all people was now called to speak words of illumination and guidance on the subject of the sufferings visited upon mankind by Hitler and his blind or criminal following', then only two interpretations were possible: *either* Heidegger was aware of the magnitude of his guilt 'when he drove our University by brute force down the road of National Socialism, and on the strength of his international reputation as a philosopher bedazzled and led astray thousands upon thousands of men and women', *or* Heidegger was and remained to this day 'staggeringly blind to reality'.

Prior to this, when it became apparent in late October/early November that the mood was turning against Heidegger, the appropriate French authorities – the advisers on higher education in the office of the *curateur* – made attempts to get Heidegger out of the firing line in Freiburg and move him to the calmer, less volatile setting of Tübingen, where the local *curateur*, the highly resourceful Germanist Capitaine Cheval, made approaches to the education authorities for Württemberg. The ground was well prepared: the acting dean, Stadelmann, had been working for some time to prepare a safety net for Heidegger. A certain degree of co-operation was also to be expected from the education authorities, since the acting director of education for Württemberg, Carlo Schmid, was concerned to ensure that the University of Tübingen acquired a certain kudos. Hence the invitation to Romano Guardini, who was offered a newly created chair in the Faculty of Philosophy. The official from the ministry with whom Cheval discussed the matter noted that the French military government would welcome Heidegger's appointment to a post at Tübingen. It was not an order, simply a suggestion. It was entirely for the Faculty to decide. The military government simply wished it to be known that it would not stand in the way of Heidegger's appointment. The proposal created a good deal of strain within the

Faculty at Tübingen in November 1945. Heidegger was the subject of lengthy (and very heated) debate, but his name was not placed at the top of the list. At the end of the day Tübingen was not prepared to play the stooge.

So the drama continued to be played out at the University of Freiburg, as the internal proceedings against Heidegger were resumed. This time he had to answer for his conduct under very different conditions and in an atmosphere that was notably more highly charged. It was *this* round of interrogation, in December 1945, that formed the background to Heidegger's account in Petzet's book *Auf einen Stern zugehen* – but with this important difference: it was the sequel to a long-drawn-out process, with all the many complications we have described; and the proceedings against Heidegger were a purely internal affair, involving only the University's own denazification commission and the Senate. His position now looked very serious. While he had been prepared to accept emeritus status in the face of the hardening opposition, he now found himself fighting for the right to teach at all.

The Faculty of Philosophy debated Heidegger's case for the first time on 1 December 1945. Following the detailed report from Gerhard Ritter, the Faculty unanimously resolved, first, to address a petition to the French military government requesting Heidegger's reinstatement; second, to petition the ministry to grant Heidegger emeritus status in accordance with his own wishes; third, 'to write to Herr Heidegger in person and tell him that we have submitted these two petitions, that we very much regret his departure, and trust that in future he will still be willing to work together with us in the Faculty when matters of particular moment arise'[224] – all this assuming, of course, that Heidegger was allowed to go on teaching. The animated debate that went on during these weeks focused primarily on Heidegger's behaviour towards Husserl, and on the telegram that Heidegger sent as rector to Reich Chancellor Adolf Hitler on 20 May 1933.

It is understandable that Heidegger's unfortunate behaviour towards his predecessor in the chair (and presumed mentor), particularly during Husserl's long illness in 1937/38, his absence at Husserl's funeral and his silence after Husserl's death now weighed very heavily against him. That was the first serious charge against him. The telegram to Hitler, in which he asked if the planned reception for the Board of the Association of German Universities could be postponed 'until such time as the much-needed realignment of the Association in accordance with the aims of *Gleichschaltung* has been accomplished', counted against him principally because of the word '*Gleichschaltung*'. The meaning and implications of the term were clear enough in May 1933, as was shown earlier.

In a series of letters addressed to the rector and to the denazification commission Heidegger sought to refute the new (or newly accentuated) accusations. In a final act of defiance he wrote to the chairman of the commission, Constantin von Dietze, on 15 December 1945, summarizing the arguments in his defence once again and putting forward a basic statement of his position:

> In 1933/34 I was no less opposed to the ideology and doctrine of National Socialism, but at that time I believed the movement could be channelled in a different intellectual direction, and I regarded such an attempt as consistent with the social and broader political tendencies of the movement. I believed that Hitler, having taken on the responsibility for the entire nation in 1933, would now transcend the Party and its doctrine, and that all men would come together on the solid ground of renewal and collective purpose to discover a responsibility for the Western world. This belief was mistaken, as I learned from the events of 30 June. But by 1933/34 it had brought me to the in-between position where I accepted the social and national (not National-Socialist) component, but rejected its intellectual and metaphysical underpinnings in the biologism of Party doctrine, because the social and national component, as I saw it, had no essential connection with the ideological doctrine of biological racialism.

He thanked the commission once again for its sympathetic and understanding attitude in correctly assessing his intentions in 1933/34. He admitted he had made many mistakes 'on the technical and staff-related side of university administration ... But I have never sacrificed the spirit and essence of science and scholarship or the

University to the Party, but have sought instead to bring about the renewal of the *universitas*.' He must leave it to the University to decide, he said, whether it wished to retain his services in any shape or form. All he asked was an assurance from the University that his thirty years of work in philosophy would be safeguarded, work

> which I believe will one day have something of value to say to the West and to the world at large. The strain of life in general and my present intellectual ordeal, together with the worry about the fate of our two sons missing in action in Russia, have sapped my energies to the point where they may only just suffice to complete some of the work that I care deeply about for the sake of the future of philosophy (From Lampe's personal papers).

Simultaneously Heidegger's thoughts turned to two friends who might be able to help his case – men whose opinions now carried some weight because they were regarded as authorities: Karl Jaspers and Archbishop Conrad Gröber. As his situation became more serious in December, Heidegger asked the commission to solicit a reference on him from Jaspers. The commission agreed, and one of its members, Oehlkers, duly wrote to Jaspers in Heidelberg, as we have already seen. The botanist Oehlkers had been friendly with Jaspers for a number of years. After the fall of the Third Reich both professors compared notes and discussed their present difficulties. On 26 May 1945, for example, Oehlkers described the highly precarious situation of Freiburg University – which had suffered heavy damage, particularly in the area where the scientific and medical institutes were situated – under the terms imposed by the French occupying power, whose future plans for the university were still veiled in obscurity. Indeed, it was not at all apparent to German observers in those weeks whether the provisional partition of Baden between the occupying powers – the southern portion of the province under French control, the north under the Americans – was to be a permanent arrangement, or whether the capital Karlsruhe was to have jurisdiction over the two universities of Freiburg and Heidelberg. And on 12 July 1945 Oehlkers gave Jaspers a detailed account of the establishment, remit and initial impressions of the denazification commission, on which he had been elected to serve as a representative of the natural sciences:

> My colleagues insisted on my collaboration, and so I was coaxed out of my reserve. For the past two weeks or so we have had meetings, hearings, discussions, etc. nearly every day, morning and afternoon. The accumulation of work is quite alarming. We are overrun with people who bombard us with an endless succession of requests, threats, warnings, words of encouragement. I know that we lay ourselves open to the charge of collaboration with the enemy. Of course I understand that. But our only concern is to make sure we have a teaching staff at the University that is 'fit to work' as *we* see it, and not as the French see it (From the personal papers of Jaspers).

At the time when Oehlkers was writing to Jaspers in these terms, Heidegger's case had not yet come before the commission. On 15 December 1945 Oehlkers briefed his friend and colleague on the Heidegger affair, emphasizing that the case was extremely complicated because it had now been referred back to the commission and new facts had emerged. As the first National Socialist rector Heidegger was 'of course seriously compromised'. But Oehlkers made no secret of the fact that the charges against Heidegger also encompassed the serious accusations against Frau Elfride Heidegger, as was indicated earlier. So what was expected of Jaspers? An assessment that went 'beyond the bare facts of his rectorship' – in other words, a comprehensive general evaluation of Heidegger's personality – with particular reference to the question of whether Heidegger had been an anti-Semite. 'He has asked us to solicit your views on this specific point.' Oehlkers never tires of pointing out that Heidegger was not 'a "Nazi" in the conventional sense of the term'. But a tragic destiny was at work during the ill-fated months of the rectorship.

> The fact is he was utterly unpolitical, and the special brand of National Socialism he had concocted for himself had nothing whatever to do with reality. In this self-created vacuum he performed his duties as rector and did untold damage to the

University, until one day he suddenly noticed the wreckage lying all around him. Only now is he beginning to understand how this came about. It's very easy to condemn his behaviour, but very hard to really understand it.

Oehlkers' wish, Heidegger's request to his one-time friend and fellow philosopher, and the quasi-official nature of Oehlkers' letter: all this placed Jaspers in a very difficult position, a dilemma from which there was no escaping. Jaspers' report of 22 December 1945 – the philosopher got down to work as soon as he received Oehlkers' letter – lay around on his desk over the weekend. Then on Christmas Eve Jaspers added a handwritten postscript to the four-page typed letter: '24/12 Since yesterday was a Sunday the letter didn't go off immediately. I thought about asking to be excused from replying, given my earlier connections with Heidegger. To reply or not to reply: in this case both go against my nature. But in the end a formal request from an official body, and especially a request from Heidegger himself, must outweigh other considerations. So I am sending the letter after all.' Jaspers may well have realized that his verdict was crucial, but also that he was tearing down the bridge between himself and Heidegger – assuming that such a bridge still existed outside his own imagination.[225]

Heidelberg, 22 December 1945

My dear Herr Oehlkers,

Your letter of 15 December arrived today. I am delighted to hear that things are working out with Gentner.[226] He may have been to see you in the meantime; he was certainly planning to in conjunction with a trip to Paris, from which he is expected back any day.

I will answer the main question in your letter straight away. Given my earlier friendship with Heidegger, it is inevitable that I should touch on personal matters, partly in order *not* to conceal a possible partiality in my judgement. You are right to call the case complicated. As with all things complicated, one must try to reduce it to simple essentials, in order not to get lost in the maze of complications. Permit me to establish one or two cardinal points:

1) I had hoped to keep quiet and say nothing, except to close friends. I have held that view ever since 1933, when I resolved to remain silent, after the terrible disappointment, in deference to my happy memories. This was made easy for me in so far as Heidegger himself, at our last meeting in 1933, either ignored awkward questions altogether or gave vague answers – particularly on the Jewish question – and thereafter ceased the regular visits he had kept up for the past decade, so that we have not met since. He carried on sending me copies of his publications to the end, but after 1937/38 he stopped acknowledging anything I sent him. I was particularly hoping that I wouldn't have to speak, that now, at last, I could keep quiet and say nothing. But since you have asked me to speak, not only officially, at the request of Herr von Dietze, but also in the light of Heidegger's own wish that my views be heard, I am left with no alternative.

2) Apart from what was public knowledge at the time, I was also privy to a number of other incidents, two of which I feel are important enough to be related here.

Acting on behalf of the National Socialist regime, Heidegger submitted a report on Baumgarten to the League of University Lecturers in Göttingen, a copy of which came to my attention many years ago. In it he writes: 'Here at any rate Baumgarten was anything but a National Socialist. In terms of family background and intellectual sympathies his roots lie in the Heidelberg circle of liberal-democratic intellectuals around Max Weber. Having failed to secure an appointment with me, he established close contact with the Jew Fraenkel, who used to teach in Göttingen and has now been dismissed from here. Through him he managed to find a place in Göttingen . . . It is too early to reach a final verdict on him, of course. He could still develop. But a decent probationary period needs to elapse before he can be permitted to join any organization of the National Socialist Party.' We have become accustomed to such horrors since then that the shock I felt on reading these words is probably hard to imagine now.

Heidegger's assistant in the Department of Philosophy, Dr Brock, was a Jew. Heidegger was not aware of this at the time of Brock's appointment. When the National Socialists began to take measures against the Jews, Brock had to leave his post. According to Brock – and I have this from his own lips at the time – Heidegger behaved impeccably towards him. He helped him to get established in England by writing him some very warm references.

In the 1920s Heidegger was not an anti-Semite. That thoroughly uncalled-for remark about 'the Jew Fraenkel' shows that by 1933 he had become an anti-Semite – in certain contexts, at least. He did not always exercise restraint in this matter. This does not rule out the possibility that in other cases, as I must assume, anti-Semitism went against his conscience and his inclinations.

3) Heidegger is a significant figure, not only in terms of the content of his philosophical world-view, but also in his ability to handle the tools of speculative thought. He has a philosophical mind whose insights are undoubtedly interesting, although my own view is that he is uncommonly uncritical and a long way removed from 'science' in any true sense. He sometimes comes across as a blend of the earnest nihilist and the mystagogue-cum-sorcerer. In the full flow of his discourse he occasionally succeeds in hitting the nerve of the philosophical enterprise in a most mysterious and marvellous way. In this, as far as I can see, he is perhaps unique among contemporary German philosophers.

It is imperative, therefore, that he be allowed to pursue his studies and writings without restriction.

4) In deciding how individual persons are to be treated we must inevitably have an eye to the overall situation in which we find ourselves today.

It is therefore essential that those who actively helped to put National Socialism in the driving seat be called to account.

Heidegger is one of the few university professors who did just that.

Countless people who were never National Socialists in their heart of hearts are currently being removed from their posts without compunction. If no restrictions of any kind are placed on Heidegger, what are his fellow academics to say when they are forced out of their jobs and made destitute, having done nothing to assist the National Socialist cause? Outstanding intellectual achievement may be legitimate grounds for allowing him to pursue that work further, but not for allowing him to remain in his post as a teacher.

In our present situation the education of the younger generation needs to be handled with the utmost responsibility and care. Total academic freedom should be our ultimate aim, but this cannot be achieved overnight. Heidegger's mode of thinking, which seems to me to be fundamentally unfree, dictatorial and uncommunicative, would have a very damaging effect on students at the present time. And the mode of thinking itself seems to me more important than the actual content of political judgements, whose aggressiveness can easily be channelled in other directions. Until such time as a genuine rebirth takes place within him, and is *seen* to be at work within him, I think it would be quite wrong to turn such a teacher loose on the young people of today, who are psychologically extremely vulnerable. First of all the young must be taught how to think for themselves.

5) I can accept to some extent the personal excuse that Heidegger was unpolitical by nature, and that the special brand of National Socialism he concocted for himself had precious little to do with the real thing. But in response to that I would first of all remind you of what Max Weber said in 1919: children who stick their fingers into the wheel of world history are going to get them broken. Secondly I would add this qualification: Heidegger undoubtedly failed to understand the true dynamics and aims of the National Socialist leadership. The very fact that he thought he could have a

will of his own proves it. But his manner of speaking and his actions do have a certain affinity with the manifestations of National Socialism – which makes his error understandable. He and Baeumler and Carl Schmitt are three very different academics who each strove for the intellectual leadership of the National Socialist movement. To no avail. They brought their very real intellectual abilities to the task, only to end up blackening the reputation of German philosophy. So I agree with you that there is a touch of the tragedy of evil about it all.

A change of heart brought about by a switch to the anti-Nazi camp has to be judged by the motives that underlie it, and to some extent these can be deduced from the timing of the change. 1934, 1938 and 1941 all signify quite different stages. To my way of thinking a change of heart that postdates 1941 is virtually meaningless – and indeed means very little unless it occurred decisively after the events of 30 June 1934.

6) Special cases sometimes call for special measures; where there is a will – and where the importance of the case warrants it – there is always a way. My recommendation is therefore as follows:

a) In consideration of his acknowledged past achievements and the likelihood that there is important work yet to come, Heidegger should be provided with a personal pension to enable him to pursue his philosophical studies and publish his works.

b) He should be suspended from teaching duties for several years, after which there should be a review of the situation based on his subsequent published work and in the light of changing academic circumstances. The question that must then be asked is whether the restoration of full academic freedom is a justifiable risk, bearing in mind that views hostile to the idea of the university, and potentially damaging to it when propounded with intellectual distinction, may well be promoted in the lecture room. Whether or not such a situation arises will depend on the course of political events and the evolution of our civic spirit.

If a special arrangement of this kind for Heidegger were not to be approved, I would regard any preferential treatment under the terms of the general dispensation as unjust.

Stated thus with a brevity that undoubtedly leaves ample scope for misunderstanding, those are my views. If you wish to communicate the contents of this letter to Heidegger, you have my permission to show him a transcript of points 1, 2 and 6, together with the last sentence under point 3 ('It is imperative . . . without restriction.').

Please excuse the schematic form of this letter, adopted in the interests of brevity. I should much prefer to discuss the matter with you in person, enlarging on specific points in response to your comments. But that isn't possible now.

You write about the hardships of the winter. I'm sure things are much worse there than they are here, even though we too are feeling the pinch. But so far we are coping. I just hope we don't have a hard frost.

My wife and I both send our warmest good wishes to you and your dear wife.

Yours sincerely,
Karl Jaspers

Since the question of Heidegger's possible anti-Semitism was discussed at length in an earlier context, we will simply summarize the main points again here. Jaspers cites the case of Baumgarten and Heidegger's report to the Göttingen office of the League of National Socialist University Lecturers, with its crude reference to 'the Jew Fraenkel . . . dismissed from here'; but he also points out that Heidegger came to the aid of his dismissed assistant, the junior lecturer Dr Brock, and 'helped him to get established in England by writing him some very warm references'. In the 1920s, he claims, Heidegger had not been an anti-Semite, although at their last meeting at his (Jaspers') home, on the occasion of the Heidelberg lecture of 30 June 1933, he had either ignored awkward questions altogether or given 'vague answers – particularly on the Jewish question'. Of especial significance is the fact that Jaspers associates Heidegger's political involvement with that of Carl Schmitt and Alfred Baeumler, thereby establishing a troika in which all three 'strove for the intellectual leadership of the National Socialist movement'. This report, which contains so many elements, albeit presented in a somewhat disorganized way, was devastating in its underlying import, both with regard to its condemnation of Heidegger's philosophical undertaking – even though Jaspers did not want his voice to be silenced – and with regard to his proposed punishment. The temporary suspension recommended by Jaspers was exactly what the French meant by '*suspendu*': neither emeritus status (with the associated right to give lectures), nor even retirement in the traditional sense, but effectively dismissal in the special form of a 'personal pension', which amounted to a kind of charitable handout, to be granted just this once as a special favour.

Jaspers' report certainly tipped the balance when the Freiburg University Senate came to make its decision, and it had an important influence on the thinking of the French military authorities. As a result, Jaspers became a figure of controversy. The dean of the faculty, Robert Heiss, informed him on 14 April 1946 that Heidegger's position was a very difficult one, and that he, Heiss, took a different view from Jaspers. In his reply of 28 May 1946 Jaspers pointed out that it was not a question of views or opinions, but of hard facts and evidence. If Heiss had seen his report, he could not have written as he had. Heiss replied on 5 July 1946 that as dean of the faculty he was well acquainted with the contents of the report. He had followed Heidegger's career for many years, had attended two of Heidegger's lecture courses himself and read up the course notes for others. At the very least it had to be said that Heidegger's views had undergone an early and radical change. He was not in any way excusing what Heidegger had done in 1933, but he had followed his subsequent career with the closest attention and could reasonably claim to know it inside out. He too was shocked and appalled by many of the things Heidegger had done in 1933. But he also knew what had come later. 'You will understand that in the light of that knowledge I do not set much store by most of the judgements made about Heidegger.

But to yours I attached the very greatest importance.' He personally believed 'that Herr Heidegger is going into a kind of exile; it may be said that he is reaping what he has sown. I cannot argue with that' (From the papers of Robert Heiss, now in the possession of the author).

In the late autumn of 1946 Hans-Georg Gadamer, Heidegger's pupil, who was looking for a chair in the Western occupation zones, travelled from Leipzig to Freiburg, interested by the possibility of succeeding Heidegger, whose case at the time looked in a very bad way. On his return to Leipzig he wrote to Jaspers on 6 October 1946, and observed that Heidegger's situation was getting worse by the day. The issue was whether he would be granted 'emeritus status – or dismissed. Probably the latter, with all its economic consequences. For that reason your report came at a tactically unfortunate moment, I fear, since you spoke only of a "pension"' (From Jaspers' personal papers). Gadamer must have read Jaspers' report during his visit to Freiburg, when he also met Heiss to discuss the possibility of succeeding Heidegger in the chair. The report and the reactions to it were not something Jaspers could shrug off just like that, and he authorized the commission to disclose the full contents of the report to Heidegger. The Heidegger affair continued to haunt him, until, as we saw at the beginning of this narrative, he attempted to re-establish contact with Heidegger in Freiburg soon after his own move to Basel.

The hope that Jaspers would save the day had thus turned out to be misplaced, and it is clear that Heidegger learned the gist of the report, and in particular the concrete proposals put forward by Jaspers, sometime between Christmas and the New Year. Since the denazification commission based its verdict very largely on Jaspers' conclusions, the only recourse left to Heidegger was a visit to his fellow countryman from Messkirch, his fatherly friend, patron and well-meaning benefactor, Archbishop Conrad Gröber: a refuge for many in those troubled weeks, for Freiburg's senior churchman was seen as a cornerstone of the Church's resistance during the period of National Socialist tyranny, and was now regarded as an unimpeachable authority, particularly *vis-à-vis* the French military government. His word carried weight: his support meant a great deal. So Heidegger slipped into the role of Prodigal Son and went round to the Archbishop's residence, after many years without contact – even though they both lived in the same city. Immediately he felt the warmth and sincerity of their shared Messkirch heritage, especially since the Archbishop's sister, Fräulein Marie, was at pains to put Heidegger at his ease. He had come cap-in-hand to Canossa: but when he heard her speaking the familiar dialect of his native town he quickly overcame his awkwardness at this, his first meeting with Gröber after so many years (and what years they had been!) A second visit soon followed, and in the last days of 1945 Gröber despatched a letter of support for Heidegger to the French military government. He also contacted the Abbé Virrion in Baden-Baden, who worked for the French military government in the department of education. Gröber was concerned that his recommendation should reach the central office in Baden-Baden as soon as possible, since the situation was looking very precarious, and a negative verdict from the University Senate could only be 'corrected' by the military government. The text of the Archbishop's letter has not survived, but its positive tenor can be inferred from a letter written by the aforementioned Abbé on 2 January 1946 and addressed to Freiburg. Here we learn that the Archbishop's letter had not yet arrived in Baden-Baden, probably because it had not yet been sent on from Freiburg. But he would attend to it the moment it arrived. He would discuss Heidegger's case with the French colonel in charge of the matter as soon as he returned from leave. 'But it will be difficult to get Heidegger reinstated at the University if the rector votes against it. However, I will do whatever I can, since you recommend the man.'[227]

Here the difficulties that would arise in the case of a negative verdict by the University Senate are openly addressed. For French policy on the universities *did* allow sufficient scope for the desire of the German universities to administer their own affairs. The military government looked upon the denazification commission as an officially authorized body, a kind of bridge between it and the University. And if the Senate voted in accordance with the commission's recommendations, it would not be easy to oppose their decision. Be that as it might, Gröber threw the full weight of his authority behind Heidegger. This needs to be emphasized, contrary to what has been stubbornly maintained by many – that the Catholic church authorities in Freiburg did everything in their power to have Heidegger removed from the University. The foremost and most influential spokesman for this view was Robert Minder, who never produced one shred of evidence to back his claim that 'it is an established fact that in the wake of the occupation the head of the Church [he means Archbishop Gröber] tried with all the means at his disposal to keep the heretic out of the University.'[228] The very opposite is true. Much later, of course, when Heidegger's case came up for review in 1949 and the question of his rehabilitation was at issue, Heidegger himself partly blamed the Catholic Church for blocking his full rehabilitation and reinstatement in his teaching post. Again, without a shred of evidence, simply on suspicion – there being little likelihood of a reconciliation between a religious doctrine founded on dogma and the thinking of Heidegger, liberated as it was from 'extra-philosophical' ties: Heidegger the perennial 'heretic'.

As the tide of opinion within the University began to turn against him, Heidegger was compelled to add a rider to his application for emeritus status at the beginning of 1946, declaring that he would refrain from all teaching until such time as the University invited him to resume.[229] The verdict reached at the meeting of the Senate on 19 January 1946 – the Heidegger affair was the only business of the day – was harsh: he was granted emeritus status, was refused permission to teach, and the commission's recommendation that the case be reviewed within a certain period was rejected. Furthermore: 'The Senate requests the Rector to inform Professor Heidegger that he is expected to maintain a low profile at public functions and gatherings of the University.'[230] The statement issued by the Faculty of Philosophy, with the unanimous support of all its members, in response to the Senate's stern judgement testified to the Faculty's unbroken goodwill towards Heidegger.

The Senate placed the Faculty's statement on record without discussion or comment. Josef Sauer, who had followed Heidegger's career for several decades, wrote disparagingly in his diary entry for 27 February 1946 of the 'absurdly muddle-headed rigmarole' from the Faculty of Philosophy, which had been greeted with howls of derision; the Faculty was falling over backwards to make itself look ridiculous. At the end of January 1946 Gerhard Ritter, a member of the denazification commission, summed up the dilemma in a letter to Hermann Heimpel, a former adherent of Heidegger's during his rectorship:

> As far as Herr Heidegger is concerned, he will no doubt be able to tell you himself how hard and successfully I have fought, against vigorous opposition in the Senate and the commission, to secure emeritus status for him rather than outright dismissal – in recognition of which the Faculty has passed a formal and unanimous vote of thanks and confidence in me. Nonetheless, Heidegger bears a heavy burden of responsibility for the way the German universities fell into line behind the Party, as he himself now recognizes. For me personally what happened in 1933 was a source of great pain and anguish; it was an intellectual disaster, which disturbed me so much that I couldn't sleep for weeks.[231]

Jaspers too, whose report did so much to influence the Senate's decision, wrote to Gerhard Ritter on 4 February 1946 and observed: 'It also occurs to me that this outcome might be the most intellectually fruitful one for Heidegger. Public appearances would land him in difficulties, and might inhibit his productivity, it seems to me. This way, too, he won't have a chance to make any gaffes, which put us all at risk at the moment.'[232]

Many people hoped that Heidegger would turn from his old ways during this period of his life, which had landed him in deep trouble (and dire financial straits), leaving him feeling ill-used and maligned, cut off from his roots in academic life. But where would he turn *to*? Did Jaspers also think that Heidegger would turn from his former ways? Brought to the verge of physical and mental breakdown by the strain, Heidegger spent part of the spring of 1946 at the Haus Baden sanatorium in Badenweiler, where he was treated for psychosomatic illness by Baron Viktor von Gebsattel. It was during this stay, incidentally, that the foundations were laid – or strengthened – for his subsequent close collaboration with a particular school of psychiatry, namely the existential-anthropological school of Ludwig Binswanger and Medard Boss, to which Gebsattel (who later became professor of psychiatry at Würzburg) also subscribed. Archbishop

Gröber, certainly, was counting on a change of heart in Heidegger. In a letter of 8 March 1946 to Father Leiber, adviser to Pope Pius XII on German political affairs, which also contained a report on the current political situation for the attention of the Pope, Gröber writes:

> The philosopher Martin Heidegger, a fellow-countryman and former pupil of mine, has been given emeritus status and is not allowed to lecture. He is at present staying at the Haus Baden sanatorium in Badenweiler, and is becoming increasingly withdrawn, as I learned yesterday from Professor Gebsattel. It was a great consolation to me when he came to see me at the start of his misfortunes and conducted himself in a most edifying way. I told him the truth, and he listened with tears in his eyes. I shall not break off relations with him, because I am hopeful of a spiritual change of heart within him.[233]

However, this was not yet the end of the Heidegger affair. At the beginning of 1946 the French military government had issued instructions for the establishment of a 'regional denazification commission', its membership to be proportionately representative of the newly licensed political parties, to which all cases would henceforth be referred – including those that had already come before the University's own internal commission. Here was yet another area bedevilled by jurisdictional uncertainties, and it was not clear to what extent the University would be represented on the new commission in order to safeguard its own interests. The Senate of Freiburg University discussed the matter on several occasions during the spring of 1946, and finally appointed Constantin von Dietze as its representative. From von Dietze's reports it is clear that the regional denazification commission was not prepared to accept any special dispensation for university personnel; on the contrary, the commission insisted on equal treatment for all and a single system of classification for the guilty, all of whom would suffer the full consequences of their actions under civil service law. Thus the view became established, despite representations to the contrary from the University, that *every* rector who had held office during the Third Reich would have to answer before the commission.

By August 1946 it was becoming clear to Heidegger that he would not be granted emeritus status, as proposed by the University in January, but would be pensioned off, which meant that he would lose his professorial status and lose for good his right to teach. Since the newly established regional commission operated under the supervision and on the instructions of the French military government, the ratification of its recommendations was a foregone conclusion. Accordingly a provisional ruling was passed by the military government on 5 October 1946, and spelled out in its final form on 28 December 1946: '*Il est interdit à M. Heidegger d'enseigner et de participer à toute activité de l'Université.*' In conformity with this decision the Baden ministry of education notified Heidegger on 11 March 1947 as follows: 'In the course of the political purge of administrative personnel a final decision was reached in your case on 28 December 1946, banning you from teaching and excluding you from all University activities. The ban on teaching is effective as of now. All salary payments will cease at the end of 1947.'[234] This financial sanction (but no other) was lifted in May 1947, when the military government approved a full retirement pension – at the same time taking the opportunity to reject explicitly any grant of emeritus status to Heidegger. The decision of 28 December 1946 remained legally binding – at a time when the *Letter on Humanism* addressed to Jean Beaufret was already on the way.

For Heidegger was already rising like a phoenix from the ashes of the funeral pyre heaped up by the French military government, destined to enter the mainstream of French intellectual thought as *the* major force in contemporary philosophy. Heidegger's thought was about to embark on its triumphal progress throughout the Romance world.

The decision taken by the military government in the autumn of 1946, which ruled out Heidegger's reinstatement at the University, did not reflect the University's own wishes. Rumours to the contrary are quite without foundation. Exactly how the French occupying power conducted itself in this and other cases is something on which the historian can throw no light while the relevant documentary records remain for all practical purposes inaccessible. But this much is certain: the French military authorities in Baden-Baden were taking a tougher line in the Heidegger case, which was to become even more intransigent as time went on – partly in response, of course, to the domestic political upheavals in France, which had an impact on policy in the occupation zones. Even Heidegger's personal library came under renewed threat of confiscation – and the threat this time was very real; the danger was only averted in 1947 by the concerted efforts of several parties. The Freiburg historian Clemens Bauer and Archbishop Gröber played a key role here, using their indirect contacts to put pressure on the military supremo in charge of French education policy in Baden-Baden, Général Raymond Schmittlein.

It was rumoured that Heidegger's books were going to be used to restock the library at the University of Mainz, recently re-established by the French occupying power. It was Franz Josef Schöningh, licensee and editor of the *Süddeutsche Zeitung* and publisher of the Catholic journal *Hochland*, a close friend of Clemens Bauer who also had good contacts with Archbishop Gröber, who intervened with Schmittlein at the request of these two men in order to prevent the threatened confiscation of Heidegger's library. Schöningh had got to know the general at a conference of German and French writers in Lahr/Baden in the summer of 1947. During his stay in Freiburg he had discussions with Archbishop Gröber and Clemens Bauer. He then wrote to Schmittlein on 6 September 1947:

> During my stay in Freiburg I learned from a number of sources that the library of the celebrated philosopher Martin Heidegger was threatened with confiscation. The persons who told me this are not personal friends of Heidegger (whom I do not know myself); indeed, they reject both his earlier philosophy and the political conclusions that Heidegger himself drew from it when the National Socialists came to power. In my own paper at the aforementioned conference I drew attention to the disastrous consequences of Heidegger's nihilism. But I also underlined the importance that this philosopher has, regrettably, acquired in France, not least through Sartre. This demonstrates clearly enough that Heidegger holds a firm place in European intellectual history, whatever one may think of him. Consequently the confiscation of his library (which I am told the University of Mainz is interested in acquiring) would cause something of a stir, and

> would, I fear, be seen as an act of vindictiveness for which the French military government would ultimately be blamed. I therefore regard it as my duty to bring this case to your attention, and to ask you to consider the possibility of intervening to prevent this step, which is so open to misunderstanding.[235]

In his paper at the Lahr conference Schöningh had argued that Sartre was 'bringing Heidegger to France like some new discovery', and that

> the terrifying nihilism of Heidegger or Sartre is seen only as the reflection of a political and social catastrophe . . . With Heidegger and Sartre the European mind has finally found that to which it has been ceaselessly drawn for centuries: the contemplation of the void. You are all familiar with Heidegger's political past. The only reason I mention it here is that it serves to illuminate the connection between nihilism and National Socialism in a particularly direct way.

He was well aware, he went on, that there were aspects of this connection that could only be explained with reference to specifically German antecedents, including Hegel, Nietzsche, Wagner and Bismarck; but that should not make his listeners lose sight of a phenomenon 'that leers out at us in every country in Europe: I mean the total absence of belief that is nihilism. This thing is the result of our common historical heritage, and therefore we must confront it and challenge it together.'[236]

Heidegger felt himself to be misunderstood, persecuted and outlawed – and all this without just cause, in his eyes. So many humiliations had been heaped upon him since the summer of 1945: parts of his house had been requisitioned for strangers; his library was under constant threat of confiscation; he was forced, as part of his punishment, to help clear the rubble from the streets of Freiburg; he was the subject of political investigations and purges at various levels. And now, to cap it all, the final humiliation: the decision by the French military government to dismiss him from his academic post and ban him from teaching altogether. Heidegger saw this as an alien dispensation imposed from without, since at that time the Germans held no political power of their own, but simply acted as the agents and instruments of the occupying power. Heidegger could not and would not accept being bracketed together with the thousands of minor officials who were being sacked from their jobs because of their political past and left without any means of support; he refused to be treated like the local primary school teacher who was dismissed from his post because he had been a member of the NSDAP and a minor Party official. One looks in vain for this kind of solidarity from him, because he saw himself as belonging to the resistance. Even in 1946 Heidegger may have been imbued with the conviction that the Germans had *not* gone under, and must persevere still through the night in order to arise – seeing things as he did in the long term, and placing his trust in the clarifying light of distant perspectives, like the distant panorama of the Swiss Alps that was constantly before his eyes at the mountain hut on the Todtnauberg. He now spent more and more of his time at his mountain retreat, and it was to this place of refuge from his troubles that he dedicated the little book he wrote in 1947, *Adventures in Thought* (Pfullingen 1954), which contains the proposition: 'The man who thinks great thoughts is liable to make great errors.' Here too, in 1946/47, he embarked on the attempt to translate the works of Lao-tzu into German. These projects clearly served as a kind of occupational therapy, now that Heidegger had been excluded from all involvement with the University.

But on legal grounds as well Heidegger refused to accept the verdict from Baden-Baden, regarding it as subject to future revision and redress. And that, for Heidegger, could only mean reinstatement in his former academic post by the University as soon as it had recovered its freedom of action with the full restoration of autonomy. Heidegger was only fifty-seven and had eleven years to go before he reached normal retirement age: what he wanted was for his chair to be kept vacant for him – in other words, a state of suspended animation. But Heidegger was expecting the impossible. To appreciate the realities of everyday university life in those critical years, it is necessary to go back to the decisions that were taken in the autumn of 1946. The problem of Heidegger's successor throws a revealing light on the legal and mental climate of the day – as we shall now see.

The search for Heidegger's successor

Heidegger himself could easily have contemplated some sort of philosophical partnership with Guardini (see Heidegger's letter of 6 August 1945 to Guardini), whom he had known for decades – as long as it meant that Guardini would get the Freiburg chair of Christian Philosophy, while he himself held the prestigious chair on which Heinrich Rickert and Edmund Husserl had conferred such distinction, a position whose occupant 'heads one of Europe's leading departments of philosophy' (as Heidegger put it in a letter to the rector's office in October 1945). The campaign to win Guardini for the chair in Freiburg was co-ordinated by the then pro-rector Franz Büchner, the man responsible for the post-war reconstruction of the University and a distinguished pathologist who looked with favour on interdisciplinary collaboration. In his letter to Guardini of 2 May 1946[237] the pro-rector unfolded his program of reconstruction. By securing the appointment of Guardini he was hoping to build up a structured interdisciplinary curriculum based on Christian principles, in which the philosophy of Guardini would enrich and enhance the study of medicine, the natural sciences and theology. He also saw a place in this for Heidegger:

> There is also the fact that Martin Heidegger lives in Freiburg, and hopefully will be lecturing again in the not-too-distant future. Although he himself has lost the Christian's secure sense of belonging, his philosophy has become an increasingly powerful expression of the yearning for that sense of belonging; he has been wrestling with the angel for so long now, trying to obtain his blessing. I know how highly he esteems both you and your work, and I could not imagine anything

finer than you two working together at the same university, engaged in a great intellectual dialogue.

This was well-intentioned but utopian. For one thing Guardini could not be induced to leave Tübingen, where he had been enticed by the clever recruiting policies of the regional director of education, Carlo Schmid, because what he really wanted – though he did not mention this in his reply, of course – was a post in Munich; and for another, this semi-public scholarly debate between Guardini and Heidegger would never have materialized, since Heidegger's star as a university philosopher was extinguished in the months that followed. In the end the chair of Christian Philosophy went to a university lecturer, Dr Max Müller.

Following the decision taken in the autumn of 1946 the vexed question of Heidegger's successor had entered an acute stage. The names of possible successors were discussed – as they were, too, in the Faculty of Philosophy at Freiburg, which resolved on 19 September 1946 to reconvene the board of appointment for Heidegger's chair, and to send informal, exploratory letters to Nicolai Hartmann, Hans-Georg Gadamer (then still in Leipzig), Gerhard Krüger, and above all Guardini. Pro-rector Büchner had anticipated this resolution by inquiring of Guardini on 3 August 1946 whether he could keep himself free to succeed Heidegger. Although no final decision had yet been reached in the denazification proceedings against Heidegger, it was clear that the outcome would not be in his favour. Büchner felt it was essential to sort out the question of a successor, since Heidegger was hardly likely to return to his teaching post – at least, not in the foreseeable future: 'I don't need to tell you what it would mean, both in practical and symbolic terms, if you were to take on the chair. Do you think, despite your initial refusal, that there is still a chance we might be able to persuade you to accept this position?' A great deal depended on Guardini's answer, he went on, for what he was trying to create at Freiburg was a new intellectual ambience that would be explicitly Christian in character. According to Büchner, what the students were looking for was clear Christian leadership. All his hopes rested on Guardini. In the course of the correspondence that passed between the two men in the autumn of 1946 it emerged that

Guardini regarded himself as totally unsuited to follow in Heidegger's footsteps. To succeed Heidegger required more than he could offer, especially since his interests lay in a totally different direction – to say nothing of the disparity in academic stature. He was not the right man for a chair in academic philosophy: 'I am not a specialist in any one area, but simply someone who looks at things and historical events and reflects on what he sees; a critic and interpreter, if you like' (letter from Guardini of 4 September 1946). Pro-rector Büchner was not to be put off so easily. He deployed all his powers of eloquence and persuasion, offering Guardini a way out of his difficulties and invoking the prospect of a radical break with the past:

> For you of all people to take over Heidegger's chair would be a symbolic act of extraordinary power, because it would signal unmistakably that Germany's universities, having gone through the misery of existential philosophy, are now awaiting the liberating word from a man who has repeatedly transcended the intellectual realm to enter the spiritual, and for whom philosophy and fundamental theology ultimately form a single whole.

The pro-rector took up Guardini's arguments against succeeding Heidegger in the chair and promptly turned them round, presenting them as good reasons why Guardini should accept the position – particularly the argument that Guardini was only a critic and interpreter. Is not all true philosophy an act of interpretation, the art of translation, the ability to reveal the hidden element in any significant phenomenon? 'Did not Heidegger do his best work when he interpreted the sufferings of his own heart cast out from the presence of God? Was it not in order to develop his own thoughts on truth that he turned to Plato, and to Hölderlin that he might speak of things sacred and holy?' The interpretation of uninterpreted signs was now a more urgent task than ever before (Letter from Büchner to Guardini of 21 September 1946).

Whatever one makes of the points raised by Büchner, and though he spoke with the tongues of angels, Guardini stood firm by his refusal: he was not, he insisted, the right man for this chair (6 October 1946). But Guardini did use the Freiburg job offer to persuade the regional directorate for culture, education and art in Tübingen (the precursor of the later ministry of education) to formally ratify his personal professorship with all its privileges and to attach a very precise job description to the chair: 'the philosophy of religion and the Christian world-view' (17 December 1946). The package deal negotiated by Guardini also included a typewriter – an almost priceless treasure in those days. Thus provided for, he could focus his gaze on his ultimate goal, namely Munich, which suddenly became an imminent prospect during these weeks: one of the first official acts of the new minister of education for Bavaria, Alois Hundhammer, was to set the wheels in motion to offer Guardini a chair at Munich (30 December 1946).

Convinced that both Guardini and Nicolai Hartmann would decline the Freiburg post when it came to decision time, Hans-Georg Gadamer travelled from Leipzig to Freiburg in the autumn of 1946 with high hopes of securing the vacant chair (as he revealed in a letter to Jaspers of 9 October 1946). The only real question mark was the attitude of the French, who for political reasons – and no doubt under the influence of the Church – might baulk at appointing a Protestant who was not a native of Baden. It seemed reasonable to suppose that the short-list of candidates would be submitted to the ministry before the end of the year, but in this Gadamer was mistaken. Guardini's refusal led to a loss of momentum in Freiburg, particularly as there was no sustained interest there in finding a permanent replacement for Heidegger in order to block the possibility of his later recall or return to his teaching post after a period of suspension. A significant faction within the Faculty of Philosophy, which was exceptionally well-disposed towards Heidegger, now adopted this line and pursued a policy of hiring temporary replacements – until further notice, as it were, until such time as the situation became stable and predictable, the mist and fog had cleared, and the political landscape of higher education had emerged into the bright light of day.

On 4 June 1947 the representative of the Faculty of Philosophy on the University Senate announced plans to secure the services of the scholar Wilhelm Szilasi, then living in Switzerland, to give a few lectures, and possibly to conduct a colloquium for students. The senators were then given a brief account of Szilasi's person and career; we are already well acquainted with him, of course, as an early pupil and friend of Heidegger's in connection with Edmund Husserl. The Senate gave the go-ahead for the necessary negotiations with the ministry and the French military government, and a few weeks later approved formal motions to engage Szilasi to give lectures and appoint him an honorary professor of Freiburg University.

Thus began the era of Wilhelm Szilasi's deputy professorship. It was a period filled with trials and tribulations, since outsiders, particularly abroad, did not always understand his special status, so that Szilasi was widely viewed as Heidegger's successor – a view that was perhaps deliberately encouraged. Almost the same age as Heidegger, Szilasi had been forced to leave Freiburg in 1933 as a victim of racial persecution, and had settled in Brissago/Locarno as a business consultant – so close to the Swiss-Italian border that the actual line of the frontier is said to have passed through the grounds of his villa. After the total collapse of Germany he was clearly living a life of comfortable independence in the safety of neutral Switzerland, making the most of the opportunities offered by this safe haven of affluence. Not the least of these was a major publishing programme for titles on cultural history ('Sammlung, Überlieferung und Auftrag') that had begun appearing under the Francke imprint in Bern back in 1945. The series was edited by Ernesto Grassi (whom we have already encountered), in association with Wilhelm Szilasi. Divided into various parts or sections, the series was a continuation of the work that Grassi had initiated earlier as honorary professor at the University of Berlin (from 1940/41), and in his role as cultural ambassador for Fascist Italy in the Reich capital: the cultivation of the intellectual tradition, the study of humanism, the reception of German philosophy in Italy. During the war years Grassi had substantial resources at his disposal and he was able, with the support of the *Duce's* regime, to circumvent the difficulties of paper rationing, if necessary by having texts printed in Italy, where an adequate supply of paper was available to him.

As we know, of course, Szilasi and Grassi had met long before their collaboration in Switzerland, where the Italian had decided to move towards the end of the war; he was even offered a teaching appointment at the University of Zurich, where the spectre of his

Fascist past rendered his position somewhat precarious.[238] As we noted earlier, the Hungarian Wilhelm Szilasi and the Italian Ernesto Grassi had been together in Freiburg from 1928 to 1933, drawn by the presence of Martin Heidegger who had been appointed to succeed Husserl. This was the second extended period of time Szilasi had spent in Freiburg, where, as always, he kept open house for guests and worked as a freelance academic. At the time of his move to Switzerland he had distanced himself from Heidegger. Now, after 1945, there would be a rapprochement, and Freiburg would become the focal point in a great tradition. There is a suggestion that Szilasi sought to re-establish contact with Georg Lukács around this time, in the hope of landing a suitable post in Budapest, but the attempt foundered on irreconcilable differences.

So Szilasi's teaching appointment in Freiburg in 1947 did not come out of the blue. He had the support of numerous friends, and his book *Macht und Ohnmacht des Geistes*, which had appeared the year before in the series edited by Grassi, had commended him to a wider public – especially the foreword dedicated 'To Ernesto Grassi', in which he invoked the tradition dear to them both:

> The tradition to which we owe our intellectual being is closely associated with the names of Husserl and Heidegger, and with those wonderful, now legendary years in Freiburg, whose like we shall not see again. The works of these two men have become an enduring legacy for mankind. Personal impressions perish with those who receive them. It is our duty to lend them permanence, for in this series of publications we have set ourselves the task of nurturing a community of thought that will keep this intellectual tradition alive and commit it to the care of future generations.

After this fulsome introduction Szilasi gives a detailed and lively account of the work of Husserl and Heidegger, emphasizing the differences between them. But he adds that all this is past history now, 'because this era belongs so wholly in the past that I know virtually nothing of Heidegger's work and writings for the last fifteen years'.

There was an obvious ambivalence about Szilasi's presence in Freiburg from 1947 onwards, in his capacity as special lecturer and honorary professor, deputizing for the 'vacant' chair of Philosophy I, reappointed from one semester to the next, variously described in the lecture list as a 'relief', 'stand-in', or 'temporary stand-in': it was easy to interpret his position as that of *de facto* successor to Heidegger – and so it was interpreted. But it was widely supposed within Freiburg University circles that Szilasi had been brought in as Heidegger's friend for the specific purpose of keeping Heidegger's philosophical influence alive among students and interested sections of the public, revitalizing it and preparing the ground for Heidegger's return: a *locum tenens*, taking Heidegger's place in the most literal sense, serving his cause, rooted in the deep soil of a friendship that went back nearly thirty years, and furnishing a safeguard against misinterpretation, deception, dishonesty, jealousy, double-dealing, and whatever else constitutes the common currency of public opinion. But the years that followed present a rather different picture of faction and discord.

Towards a rehabilitation of Heidegger

On 26 September 1949 Heidegger celebrated his sixtieth birthday, reaching the age when it is customary for scholars to be presented with a *Festschrift*. With this birthday in prospect, moves were made to breach the wall of silence that surrounded Heidegger. Hans-Georg Gadamer, his one-time pupil, did his best to organize a *Festschrift* with contributions from distinguished philosophers who were associated in any way with Heidegger's thought. The uncontentious working title, 'A *Festschrift* in honour of the philosopher Martin Heidegger', would have served to convey the aims of the book without the need for further words of dedication.[239] Gadamer had to handle the project with the utmost delicacy, meeting with guarded acceptances, conditional willingness, and then in the end refusals. At first it looked as if the philosophical community was standing aloof. But the project eventually came to fruition, albeit a little late, and 1950 saw the publication of the volume *Common concerns. For Martin Heidegger on his 60th birthday* (Frankfurt a.M.). A parallel publication, edited by Wilhelm Szilasi, the temporary occupant of Heidegger's chair, appeared on time: *Martin Heidegger's influence on the sciences* (Bern 1949). Friends of Heidegger's in linguistics and philology honoured him in a separate commemorative volume of their own.[240] The ice had been broken: the academic world had brought Heidegger out of the closet and nailed its colours to the mast. The shadows were receding, even if 'the Heidegger affair' remained unresolved. Also significant was the fact that the reactions of the media on 26 September 1949 were largely positive, expressing astonishment that Heidegger should still be condemned to silence.

At the beginning of 1949, when the Basic Law was soon to be enacted and the establishment of the Federal Republic of Germany was now an imminent prospect; when the West Germans were about to acquire a significant measure of national autonomy again and the occupying powers visibly began to relax their grip on the system of law and justice; when the various denazification proceedings had started to become tiresome – enough was surely enough! – and the grass had begun to grow over the recent past: then perhaps the time had come to lift the ban on Heidegger and clarify the legal basis of his position at Freiburg University. The omens were favourable, since the rector appointed for the academic year 1949/50 was the historian Gerd Tellenbach. This meant that the Faculty of Philosophy, Heidegger's own faculty, was now represented at the highest level of the University's administration. The dean of the Faculty during this rectorship was Clemens Bauer. These two colleagues had already displayed a sympathetic understanding of Heidegger's difficulties during the preceding years. Bauer's personal intervention to save Heidegger's library has already been noted; Tellenbach, for his part, had shielded Heidegger when he was threatened with further restrictions on the use of his home and the confiscation of his furniture, and again when the labour office sought to conscript him for compulsory work service. In the summer of 1947 that was a very real danger.[241]

Now, at the beginning of 1949, there seemed to be grounds for optimism. On 9 January 1949 Heidegger sent Tellenbach, the rector-designate, a brief account under the title 'My relationship with the University', in which he summarized the events since 1945. He concluded: 'If the Faculty is now thinking to put an end to this state of affairs, then the only acceptable way forward, after what has occurred, is for the University to persuade the military government to lift its teaching ban, so that my application for emeritus status may be properly processed.' This plan of action proposed by Heidegger himself should be noted, because it was not long before tempers began to fray and campaigns were set in motion behind the scenes to discredit Freiburg University for its alleged failure to rehabilitate Heidegger adequately. As far as Heidegger was concerned, then, his legal status, his 'relationship with the University', was clear enough on 9 January 1949: he was subject to a teaching ban that must be lifted so that he could formally apply for emeritus status. This summary of the legal position needs to be kept in mind. For the moment it was not affected by the proceedings of the denazification tribunal before which Heidegger, like all former members of the Party, had to appear. In March 1949 he was classified by the state commissioner for denazification proceedings as a 'nominal Party member', against whom no punitive action was to be taken.[242]

The Faculty was prepared to submit an application for the conferral of emeritus status on Heidegger, fully accepting that there could be no substantive revision of the decision taken by the Freiburg University Senate in January 1946, but only a formal revision in the sense that a grant of emeritus status would indirectly constitute a revocation of the teaching ban. It would have been a big mistake to try and reinstate Heidegger in his old chair at this point. After Rector Tellenbach had sounded out the French liaison officer on the one hand and the regional government of southern Baden on the other, the Faculty of Philosophy submitted an application to the Senate in May 1949 requesting that Heidegger be retired with all the rights and privileges of an emeritus professor.[243] The submission pointed out amongst other things that because of his international standing as a philosopher Heidegger's relationship with the University had not been regulated through the normal channels, but had been subject to a special review process in which the final decision rested with the French military governor. The Faculty now deemed it appropriate to take steps within the University to effect his rehabilitation. New circumstances had arisen in the meantime:

> The continuing world-wide interest in Heidegger's philosophy and its various developments makes it desirable that Heidegger should be allowed to speak for himself again. The conspicuous restraint shown by Herr Heidegger has also prevented him from repeating the criticisms he voiced earlier in his lectures, as many have testified, against certain pernicious tendencies of the times, and has meant that the latest developments in his thought are entirely unknown.

It was simply unacceptable, in a state founded on freedom of speech, that a man of Heidegger's stature should be denied a voice indefinitely – which is what a continuation of the teaching ban would amount to.

The Faculty felt that it would be a real gain for the entire University if Heidegger, whose intellectual significance would be underlined in the public eye by his forthcoming sixtieth birthday, were allowed to enter the University again and speak his mind in the modest role of an emeritus professor: no longer having any say in the conduct of University policy, but able to articulate in a fitting manner the thoughts that he had not been allowed to express since 1934.

The discussion initiated by this submission in the Senate, the platform for debate at the University of Freiburg, touched on old wounds. As the record of the Senate meetings in May 1949 reveals, there was considerable opposition to the proposals.[244] After two lengthy meetings of the Senate on 4 and 18 May, the Faculty's application was finally granted, albeit by a slender majority (7:5). As the number of dissenting voices indicates, the air had by no means cleared. Those opposed to the motion expressed serious doubts about the quality of Heidegger's philosophical work. It was claimed that the Faculty was vastly overrating Heidegger's intellectual importance; he was at best a vogue figure, at worst a charlatan, whose teachings were dangerous and rightly banned. The Faculty of Philosophy defended itself in a statement reiterating the academic argument: as the one Faculty professionally qualified to judge of these matters, its views had not been given the weight they deserved. Heidegger's importance as a philosopher was the *only* factor influencing its deliberations:

> Is Heidegger's contribution to the philosophical debate so crucially important, alongside those of Hegel, Kierkegaard, Nietzsche, Dilthey, Husserl and other outstanding thinkers, that our idea of the university and of science and scholarship demands that he be allowed to give appropriate expression to his thoughts once again, irrespective of whether one agrees with him or disagrees with him? For its part, the Faculty has answered this question in the affirmative, and after carefully weighing all its misgivings has decided that it does *not* wish to bid farewell to Martin Heidegger for good.

It was announced that the Faculty would be collecting references from distinguished outside scholars in support of its application, and specifically from Romano Guardini (Munich), Karl Jaspers (Basel),

Nicolai Hartmann (Göttingen), Charles Bayer (Paris), Emil Staiger (Zurich) and Werner Heisenberg (Göttingen) – a truly international field. The Senate agreed to this arrangement.

One of the leading players in this act of the rehabilitation drama was Max Müller, holder of the chair of Christian Philosophy, who saw himself as a disciple of Heidegger, even though 'the old man of the mountain' (Müller's own phrase) had not shown him any particular goodwill in his day. Amongst other things, Müller took on the role of corresponding with Guardini, in the hope of eliciting a report from him in Heidegger's favour. That was no easy task, since Guardini, now installed at Munich, would much rather have kept quiet, despite the fact that – or precisely because – he was thoroughly familiar with the Heidegger case. In a letter of 11 June 1949 Müller told Guardini that Heidegger longed with all his heart to be able to enter once again as an emeritus into the university 'whose pride and glory he has so long been'.

> Regardless of whether we share his views and regard them as genuine insights or whether we oppose them, we believe his contribution to the philosophical debate to be so crucial and indispensable (particularly in view of the many misunderstandings surrounding him) that we regard a further continuance of the teaching ban as unjustifiable. We are not concerned here with political judgements or revisions, but *simply and solely with the question of whether Heidegger's extraordinary intellectual standing does not require that he be accorded the opportunity he desires, subject to the restrictions of emeritus status, to work again within the setting of the University*.

Müller said that the more individual and personal the report turned out, the better pleased the Faculty would be. He also added that Heidegger knew nothing of this campaign on his behalf.[245] In his reply of 1 July 1949 Guardini declared that he did not feel it was his place to pass judgement on 'how and in what way the teaching ban imposed on Martin Heidegger is lifted'. On a personal level, Heidegger had been a very important figure in his life for over thirty years. As far as intellectual standing was concerned, he was of the opinion that Heidegger 'is the most potent force in German philosophy today', and he hoped he would have an opportunity to document that claim publicly. It was imperative that Heidegger should once again have complete freedom to develop his views, through both the spoken and the written word. 'So despite any political or philosophical reservations I may have, I would gladly do whatever I can to help his cause.' As an outsider, however, it was not for him to say what Heidegger's formal position should be at Freiburg University. When Müller continued to press him, Guardini finally agreed to let his letter be passed on to the Senate. The letters from the other referees were also positive, particularly the one from Karl Jaspers, who used this opportunity to mitigate the severity of the highly damaging report he had written in December 1945.

By the summer of 1949, therefore, the University had paved the way for a development that gave Heidegger what he wanted: the grant of emeritus status following a revocation of the teaching ban. But initially the application for emeritus status ran into formal legal difficulties. After protracted skirmishes between legal experts it was established that under current civil service law a grant of emeritus status could not be made before the statutory age of 62, and Heidegger would not attain this age until September 1951. It was necessary, therefore, to resort to *ad hoc* arrangements. Thanks to some tough negotiating by Rector Tellenbach with the government ministries concerned, the following compromise was reached: if Heidegger so requested, the teaching ban could be lifted and he could be retired on a full pension, with the understanding that when he reached the age of sixty-two he would be given emeritus status. The difference in salary up until that point would be made good through teaching assignments. The ministry of state under the direction of the Baden state president Leo Wohleb gave a formal guarantee that Heidegger's emeritus status would become effective as of 26 September 1951.

In the course of the aforementioned discussion about the difficulties of the case under civil service law the Baden state president Wohleb, a highly cultivated man who was deeply devoted to the University of Freiburg, had mooted the possibility of Heidegger's reappointment to the chair – unaware, clearly, of the weight of opinion and the balance of forces within the University Senate.

The state president's benevolent attitude was again attributable to the influence of Max Müller, who was Wohleb's *éminence grise* in matters of academic policy. As the negotiations were drawing to a close at the end of March 1950, a minority application was submitted by the Faculty of Philosophy on 1 April requesting Heidegger's full reinstatement (i.e., his restoration to his former teaching post) – a move that caused a considerable stir, clouding, not to say poisoning, the atmosphere. Heidegger welcomed this initiative. As he remarked in a letter to Rector Tellenbach on 6 April 1950: if the Faculty was now making application for his full reinstatement, then he was bound to say that he regarded this option – which had not yet been discussed – 'as the more appropriate one in the circumstances, particularly after I have been subjected for five years to "punitive measures" that go far beyond what the University's denazification commission approved in 1945'. But this was too much for the rector and dean, who resolutely opposed such a step, realists that they were, knowing that feelings would run dangerously high within the University if this petition were granted. This minority application, which had obviously been discussed in advance with Heidegger, prompted the rector to issue him with an ultimatum: Heidegger must make up his mind one way or the other, either to follow the course of action suggested by the rector, or to support the application submitted by a minority within the Faculty of Philosophy. After intensive negotiations Heidegger gave in, and agreed to follow the course recommended by the rector and dean, accepting retirement with a guaranteed grant of emeritus status to follow – particularly as this would mean the immediate lifting of the teaching ban. And so it was decided. Martin Heidegger was officially cleared to begin lecturing again in the winter semester of 1950/51: he had come in from the wilderness and been rehabilitated. Of course, Heidegger himself could never see this as anything more than a third-rate form of rehabilitation, and he bore a grudge against those responsible for the rest of his life.

The public had rehabilitated him in a different way. In small, elitist gatherings the faithful flocked to hear Heidegger lecture – beginning with the 'Club of Bremen' in 1949. 'Who is Zarathustra?' and 'The principle of sufficient reason' were among the topics first presented there, and later repeated on 25/26 March 1950, just as the final decision on his case was being taken at Freiburg University, in the fashionable, not to say exclusive, surroundings of the 'Bühler Höhe' spa assembly rooms, where a very mixed audience gathered to hear the celebrated philosopher.[246] However, the real breakthrough for Heidegger came in the summer of 1950, when the Bavarian Academy of Fine Arts sponsored his lecture 'On things' ['Über das Ding']. It caused a political furore, but that was seen as the work of envy. When Heidegger gave his first public lecture again in Freiburg in the summer semester of 1952 – a moment of quiet triumph: in the preceding semesters he had only given classes – the students overflowed the lecture hall. What now emerged was a pent-up thirst for knowledge, coupled with curiosity and a longing to hear the word of Heidegger the thinker. When Heidegger posed the question 'What does it mean to think?', one simply had to be there to hear the great man's voice.[247] From 1953 to 1957 the Bavarian Academy of Fine Arts continued to play host to Heidegger, in lecture cycles prepared by a prestigious intellectual elite that also included Werner Heisenberg, Carl Friedrich von Weizsäcker, Friedrich Georg Jünger and Ernst Jünger, Carl J. Burckhardt and others. The most conspicuously successful was the lecture series on 'The arts in the age of technology' (1953). As we learn from Petzet (1983, p. 81): 'When Heidegger closed with the now-famous proposition "Asking questions is the piety of thought", a never-ending storm of applause erupted from a thousand throats. I had the feeling then that the ring of mistrust and malice surrounding the master – and friend – had finally been broken. It was perhaps his greatest public triumph.' Formal honours also came his way. In 1959, on his seventieth birthday, he was made an honorary citizen of his native town of Messkirch, and in 1960 the *Land* of Baden-Württemberg awarded him the Hebel Prize, previously bestowed upon Albert Schweitzer and Carl J. Burckhardt.

The later works took shape. Hitherto little travelled, Heidegger now visited parts of Europe he had known only through literature. Todtnauberg became a place of pilgrimage, and the tranquillity of the Black Forest mountain-top was not infrequently disturbed by visitors – some who came by invitation, others uninvited. The debate about the political past continued to flare up: but what of it? As in France, where Jean Beaufret and others saw to it that no stain should attach to Heidegger's name, so in Germany the critical voices were silenced. There were always friends ready to throw themselves into the breach – if indeed the walls ever were breached. When the thorny issue was aired in the pages of *Der Spiegel* in 1966, Erhart Kästner, looking far ahead and thinking of the long-term strategy, pressed Heidegger to give an interview for the magazine. The interview duly took place, after careful preparation, in the late summer of 1966 – the transcript stored away for publication only after Heidegger's death.[248]

The following year he had an opportunity to meet the Jewish poet Paul Celan, who had been invited to read from his poems before a huge audience gathered at Freiburg University at the end of the summer semester (24 July 1967). 'I have long cherished the desire to meet Paul Celan. He has gone further than anyone else – and holds back more than anyone. I know all his stuff, and I know too about the serious crisis that he managed to pull himself out of by his own efforts – in so far as that is possible for a man.'[249] It was a difficult meeting; the past lay heavy between them, preventing them from getting close, at least to begin with. But then, contrary to all expectation, Celan accepted Heidegger's invitation to the mountain hut at Todtnauberg: the next day, 25 July 1967, yielded the conversation that broke the ice – but failed to bring it into the clear light of day. Celan's disappointment can be heard in the lines he wrote in the visitors' book at the hut: 'Recorded in the visitors' book, looking out at the star over the well, in the hope of a word to come in my heart. 25 July 1967/Paul Celan'.[250]

All the same, back in Frankfurt, on 1 August 1967, Celan wrote the poem '*Todtnauberg*', which appeared initially in 1968 in a limited collector's edition.

> ARNICA and EYEBRIGHT,
> the drink from the well with the
> star-block on top,
>
> in the hut

the line written
in the book
– whose name did it
record before mine? –
the line that tells
of a hope, today,
of the word
of a thinker that will
come, come at once
in my heart,

grass beneath the trees, unlevelled,

orchis upon orchis,
standing alone,

coarse things, later, as we drive,
very clear,

the man
who drives us,
who hears it too,

the half-trodden
log paths
in the high moor,

dampness,
all around.[251]

Epilogue

Heidegger lived to an advanced age, but the later years took their physical toll in illness and frailty – the usual concomitants of old age. The death of friends, relatives, people who were close to him, was a constant reminder of the ever-present reality of final departure. Talk of death, philosophical talk included, was quite different from being in the physical presence of the dying, as he had once remarked to the Tübingen philosopher Walter Schulz[252], in conversation at the 'Stag Inn' in the grounds of the former Cistercian monastery of Bebenhausen. He went on to explain that his analysis of death in *Being and Time* had been written for the medical profession. Indeed: nowhere, perhaps, did Heidegger see the limits of 'practical philosophy' more sharply drawn than in the phenomenon of 'death', not least because it is here that the issue of 'faith and philosophy' presents itself – and no longer as a purely theoretical issue, as Heidegger had written to Jaspers on 1 March 1927, when Heidegger's mother was nearing death. Perhaps his whole life had been but a preparation for his own death. How else are we to understand the verse from the first chorus of Sophocles' *Antigone*, as rendered by Heidegger:

> Wandering hither and thither
> Not knowing where to turn
> He comes upon the void.
> But from Death's dread pursuit
> Nothing can ever save him,
> Though he have thus far 'scaped
> The pangs of lingering affliction.[253]

Heidegger died on 26 May 1976 in Freiburg. The funeral took place on 28 May. He wanted to be buried in his native soil, in the place where he was really at home, where the remembrance of his ancestors was rooted, where the skies look down upon a land that is free and open and radiant, whose harsh austerity conceals a kind of gaiety. He wanted to return to the place whence he had started out, the land of his fathers, the source of his being. He wanted to go home, quitting this world full of sound and fury and desolation to enter into the collective inheritance of his native land, fashioned by the forces of heaven and earth over the centuries: like the church of St Martin in Messkirch, which stands as a symbol of this enduring constancy. He wanted to bring back to this consecrated Christian homeland the inheritance that he had increased, to do that which he had been commanded to do: 'To help bear the sufferings of a godless age in his thought, and for the rest to interpret the way of the times and the world as a way that leads thither' – in other words, 'to wait with patient expectation upon the epiphany of the divine God', as the Catholic priest and theologian Bernhard Welte, his fellow countryman and companion in thought, put it in his funeral address for Heidegger in the cemetery chapel at Messkirch.[254] The words of the priest, taken from the prophet Jeremiah – 'But the Lord said unto me, Say not, I am a child: for thou shalt go to all that I shall send thee, and whatsoever I command thee thou shalt speak' (Jeremiah 1, verse 7) – were an attempt, by no means easy, to answer the question of whether Heidegger had returned to his native soil as a Christian, and a Catholic Christian to boot. The plain headstone is decorated with a star, but the sign of the cross, carved on the neighbouring headstones of his parents and brother, appears to be touching the philosopher's grave as well. Was it befitting, asked Bernhard Welte, to give Martin Heidegger a Christian burial:

> Does it befit the message of Christianity, does it befit the intellectual path that Heidegger followed? Certainly it was his wish. In fact he never severed his links with the fellowship of believers. He went his own way, to be sure, the way he doubtless had to go, obeying his inner call; and we might hesitate to call his way 'Christian' in the normal sense of the term. But it was the way of one who was perhaps the greatest seeker after truth this century has known. He sought the divine God and His glory with patient expectation, hearkening

> unto the message. He sought Him also in the preaching of Jesus. So I think it is meet that we should speak the words of comfort from the Gospel over the grave of this great seeker after truth, and the prayers of the psalmist, particularly from the psalm 'De profundis', and the greatest of all prayers, the one that Jesus taught us.

The language of the liturgical forms was Christian; the officiating priest, Heidegger's nephew Heinrich, handled them with great care, in accordance with his uncle's wishes, and mindful of his closeness to, and remoteness from, the Catholic Church. The free prayers in the cemetery chapel were followed by the prescribed texts and rites of the Catholic burial liturgy. Was this the return of the Prodigal Son to the bosom of the Church? The folk writer Albert Krautheimer, who for more than a decade was parish priest of Bietingen, next door to Messkirch, used to say that the church authorities would love to see Martin walk through the main door of St Martin's in Messkirch dressed in the robes of a penitent; in fact, he had long since entered the church by the vestry door, just as he had as a young lad when his father was sexton. And Krautheimer knew what he was talking about. He it was, after all, who had brought Heidegger's manuscripts safely through the turmoil and uncertainty of the war and the post-war period, hiding them away inside the stout, massive church tower of Bietingen and preserving them for the duration.

Afterword to the second, revised edition

When Alexander Schwan added a lengthy postcript to the 1989 reprint of his dissertation 'Political philosophy in Heidegger's thought' (originally published in 1965), it was forcefully entitled 'A plea for an internal perspective on Heidegger', and advocated 'a discussion of Heidegger and his thought in terms of his own assumptions and arguments, particularly with regard to the relationship between "philosophy" and "politics" in Heidegger.'[255] Alexander Schwan continued to work at this self-imposed task right up until the last weeks of his life (he died on 30 November 1989, not long after his sixtieth birthday), and in my view has given us the most complete analysis we have of Heidegger's political thought; he has taken the true measure of it, pointing up its possibilities and limitations, and above all he has laid bare Heidegger's inability to participate with any moral authority in the creation of a liberal political democracy.

It is a dismal enough record, but one that we have to accept. Otherwise there would be no way of understanding Heidegger's philosophical calling to a line of thought that extends back beyond the Western tradition, back to a different beginning that is essentially located among the early Greeks – a beginning that has 'irrupted into our future', that 'stands there as the distant dispensation over us, waiting to claim its greatness again'. This statement from the rectorship address never lost its validity for Heidegger, since he resolutely submitted to this dispensation 'in order to regain the greatness of the beginning'. He had long since fixed upon his own calling: 'But if we submit to the distant dispensation of the beginning, then science and knowledge must become the fundamental event in our intellectual-ethnic existence.' There are no signs here as yet of that intercultural critique in which the late Heidegger is said by a growing number of recent commentators to have been engaged: his rejection of Eurocentrism, his destruction of the philosophical foundations.

What this means is that the national revolution brought about by the movement under Adolf Hitler was the crucial historic event in the history of Being, an event in the sense of the dis-closure of Being, which at once reveals and conceals itself in what is 'founded out there' [*Dagründung*] – Adolf Hitler, in other words, as the executor destined and despatched by Being to fulfil a historical mission.

But the Führer declined to assume his appointed role, whereupon Heidegger took querulous issue with the 'movement' in his *Contributions to philosophy: Of the event*. It was a very private and personal reckoning, it has to be said; this motley collection of sketches, plans, outlines, notes on his reading and attempted definitions, written down between 1936 and 1938, was intended for his own consumption only. Whether these texts, published in 1989 as Volume 65 of the *Collected Works*, can 'rightly be regarded as Heidegger's second main work' after *Being and Time* (according to the blurb on the jacket) is a matter of opinion. At all events, the turning-away from the 'inward truth and greatness of National Socialism' (*Introduction to Metaphysics*, 1935) did not take place. On the contrary, Heidegger remained a loyal member of the Party.

In May 1938 local Party officials in Freiburg gave Heidegger a clean bill of health in the questionnaire they were required to submit to NSDAP headquarters in Karlsruhe. Perhaps he could afford to give a little more generously to Party funds, but otherwise he was 'an exemplary Party member'; although no longer publicly active in the cause, he was loyal to state and Party, and above all he was fiercely opposed to Catholicism – a point that was particularly stressed.[256] Heidegger's record on academic policy and appointments was fully in keeping with this picture: the 'political' references he wrote for Max Müller and Gustav Siewerth, two Christian candidates for lectureships in philosophy, fall precisely within this period. Under National Socialism both men remained barred from a university career. When the chips were down, the Party could always count on Martin Heidegger.

In private, meanwhile, Heidegger made no secret of his disappointments, particularly with regard to the failure of his rectorship and the rejection of his academic vision. In the autumn of 1937 he spoke of his feelings to a group of intimates at Freiburg, professors and lecturers from the faculties of medicine and natural sciences/mathematics who were taking part in a study group on the theme 'Science under threat'. He was caustic in his criticism, if somewhat less than precise: 'During my rectorship I made many serious mistakes. But the two biggest mistakes were: (1) that I didn't reckon with the mean-mindedness of my so-called colleagues and the spineless treachery of the student body; (2) that I didn't realize it is pointless to approach a government ministry with creative demands and ambitious goals. Which is why this ministry preferred to operate through underhand machinations involving both students and staff – here and elsewhere.' And his response, now, in 1937:

> Resignation? No. Blind acceptance? No. Toeing the line? No. The only thing is to build for the future . . ., to stay on and do whatever I can to influence individuals. Not by way of preparing for the new University (there's no point in that any more), but in order to preserve traditions, to offer models and examples, to plant the seeds of new ambitions here and there in key individuals – somewhere, sometime, for somebody. That is neither resignation nor an evasion, but the imperative implicit in the essential philosophical task of the second beginning. It is a question of creating knowledge – but only when the need for truth has been discovered; and that requires an awareness that Being has been forgotten and truth destroyed (from the unpublished text of Heidegger's talk).

With this Heidegger had reverted to his basic philosophical premise. To what extent he was betrayed during his rectorship by his students or student functionaries must remain an open question for the present. I am currently working on a detailed study of Heidegger's relations with the changing student leadership at Freiburg between October 1933 and February 1934. The story revolves around a number of interconnected episodes, many of them having to do with the controversial suspension of the Catholic student fraternity 'Ripuaria' (briefly referred to on p. 245 of my book). The background to these events is particularly important, since the members of this fraternity pointedly boycotted the election on 12 November 1933 – after Heidegger had thrown the full weight of his authority behind the referendum.

Once again we can only lament the fact that Heidegger's addresses and lectures from 1933 and the months that followed have not yet been published, forcing us to rely on selective extracts from these texts.[257]

The considerable reaction that my book has provoked on the subject of Heidegger's involvement with National Socialism has focused primarily on the case of Hermann Staudinger, or 'Operation Sternheim' as it was code-named by the Gestapo. And rightly so, I am bound to say, even though Heidegger's conduct in informing against a colleague remains a mystery to me to this day. Touching in the extreme is Silvio Vietta's heroic attempt[258] to remove Heidegger from the firing-line by his portrayal of the world-renowned chemist Staudinger as the representative of a utilitarian, industrialized science, whom Heidegger was bent on opposing for that very reason (and also bent on 'eliminating' or 'neutralizing', presumably). Anyone with any vestige of common sense can only shake their head in wonderment. I remain convinced that the explanation for his denunciation of Staudinger is ultimately to be found only in the realms of depth psychology, and this is a question to which we must return at the appropriate time. For in this as in other matters, it is not yet possible to reveal everything that is known. All in good time.

Significant portions of my book are based on the correspondence between Jaspers and Heidegger, which has since been published after many years of editorial preparation[259] and is now open to general interpretation. I have already made use of the most important elements, and essentially I have nothing further to add, particularly since the editors of the correspondence have provided a detailed critical commentary. Letters are remarkably immediate sources; although in many cases they are conscious constructs, the underlying personal substance still shows through, giving a direct insight into the writer's state of mind. This is even more true for letters that have been written from a position of personal involvement. Therein lies the peculiar importance of the correspondence between Heidegger and Elisabeth Blochmann, published in 1989[260]: what we have here is a dialogue shot through with a special intimacy that in 1929 began to grow beyond the merely formal to become something altogether deeper. Heidegger confided in his correspondent about his mental and spiritual development, constantly referring back to the times when they had met. The Beuron motif, for example, which is touched upon at several points in my book, comes across with greater clarity and intensity in the letters. In fact it emerges as something of a Beuron syndrome – as when, for example, Heidegger introduces Elisabeth Blochmann to Beuron in the autumn of 1929 and the place becomes a kind of Indian summer to 'the summer days we spent together'. In his letter of 12 September 1929 he encapsulates the 'Beuron experience' in prose of almost classical measure, whose meaning nevertheless remains obscure because everything centres on the *mood* into which things past are here transmuted:

> The past of human existence as a whole is not a nothing, but that to which we always return when we have put down deep roots. But this return is not a passive acceptance [Übernehmen] of what has been, but its transmutation. So we can only abhor contemporary Catholicism and all that goes with it, and Protestantism no less so: and yet 'Beuron' – to use the name as a kind of shorthand – will unfold as the seed corn of something essential. This is already clear from your feelings about compline, which *had* to give you more than you could ever get from the High Mass. To say that we walk daily into the night is a truism at best for modern man, who for the most part sees the night simply as an extension of the day as he understands it, as a continuation of the busy, intoxicating round. But compline embodies still the elemental power of the night as a mythical and metaphysical presence, which we must constantly break through in order truly to exist. For the good is only the good side of evil. Modern man is more than accomplished in the organization of every conceivable thing, but he is no longer capable of composing himself for the night. We *appear* to *be* something and to achieve something in a state of 'motion'; but when rest and leisure come our way, we simply do not know what to do with ourselves. So compline has come

> to symbolize for you the projection [Hineingehaltensein] of existence into the night, and the inner necessity of being prepared for it each and every day.

It is a momentous passage – momentous and intemperate. Here speaks the philosopher who only a few weeks before had achieved a resounding success with his inaugural lecture at Freiburg on the theme 'What is metaphysics?' He has set himself up to pass judgement on everything under the sun. Or is that going too far?

It was only a couple of weeks later that Heidegger wrote the highly damaging letter to Viktor Schwoerer (2 October 1929) which Ulrich Sieg has subsequently published and analysed in *Die Zeit*.[261] Heidegger had applied for a grant on behalf of his pupil Eduard Baumgarten to the Emergency Association for German Science, whose vice-president was the Freiburg academic Viktor Schwoerer, and he followed up the official application with a personal letter in which he emphasizes the urgency and importance of his request in uncompromising language:

> . . . what is at stake here is nothing less than the need to recognize without delay that we face a choice between sustaining our *German* intellectual life through a renewed infusion of genuine, native teachers and educators, or abandoning it once and for all to the growing Jewish influence – in both the wider and the narrower sense. We shall get back on the right track only if we are able to promote the careers of a new generation of teachers without harassment and unhelpful confrontation.

It must be said that Heidegger was in good company here. This kind of attitude was widespread during those years; indeed, it was almost a hallmark of German cultural criticism. So to that extent Heidegger's comments are not particularly surprising. What does set this incident apart is the categorical insistence of his views about where *German* intellectual life was headed: either into a state of groundless alienation, or onto the firm ground of ethnic identity (there is a direct link here with the language and ideas of the Leipzig proclamation of 11 November 1933, where the 'idolization of groundless and powerless thinking' is to be superseded by an 'ethnic and national science'). There is a piquant irony here. In 1929 Eduard Baumgarten is seen as the great white hope of German intellectual life, a bulwark against the rising tide of Jewish influence. But in the unfavourable report that Heidegger addressed to the League of National Socialist University Lecturers in December 1933 the renegade Baumgarten, his former pupil, is branded as a Jewish protégé. So there seems to be a continuity in Heidegger's mental disposition here, from the 1920s through the climactic days of the early 1930s to Germany's year of destiny in 1933.

In my book I drew attention to Heidegger's growing *rapprochement* with organized National Socialism, which took place by 1932 at the latest. This conclusion is based in part on a reading of the letter to Jaspers of 8 December 1932. In his letters to Elisabeth Blochmann, however, Heidegger gives us an even deeper insight into his mental and spiritual state. On 19 January 1933, for example, he writes from Freiburg and paints a picture of his mental development over the preceding weeks, apologizing for the delay in responding to her Christmas greetings:

> The delay is solely due to the events of recent weeks, when I was overtaken by a great storm which I faced out with all sails set. Much of the old rigging was torn to shreds in the process. But there's no point trying to patch and mend it . . . Up there [Heidegger is referring to his mountain hut: author's note] the seeds germinate and grow, while down here the ripe fruits fall. But I need a lot more seed and germination, and I leave that to the March sun and its storms. It is strangely unsettling, this business of waiting for it to happen, of being unable to force the pace.

The storm metaphor has a familiar ring, taking us straight to the closing quotation of Heidegger's rectorship address of 27 May 1933 – 'All great things stand fast in the storm' – while the quasi-religious, Pentecostal allusion is likewise unmistakable. Suffice it to say that the numerous letters written by Heidegger to Elisabeth Blochmann in 1933 are a very important documentary source, adding significantly to our knowledge of concrete particulars and enabling us to trace the broad outline of Heidegger's academic policy. By and large the letters confirm what we already know and bring it into sharper focus. This applies, for example, to the 'academic summer camp' of 1933, to which Heidegger attributes so much importance in his apologia.

The personal and intellectual relationship between Edmund Husserl and Martin Heidegger forms one of the key elements in my biographical study. I would like to return briefly to April 1933 – bearing in mind the events of 1929 that have just been touched upon – and recall the respective parts played by Heidegger and his wife Elfride when they expressed their sympathy, as individuals and as a family, with the fate suffered by the patriotically-minded Husserls under the terms of the 'Reich Law on the re-establishment of a permanent civil service'. Reprinted below is the text of the long-lost letter that Frau Heidegger wrote to Frau Husserl on 29 April. A copy of the letter (in Frau Heidegger's hand) was sent to me via Jean-Michel Palmier, author of the Afterword to the French edition of my Heidegger book. It was discovered among the papers of Frédéric de Torwanicki, who called on the Heidegger family in the autumn of 1945 when he was serving as a French officer, and subsequently published an account of his visit in Sartre's journal *Les Temps Modernes*.

29 April 1933

My dear Frau Husserl,

On behalf of my husband and myself I feel I must write a few lines to you and your husband in these difficult weeks. We wanted to tell you both that today, as always, we remember with undiminished gratitude all that you have done for us in the past. Although my husband's philosophical interests took him in a different direction, he will never forget how much of value for his own undertaking he learned from your husband's teaching. And I will never forget the kindness and friendship that you yourself showed us in the difficult years after the War. It has pained me greatly that I have not been able to show you my gratitude for the last few years – although I never really understood what tissue of misunderstandings could have led you to see in us only the people who had let you both down. To all that, of course, must be added profound gratitude for the spirit of sacrifice displayed by yours sons, and we are merely following the lead given by this new law (which is harsh but sensible from the German point of view)* when we profess our total support and sincere regard for those who have given proof, by their actions, of their allegiance to our German nation in its hour of greatest need. So we were all the more shocked to read the recent press reports about your son at Kiel. We trust that it is all an unfortunate mistake by some over-zealous junior official, carried away in the general excitement of the last few weeks – like the distressing miscarriages of injustice that occurred during the revolutionary weeks of 1918.

I beg you, my dear Frau Husserl, to accept these few lines for what they are: an expression of sincere and undiminished gratitude.

Yours sincerely
Elfride Heidegger

Here the writer unburdens herself as one wife to another, one mother to another, her expressions of gratitude wrung from her by a guilty conscience during those watershed days of April 1933, when Heidegger took over the rectorship and became jointly responsible for implementing the anti-Jewish measures. The Reich Law in question, which was primarily directed against non-Aryans and was 'sensible from the German point of view', may have been harsh, but it did admit of exceptions: Jewish war veterans were exempted from its provisions. Hence the shock felt by the Heidegger family when the former officer and war veteran Gerhart Husserl, who had been seriously wounded on several occasions in the service of his country, was caught up in the flurry of dismissals. Had not the Husserls, as Jews, 'given proof, by their actions, of their allegiance to our German nation in its hour of greatest need'? In the final analysis, of course, they were excluded

* The words in brackets have been crossed out in the original, presumably by Elfride Heidegger herself, who felt this particular passage to be inopportune in the autumn of 1945.

from the fellowship of the German people, permitted only to play a walk-on part. The bottom line was that German intellectual life was now no longer abandoned to the growing Jewish influence. Now at last the German spirit would be sustained by an infusion of genuine native teachers and educators. The National Socialist revolution, in which over-zealous junior officials were bound to overstep the mark occasionally – omelettes, broken eggs and all that – 'brings with it the total transformation of our German being'. The Germans had renounced 'the idolization of groundless and powerless thinking'. 'Ethnic and national science', meaning German science, or to be more precise, German philosophy, had finally broken free from the spell of Jewish influence – in the shape of Edmund Husserl and his like – and Heidegger was 'certain of this: that the clear hardness and workmanlike assurance of an unyielding, simple questioning of the essence of Being is now coming back'. I have chosen to read the letter of 29 April 1933, informed as it is by Heidegger's thinking, in conjunction with the words he spoke in the Leipzig proclamation of November 1933 and the letter he wrote to Viktor Schwoerer in October 1929, because it is only through such a juxtaposition that we get a clear picture of his mentality.

The interest aroused by my book – if I may venture to claim as much – rests primarily on its close examination of Heidegger's origins and early educational experience. The structural principle of the book, which is taken from that crucial passage in Heidegger's letter to Jaspers of 1 July 1935, has been understood and accepted, not least in terms of what it tells us about Heidegger's own perception of himself. As can be seen from the more recent additions to my bibliography, the issue of his Catholic origins has continued to exercise me. Much remains to be explained. Theodore Kisiel's forthcoming book, *The Genesis of Heidegger's 'Being and Time'*,[262] will throw a good deal of new light on the early stages of Heidegger's career, particularly the curious interlude between 1917 and 1919 – the high point of his religious-philosophical quest, when he turned his back on the 'system of Catholicism' and espoused a broadly Protestant band of Christianity. We now know that Heidegger's Catholic marriage in March 1917 was annulled only a few weeks later, under the terms of Catholic canon law as it then was, by a Protestant wedding in

Wiesbaden conducted by Pastor Lieber. All this amounted to a very significant confessional volte-face.

But we also know (and Kisiel's book goes into some detail on the matter) what Heidegger, in his search for a phenomenology of religion, subsumed under the heading 'the system of Catholicism', namely the very antithesis of religious immanence. Throughout the period 1917-18 Heidegger's thinking centred on the direct immediacy of religious experience and the irrational element in mysticism, from St Bernard of Clairvaux to the Spanish mystics of the sixteenth century. This preoccupation led him to the realization that 'a dogmatic, casuistic pseudo-philosophy' derived from a 'particular religious system', as exemplified in Catholicism, was by no means the closest thing to religion and the religious: indeed, it served only to mask and conceal them, since the problem of religious immanence (for example) is not philosophically addressed. Our capacity for experience is restricted in and by the system, and in the end it becomes completely deadened, because the original awareness is missing. For this reason the system of Catholicism (wrote Heidegger in 1918) 'had to pass through a convoluted, artificial, dogmatic cordon of propositions and proofs whose theoretical basis remained totally unclear, until it emerged on the other side as a canon-law code with police powers that overwhelmed and oppressed the individual'. (This and the following quotations are taken from some unpublished notes of Heidegger's – essentially marginal notes on his extensive readings in the philosophy of religion – which Kisiel has had an opportunity to study.)

Heidegger was in no doubt that one of the chief threats to the immediacy of religious life was scholasticism, which was so preoccupied with theology and dogma that it had forgotten religion, thereby bringing to its culmination a process that had begun in the days of the early Church: 'the theorizing, dogmatizing influence of canon-law institutions and ordinances'. In this context Heidegger viewed medieval mysticism as a natural counter-movement, and in particular he saw Martin Luther as the man who brought those mystical beginnings to full fruition by ushering in 'an *original* form of piety'. In these notes Heidegger's struggle with the definitions and meanings of faith and grace becomes abundantly clear: the Catholic faith as a 'holding to be true' versus the Reformation's 'trusting in the promise of salvation'. Or the Catholic theological axiom on the relationship between grace and man: *gratia praesupponit naturam*.

So it is all the more significant that Heidegger never broke free from Catholicism, the faith of his birth. This religious, confessional inheritance remained a thorn in his flesh, ineradicable and incurable. I for my part remain convinced that this is one of the fundamental keys to an understanding of Heidegger's life, even though I have been repeatedly attacked on this issue and grotesquely misrepresented – as if, for example, I had argued that Heidegger was a traitor to his Catholic origins, and that if only he had stuck to his original path he would not have gone astray politically and become embroiled in these unfortunate events. No one who reads my book properly could seriously subscribe to such an interpretation. Nor have I been at all deterred by the vehement attacks that Heidegger launched against Catholicism in general and 'Jesuitism' in particular – in his letters to Elisabeth Blochmann, for example. When, in his *Contributions to philosophy*, Heidegger argues and writes against the Christian God, finally pronouncing Him dead from the judgement seat of philosophy ('The death of this God spells the end of all theisms'), we are dealing with the unavailing struggles of a man tormented by his faith in God. It does not surprise me in the least that Heidegger avoided any mention of those early works of his that appeared up until 1912 in the pages of the Catholic journal *Der Akademiker*. A fuller account of these matters may be found in my introduction to the recent German–English edition of those early texts.[263]

Merzhausen, December 1991 HUGO OTT

NOTES

1 15 June 1988, pp.38–47.

2 I refer to the sequence of three related articles (1983, 1984a and 1984b) devoted specifically to this theme.

3 Martin Heidegger 1983.

4 Conducted in 1966, on the understanding that it would not be published until after Heidegger's death. The interview duly appeared on 31 May 1976.

5 Petzet 1983.

6 In addition to a lecture on Heidegger's rectorship year 1933/34, I also gave a lecture dealing with the period after 1945, subsequently published under the title 'Martin Heidegger und die Universität Freiburg nach 1945' (Ott 1985).

7 Cf. Ott 1984c and 1986.

8 See Antje Bultmann Lemke, 'Der unveröffentlichte Nachlass von Rudolf Bultmann', in: Bernd Jaspert (ed.), *Rudolf Bultmanns Werk und Wirkung*, Darmstadt 1984, pp.194–207, and specifically p.203.

9 As quoted in Bultmann Lemke, *loc. cit.* (see note 8), p.202.

10 *Heidegger. The Man and the Thinker*, Precedent 1981.

11 See especially 'The Missing Link in the Early Heidegger', in: Joseph Kockelmans (ed.), *Hermeneutic Phenomenology: Lectures and Essays*, Washington D.C. 1988, pp.1–40; 'War der frühe Heidegger tatsächlich ein "christlicher Theologe"?', in: Gethmann-Siefert and Meist 1988. I am extremely grateful to my colleague, Theodore Kisiel, for giving me access to the manuscripts of these essays at an early stage, thereby furnishing me with a number of valuable indications.

12 Cf. Elisabeth Young-Bruehl, *Hannah Arendt. For Love of the World*, New Haven/London 1982.

13 *Husserl-Chronik. Denk- und Lebensweg Edmund Husserls*, Den Haag 1977.

14 *Martin Heidegger*, Stuttgart 1976.

15 *Philosophische Autobiographie*, enlarged edition, Munich 1977. *Notizen zu Martin Heidegger*, ed. Hans Saner, Munich and Zurich 1978.

16 *Philosophische Lehrjahre*, 1977.

17 Neske 1977.

18 A second edition appeared in 1983, enlarged by an extended postscript.

19 28/29 November 1987 (issue No.277), under the title 'Wege und Abwege. Zu Victor Farias' Kritischer Heidegger-Studie'.

20 In *Le débat*, No.49, March-April 1988, pp.185–191, under the title 'Le Débat du Débat – Chemins et fourvoiements'.

21 3/4 November 1984 (issue No.257).

22 Volume 53 of the Complete Works: *Hölderlins Hymne 'Der Ister'*, ed. Walter Biemel 1984.

23 As quoted in Pöggeler 1988, p.41.

24 From the personal papers of Romano Guardini in the Bavarian State Library, Munich.

25 This letter and the following quotations are from Rudolf Stadelmann's personal papers in the Federal Archive in Koblenz (R. 183). For further information on Rudolf Stadelmann, who died in

1949 at the early age of 47, see also *Rudolf Stadelmann zum Gedächtnis*, a commemorative address by Eduard Spranger (Tübingen 1950), and the affectionate memoir by Hermann Heimpel in *Historische Zeitschrift* 172, 1951, pp.285–307.

26 Published in 1934 as Volume 47 in the series 'Philosophie und Geschichte' by J.C.B. Mohr in Tübingen. A much toned-down version may be found in Stadelmann 1942, pp.5–31.

27 Hitherto unpublished. Nor are there any plans, so far as I can see, to include the text in the Complete Works; which is a pity, since Heidegger very probably has something to say there about the earlier lectures in the series (including the notorious lecture given by Erik Wolf on 7 December 1933 entitled 'True law in the National Socialist state', *Freiburger Universitätsreden*, Volume 13, Freiburg i. Br. 1934).

28 Cf. Ott 1985.

29 The relevant documents (from the personal papers of Clemens Bauer) are now in my possession. Edgar Morin made a name for himself in 1946 with the book *L'an zéro de l'Allemagne*. He later renounced his allegiance to the Communist Party. Today he is director of research at the 'Centre National de la Recherche Scientifique' in Paris.

30 The original document is in my possession.

31 This is particularly true of France. To be more specific, the controversy that has been raging in that country since the appearance of Victor Farias' book *Heidegger et le nazisme* in 1987 deserves a study to itself.

32 René Schickele, *Werke in drei Bänden*, Cologne and Berlin 1959, p.1040.

33 From the personal papers of Karl Jaspers in the German Literary Archive in Marbach a.N.

34 The rectorship address appeared as Volume II in the series *Freiburger Universitätsreden*, and simultaneously under the imprint of G. W. Korn (Breslau). A second edition was published by Korn in 1934. A slightly revised version was reissued in 1983 in a single volume that also contained Martin Heidegger's apologia, edited by Hermann Heidegger and published by Vittorio Klostermann (Heidegger 1983).

35 *Die Zeit*, No. 39, 24 September 1953.

36 *Hölderlins Hymne: Wie wenn am Feiertage*, Frankfurt a.M. 1941 (published in Halle a.d.S.).

37 From the personal papers of Karl Jaspers in the German Literary Archive in Marbach a.N.

38 For a fuller account see Petzet 1983, p.74 ff.

39 Jaspers 1977, p.102.

40 The full text is quoted below, p. 324 ff.

41 *Nietzsche: Der Wille zur Macht als Kunst*, Volume 43 of the Complete Works, p.26 f.

42 The first published edition of the text (*Über den Humanismus*) appeared in Germany in 1949 under the imprint of Vittorio Klostermann (Frankfurt a.M.).

43 Cf. Ott 1984c.

44 Vittorio Klostermann, Frankfurt a.M. (first edition not available in bookshops).

45 The relevant documents may be found in the Archiepiscopal Diocesan Archive in Freiburg, filed under 'Pfarrei Messkirch, Mesnerdienste'.

46 These personal details were supplied by Martin Heidegger himself in 1933 in the declaration that civil servants were required to submit as proof of their Aryan origins (Central State Archive in Stuttgart, filed under 'Kultusministerium Baden-Württemberg EA III/1').

47 Gustav Kempf has left us a history of the village of Göggingen.

48 'Ein Geburtstagsbrief des Bruders' ['A birthday letter from Heidegger's

brother'], in: *Martin Heidegger zum 80. Geburtstag von seiner Heimatstadt Messkirch*, Frankfurt a.M. 1969, p.58 ff.

49 On Gröber see also my biographical entries in: *Badische Biographien*, Vol. 1, Stuttgart 1982, pp.144–148, and in: *Zeitgeschichte in Lebensbildern*, Vol. 6, Mainz 1984, pp.65–75.

50 Cf. Ott 1985.

51 Minutes of the meetings of the Heidelberg Academy of Sciences, 1957/58, p.20 f.

52 Remarks made by Heidegger on the occasion of the 80th birthday of the publisher Hermann Niemeyer. See *Hermann Niemeyer zum 80. Geburtstag am* 16. April 1963, Tübingen 1963 (printed privately), p.28.

53 Cf. my article 'Dr Max Josef Metzger', *Freiburger Diözesan-Archiv*, No.106, 1986, p.187 ff.

54 This letter, the original of which is preserved in the Konradihaus in Constance, was brought to my attention by the director of studies Lothar Samson (Constance).

55 'Necrologium Friburgense 1946–1950', *Freiburger Diözesan-Archiv*, No.71, 1951, p.221 ff.

56 'Martin Heidegger als Mitschüler', in: Ernst Ziegler (ed.), *Kunst und Kultur um den Bodensee*, Sigmaringen 1986, pp.343–360.

57 Cf. Haeffner 1981, p.361. I am indebted to my colleague Gerd Haeffner for enlarging on this episode in a letter.

58 'Aus einem Gespräch zur Sprache. Zwischen einem Japaner und einem Fragenden' ['From a conversation on language. Between a Japanese and an inquiring spirit'], in: Martin Heidegger, *Unterwegs zur Sprache*, Pfullingen 1959, p.96.

59 Cf. *Armin Kausen. Ein Buch des Andenkens an seine Persönlichkeit, sein Leben und sein Wirken*, Munich 1928.

60 Martin Heidegger published in this journal until 1912.

61 On the problem of German modernism, see also Oskar Köhler, *Bewusstseinsstörungen im Katholizismus*, Frankfurt a.M. 1972.

62 Report dated 2 April 1911, filed in the Archiepiscopal Diocesan Archive in Freiburg (B2 – 32/174).

63 Cf. Odilo Engels, 'Heinrich Finke (1855–1938)' in: *Badische Biographien*, New Series Vol. 2, Stuttgart 1987, pp.87–89.

64 See below, p.75 ff.

65 Baeumker (1853–1924), a professor at Strasbourg since 1903, was one of the leading representatives of neoscholasticism, and together with Georg von Hertling, whom he succeeded at Munich in 1912, he was a vigorous champion of educational policies designed to further the careers of young Catholics – notably through the work of the Görres Society, which had been founded in 1876 at the time of the *Kulturkampf*. He was also the teacher of Martin Grabmann, with whom the young Heidegger had some professional contact.

66 Cf. Klaus Ganzer, 'Die Theologische Fakultät der Universität Würzburg im theologischen und kirchenpolitischen Spannungsfeld der zweiten Hälfte des 19. Jahrhunderts', in: *Vierhundert Jahre Universität Würzburg. Eine Festschrift*, Neustadt an der Aisch 1982, pp.317–373, particularly p.361 ff.

67 Martin Heidegger, *Hölderlins Hymnen: 'Germanien' und 'Der Rhein'*, Complete Works Vol. 39, Frankfurt a.M. 1980, p.145 f.

68 See for example Karl Lehmann, 'Metaphysik, Transzendentalphilosophie und Phänomenologie in den ersten Schriften Martin Heideggers (1912–1916)', *Philosophisches Jahrbuch der Görres-Gesellschaft* 71, 1963/64, pp.331–357, which looks at Heidegger from the perspective of Catholic theology. For a Marxist interpretation, see Wolf-Dieter von Gudopp, *Der junge Heidegger. Realität und Wahrheit in der Vorgeschichte von*

'Sein und Zeit', Berlin/Frankfurt a.M. 1983.

69 From the personal papers of Josef Sauer, now in the possession of his nephew, Canon Josef Sauer of Freiburg, to whom my thanks are due for making them available to me.

70 Martin Heidegger, *Frühe Schriften*, Complete Works Vol. 1, Frankfurt a.M. 1978, pp.412–433.

71 For the background to this and the following discussion see Ott 1986, where full source references will be found.

72 These particulars are taken from the grant award files in the Archiepiscopal Diocesan Archive in Freiburg.

73 See also my article 'Constantin von Schaezler (1827–1880) und Olga von Leonrod geb. von Schaezler (1828–1901). Ein Beitrag zum Spannungsverhältnis der Konfessionen im 19. Jahrhundert', in: *Historia oeconomica et socialis. Festschrift für Wolfgang Zorn zum 65. Geburtstag*, Stuttgart 1987 (VSWG, Supplement No. 84), pp.308–315.

74 The résumé was published for the first time in Ott 1984c.

75 *Meister Eckarts Predigten*, Vol. 2, edited and translated by Josef Quint, Stuttgart 1971, p.132 ff.

76 Cf. Ott 1988c.

77 On Kraus (1840–1901) and his difficulties with the Church ministry, see Hubert Schiel's summary in the *Lexikon für Theologie und Kirche*, 2nd edition, Freiburg i. Br. 1961, Vol. 6, column 596.

78 Schneeberger 1962, No. 176.

79 We owe our insight into this correspondence, now preserved in the Husserl Archive at the Catholic University of Louvain in Belgium, to the untiring efforts of those North American scholars who have subjected the career of the early Heidegger (leading up to *Being and Time*) to the most intensive scrutiny (see Sheehan 1981 and Kisiel 1988).

80 Ochwadt/Tecklenborg 1981.

81 This account of events is based on the diary of Engelbert Krebs.

82 Ochwadt/Tecklenborg 1981, p.93 f.

83 I am indebted for this information to the diary notes of Bernhard Welte (1957), to which I was given access.

84 Caspar 1980. Caspar has misread one important passage. Instead of 'Meine Frau, die Sie erst besucht hat' ['My wife, who has just visited you'], the passage reads: 'Meine Frau, die Sie recht berichtet hat' ['My wife, who has informed you correctly'].

85 For a fuller discussion see Ott 1987a. The Latin sentence that ends the diary entry may be rendered as follows: 'I praise you, O Father, that you have hidden these things from the high and mighty, revealing them to the humble and lowly. Thus it has pleased you, O Father.'

86 Ochwadt/Tecklenborg 1981, p.157 ff.

87 The best brief account of Rudolf Otto (1869–1937) is that by G. Wünsch in: *Die Religion in Geschichte und Gegenwart*, Vol. 4, Tübingen 1960, column 1749 f. Otto's book *Das Heilige. Über das Irrationale in der Idee des Göttlichen und sein Verhältnis zum Rationalen*, published in 1917, was studied closely by Husserl, who discussed it at length with Heidegger. See bibliography in Rudolf Otto, *Aufsätze zur Ethik*, ed. Jakob Stewart Boozer, Munich 1981.

88 Cf. Joseph Ziegler, 'In memoriam Dr Peter Katz (Cambridge)', *Theologische Literaturzeitung* 1962, No.10, p.793 ff.

89 For a fuller account of Szilasi see Ott 1988d.

90 Löwith 1986, p.45.

91 The account that follows is largely based on Kisiel 1988a. I am also indebted for a number of important insights to the work of Karl Schuhmann (Utrecht), who made available to me the continuation of his *Husserl-Chronik* in manuscript form with extracts from Husserl's correspondence.

92 As reported by Karl Schuhmann.

93 Kisiel 1988a, p.7.

94 As reported by Karl Schuhmann.

95 From the personal papers of Karl Jaspers in the German Literary Archive in Marbach. Likewise the extracts from letters that follow, for which individual source references will not (therefore) be given.

96 As reported by Karl Schuhmann.

97 Cf. Bultmann Lemke, Antje, 'Der unveröffentlichte Nachlass von Rudolf Bultmann' in: Bernd Jaspert (ed.), *Rudolf Bultmanns Werk und Wirkung*, Darmstadt 1984, p.202.

98 Mörchen 1984, p.234.

99 Hessen State Archive, Marburg, Inventory No.307 d, Accession 1966/10. Most of the text – with a few misreadings – is published in Kisiel 1988a.

100 *Ibid.* The substantive portion of the Faculty's recommendation appears in English translation in Kisiel's text.

101 As reported by Karl Schuhmann. See also Schuhmann 1978.

102 Arnold von Buggenhagen, *Philosophische Autobiographie*, Meisenheim am Glan 1975, p.134. Max Scheler died in May 1928.

103 The title on the printed invitation card, signed by the president G. Gentile and the director G. Gabetti, reads: 'Hölderlin e l'essenza della poesia'.

104 From the personal papers of Karl Jaspers in the German Literary Archive in Marbach.

105 Löwith 1986, p.57.

106 Cf. *Klassiker in finsteren Zeiten* 1933–1945. Catalogue for the exhibition by the German Literary Archive in the Schiller-Nationalmuseum, Marbach a. Neckar, ed. Bernhard Zeller, Vol. 1, Stuttgart 1983, pp.344–365 ('Zwiesprache von Dichten und Denken. Hölderlin bei Martin Heidegger und Max Kommerell'). See especially p.351 f.

107 See Schickele's diaries, published in *Werke in drei Bänden*, Cologne/Berlin 1959, p.1040.

108 I am indebted to Karl Schuhmann (Utrecht) for bringing this and other letters to my attention.

109 Cf. Ott 1985.

110 From the personal papers of Karl Jaspers.

111 From the personal papers of Herbert Marcuse (University Library, Frankfurt a.M.).

112 Cf. Ott 1983, 1984a, 1984b. Much light is thrown on the events narrated here by Sauer's meticulously kept diary, which was the central source for my initial investigations into Heidegger's rectorship.

113 Filed at the Central State Archive in Stuttgart, under 'Kultusministerium Baden-Württemberg EA III/l Univ. Freiburg, Heidegger, Martin'.

114 Schwan 1965 remains the definitive study of this question.

115 Towards the end of 1932 Sommerfeldt had published a sketch at Göring's direction entitled '*Göring, was fällt Ihnen ein*!' ['Göring, what are you thinking of?']. The third, updated edition appeared under the new title, and soon reached 350,000 copies.

116 Cf. Marten 1988, p.90. Marten omits the name of the family. But I know for a certainty that it was the art historian Hans Jantzen, whom Heidegger knew very well and was friendly with during his (Jantzen's) time at Freiburg.

117 Cf. M.H. Sommerfeldt, *Ich war dabei. Die Verschwörung der Dämonen 1933–1939*, Darmstadt 1949, p.22.

118 Sauer's diary, entry for 28 May 1933.

119 Heidegger 1983, p.31.

120 Martin/Schramm 1986, p.28. The incident that follows is referred to there in footnote 51.

121 Freiburg University Archive, V/l (under 'Generalia/Vereine').

122 'Die Universität im Neuen Reich'

['The University in the New Reich'], *Der Heidelberger Student*, Supplement for Ruperto Carola, 13 July 1933 (as quoted in Mussgnug 1985, p.491).

123 Some useful introductory work on this has been done by Dolf Sternberger (1984).

124 Winfried Franzen, who has written on Heidegger's place in the history of philosophy and also contributed a biographical sketch, published a study in 1988 under the title 'Die Sehnsucht nach Härte und Schwere' ['The yearning for hardness and rigour']. Based on a close reading of Heidegger's 1929/30 lecture on 'The basic concepts of metaphysics' he argues plausibly that the yearning for hardness and rigour which is discernible there may well have predisposed Heidegger to become politically involved with the National Socialists. At the very least, argues Franzen, a study of the language and content reveals a strong affinity with National Socialist ideology.

125 Rudolph Berlinger, from an address given at the funeral of Karl Ulmer, 29 May 1981 (printed privately).

126 *Introduction to Metaphysics* (1953 edition), p.28 f. The 'pincer' analogy has been taken up by latter-day environmentalists in their analysis of the arms race. See for example Hanspeter Padrutt, *Der epochale Winter. Zeitgemässe Betrachtungen*, Zurich 1984, p.199 ff. 'Heidegger as the star political witness for today's worldwide missile confrontation'. These writers forget, of course, that Germany built the first rockets to be used as weapons of mass destruction (i.e. the V1 and V2) during the Second World War.

127 Proclamation in the *Freiburger Studentenzeitung*; cf. Schneeberger 1962, No. 114.

128 Petzet 1983, p.52.

129 For a fuller account of this episode see Ott 1988b, p.73 f.

130 Enrico Castelli, *Il tempo invertebrato*, Padua 1969, p.51, note 4.

131 Letter of 14 April 1928. I am indebted to Professor Karl Schuhmann of the University of Utrecht for the opportunity to study Heidegger's correspondence with Julius Stenzel.

132 Letter to the editor of the *Frankfurter Allgemeine Zeitung*, 14 April 1984.

133 Pöggeler 1988, p.31 f.

134 *Zeitschrift für Sozialforschung* 3, 1934, pp.193 and 194.

135 Freiburg University Archive, II/2-63. Circular from the rector to all heads of department.

136 Both the place of this meeting in Berlin and the alleged attack on Heidegger are the products of pure misunderstanding, as Karl Schuhmann (1978) has painstakingly established.

137 Golo Mann, *Erinnerungen und Gedanken*, Frankfurt a.M. 1986, p.324.

138 Most recently repeated in Léopoldine Weizmann, 'Heidegger était-il nazi?', *Études*, May 1988, p.637 ff.: 'Heidegger interdit alors à Husserl l'accès à l'université parce qu'il était juif.'

139 This particular strand is preserved in Schuhmann's *Husserl-Chronik* (1977).

140 Ott 1988c.

141 *Pfänder-Studien*, ed. Herbert Spiegelberg and Eberhard Avé-Lallemant, Den Haag 1982, p.342 ff.

142 Jaspers 1977, pp.97 and 103.

143 *Frankfurter Allgemeine Zeitung*, 22 October 1984.

144 See for example Willms 1977, p.16 f.

145 For a general account of the whole episode see also Hans-Wolfgang Strätz, 'Die studentische Aktion wider den undeutschen Geist im Frühjahr 1933', *Vierteljahreshefte für Zeitgeschichte* 16, 1968, pp.347-372.

146 This is the version given in Jaspers' report on Heidegger of 22 December 1945, which will be examined in detail below. Hans-Joachim Dahms (1987, p.182) offers a somewhat different version, taken from the file on

Baumgarten in the Göttingen University Archive: 'In terms of his family background and intellectual sympathies Dr Baumgarten is rooted in the Heidelberg circle of liberal-democratic intellectuals around Max Weber. During his time here he was anything but a National Socialist . . . When Baumgarten failed to secure an appointment with me he began to associate regularly with the Jew Fraenkel, who used to teach in Göttingen and has now been dismissed from here. I imagine this is how Baumgarten got himself fixed up in Göttingen – which would also explain his current connections with the place. For the present there can be no question, in my view, of admitting him either to the ranks of the SA or to the teaching body.'

147 So writes Jürgen Busche in the *FAZ* for 30 April 1983 ('Der Standpunkt Martin Heideggers'). See also the subsequent letter to the editor from Wilhelm Schoeppe ('Heidegger und Baumgarten') in the *FAZ* for 28 May 1983.

148 Political Archive of the Foreign Office, R III/218/3a. Further reference will be made below to other documents from this source.

149 Cf. Markus Mattmüller, *Leonhard Ragaz und der religiöse Sozialismus*, Vol. I, Basel 1957, Vol. II, Zurich 1968. The references to Hermann and Dora Staudinger will be found in Vol. II.

150 Cf. Claus Priesner, 'Hermann Staudinger und die makromolekulare Chemie in Freiburg. Dokumente zur Hochschulpolitik 1925–1955', *Chemie in unserer Zeit*, Vol. 21, 1987, pp.151–160.

151 On Bühl and his like, see Alan D. Beyerchen, *Wissenschaftler unter Hitler. Physiker im Dritten Reich*, Cologne 1980, *passim*.

152 From the Freiburg University Archive.

153 From the personal papers of Rudolf Stadelmann in the Federal Archive, Koblenz.

154 Neske 1977, p.53.

155 Preserved in the Federal Archive in Koblenz (R 183). In addition to the circulars the collection also contains the key letters from Heidegger to Stadelmann of 11 and 23 October 1933, together with Stadelmann's long letter to Heidegger of 16 October 1933. This correspondence is particularly revealing for an understanding of the allegiance/leadership syndrome.

156 Cf. *Semper Apertus* 1985, Index.

157 Cf. Werner Walz, Elisabeth Glatt, Eduard Seidler, *Radiologie in Freiburg 1895–1980*, Freiburg i. Br. 1980, pp.49 and 54 f.

158 Rudolf Stadelmann, 'Vom geschichtlichen Wesen der deutschen Revolutionen' ['The historical character of the German revolutions'], *Zeitwende* X, 1934, pp.109–116. This article was very probably based on lectures Stadelmann gave in the autumn of 1933.

159 Stadelmann 1942, p.17.

160 Heidegger 1983, p.38.

161 Hamburg 1933.

162 The letter is published in *Telos* No. 72, 1987, p.132. This issue is devoted to the work of Carl Schmitt. I am grateful to Dr Johannes Gross for drawing it to my attention.

163 Hollerbach 1986. The following account is based on Ott 1984a, where detailed source references may be found.

164 *Freiburger Universitätsreden* Vol. 13. This lecture was one of a series of public lectures given in the winter semester of 1933/34 under the title 'The role of intellectual life in the National Socialist state'.

165 *Archiv für Rechts- und Sozialphilosophie* 28, 1934/35, pp.348–363.

166 Martin Heidegger, *Hölderlins Hymne 'Wie wenn am Feiertage . . .'*, Halle a.d.S. 1941, p.31 f. The following quotations will be found there and on p.30.

167 On Krieck, see Gerhard Müller's exhaustive Freiburg dissertation

Die Wissenschaftslehre Ernst Kriecks, Motive und Strukturen einer gescheiterten nationalsozialistischen Wissenschaftsreform, Weinheim 1976. Löwith furnishes an interesting insight into the character of Erich Jaensch. He relates how Romano Guardini gave a lecture on Pascal in Marburg in the summer semester of 1933: 'The National Socialist psychologist Jaensch, who had by then invented the "counter-type" to the German character, was incensed by this lecture, and declared it was a disgrace that the University should have had to listen to a lecture on a Frenchman given by a foreign academic (Guardini was Italian by birth) at the present time.' (1986, p.76 f.)

168 Alfred Baeumler, who as a member of the Fighting League for German Culture had already been associated with Alfred Rosenberg before Hitler's seizure of power, and was chosen by him to head the Department of Science, had been appointed to a chair in 'political education' at the University of Berlin in the summer semester of 1933 – as we saw above.

169 Cf. Poliakov/Wulf 1983, p.548. Also discussed in Christoph von Wolzogen, '"Es gibt". Heidegger und Natorps "Praktische Philosophie"', in: Gethmann-Siefert and Pöggeler 1988, p.330 f.

170 Central State Archive, Merseburg Office, Rep. 76 Va Sekt. 1, Tit. IV, No.71. The most important pages are 42–73, 476–487 and 499–505. I am most grateful to the Merseburg Office of the Central State Archive for sending me photocopies. I am also indebted to my colleague, Hans-Martin Gerlach (Martin-Luther University, Halle-Wittenberg) for putting me in touch with this source, and indeed for all his helpful advice. Farias was the first to narrate these events (1987, p.215 ff.), though the attacks on Heidegger by the Krieck-Jaensch faction receive only very cursory mention.

171 Published for the first time in Christoph von Wolzogen, *loc. cit.*, (see footnote 169), p.331.

172 Cf. the memorable chapter 'Hans Lipps' in Gadamer 1977, pp.161–165.

173 Heidegger had given a course of lectures in the summer semester of 1935 entitled 'Schelling: On the nature of human freedom'.

174 How divergent the assessments of Heidegger's character by Party agencies were is demonstrated by a questionnaire preserved in the archive at the Quai d'Orsay (Paris), which was filled in by the local Party leadership in Freiburg and sent to Party district headquarters in Karlsruhe (to be used by the 'Textbook Censorship Office' as a source of information about the authors concerned). In the general summary of Heidegger's character presented here the main emphasis is on his fierce opposition to Catholicism. This document from the spring of 1938 also suggests that at that time Heidegger was still very much *persona grata*, at least in local Party circles. (An initial account of this document – which I have not seen myself – is given by Jacques Le Rider in an article entitled 'Le dossier d'un nazi "ordinaire"' that appeared in *Le Monde* for 14 October 1988.)

175 Haecker 1933.

176 Ochwadt/Tecklenborg 1981, p.109.

177 Haecker 1933, p.17.

178 Cf. Schneeberger 1962, No.129.

179 *Ibid.*, Nos.116 and 129.

180 Haecker's book was something of a popular success – in marked contrast to Heidegger's rectorship address.

181 This lecture has not yet received the critical attention it deserves. The following passages are quoted from the text published in Tübingen in 1953 (4th edition 1976).

182 The events in question, reconstructed from archive sources, are recounted in Remigius Baeumer,

'Die Theologische Fakultät Freiburg und das Dritte Reich', *Freiburger Diözesan-Archiv*, No.103, 1983, pp.265–289, particularly p.285 f.

183 Haecker's book *Christ und Geschichte* had been published that year in Leipzig, testifying unequivocally to the belief that the Christian God alone is the God of history.

184 Cf. Ott 1988a.

185 *S. th.* I q. 84 a. 7.

186 From the files on doctoral candidates, Freiburg University Archive.

187 Cf. Franz Pöggeler, 'Gedenkworte für Gustav Siewerth', *Jahres-und Tagungsbericht der Görres-Gesellschaft* 1964, p.55 ff.

188 Cf. Martin and Schramm 1986.

189 Freiburg University Archive.

190 A full report – including transcripts of the speeches – will be found in the *Freiburger Zeitung* for 30 May 1930 (No.147).

191 From the personal papers of Clemens Bauer – now in the possession of the author.

192 *Ibid.*: Müller's application for a lectureship of 1 June 1945. In an appendix to this application Müller relates the whole episode in great detail – speaking more openly than in the later 'interview' (*loc. cit.*, footnote 188), where his treatment of Heidegger is more indulgent.

193 Cf. Ott 1988a.

194 I had a long and very informative conversation with Professor Heinz Bollinger, centring on the letter mentioned in note 195. My account here is based in part on what he told me.

195 From the personal papers of Martin and Irmgard Honecker. I am grateful to Dr Raimund Honecker and Frau Jansen, née Honecker, for permission to make use of this material. The letter from Frau Honecker is a key document for our understanding of events behind the scenes.

196 Heidegger quotes the directive (Z.D. 165/34, issue No.7514): 'The essay by Martin Heidegger, "Plato's theory of truth", which appears in the forthcoming issue of the *Jahrbuch für geistige Überlieferung* published by Helmut Küpper Verlag of Berlin, is not to be reviewed or referred to. No mention is to be made of Heidegger's work as a contributor to Volume II of this Yearbook – which for the rest may be reviewed quite freely.'

197 *Nationalsozialistische Monatshefte*, No.140, November 1941, pp.926–932.

198 'Professor Martin Heidegger gave us a learned address' (annual report of the archabbey of St Martin, Beuron).

199 *Jahrbuch für geistige Überlieferung*, Vol. II, p.122 f.

200 All these documents are preserved in the Federal Archive in Koblenz (NS 15/209).

201 See above, p.210 f.

202 See also my essay 'Alfred Rosenbergs Grosskundgebung auf dem Freiburger Münsterplatz am 16. Oktober 1937', in: *Freiburger Diözesan-Archiv*, No.107, 1987, pp.303–319.

203 *Der Alemanne*, evening edition of 10 June 1938.

204 All these documents are preserved among the personal papers of Clemens Bauer.

205 Letter No.132 in: Schwabe and Reichardt 1984.

206 *Frankfurter Allgemeine Zeitung*, 25 July 1953.

207 Pöggeler 1983, postscript, p.340 f.

208 Heidegger is clearly alluding to the works of Fritz Joachim von Rintelen and areas of study such as 'philosophia perennis'.

209 Translated and published by Heidegger in *Introduction to Metaphysics*, Tübingen 1953, p.112 f.

210 In his important book on Heidegger the American Jesuit William F. Richardson has drawn up a catalogue – authorized by Heidegger himself – of all the lectures and seminars given

or projected by Heidegger (1963, p.670 f.).

211 Petzet 1983, p.52.

212 As quoted in the text of Heidegger's lecture in the Complete Works, Vol. 53, p.4.

213 F. Hölderlin, *Sämtliche Werke*, ed. F. Beissner (Grosse Stuttgarter Ausgabe, Vol. 4, 1., Stuttgart 1961, p.309).

214 *Hölderlins Hymne 'Der Ister'*, *loc. cit.* (see note 22), pp.98 and 106.

215 Moehling 1972.

216 It is recorded in the Freiburg Municipal Archive (C5/402 c) and related in more detail in Ott 1985.

217 Extracts from the letter are given in English translation in Moehling (1972).

218 Petzet 1983, p.52.

219 From the personal papers of Romano Guardini (Bavarian State Library, Munich).

220 From the personal papers of Adolf Lampe (Archiv für Christlich-Demokratische Politik, Konrad-Adenauer-Stiftung St Augustin).

221 Ochwadt/Tecklenborg 1981, p.125 f.

222 From the personal papers of Clemens Bauer.

223 *Ibid.*

224 Minutes of the Faculty of Philosophy, University of Freiburg (permanent record of proceedings).

225 The original text of Jaspers' report came into my possession in a somewhat roundabout way via the personal papers of Robert Heiss, who was dean of the Faculty of Philosophy in 1946. Heiss kept this report (together with his own correspondence with Jaspers) in his private files.

226 The reference here is to the physicist Wolfgang Gentner, who was invited to take over the chair of physics at Freiburg vacated by the dismissal of Steinke, the local leader of the League of NS University Lecturers.

227 From the personal papers of Gröber preserved in the Archiepiscopal Archive in Freiburg (Gröber/67).

228 Robert Minder, *Hölderlin unter den Deutschen und andere Aufsätze zur deutschen Literatur*, Frankfurt a.M. 1968, p.140.

229 Record of proceedings of the Faculty of Philosophy, 6 January 1946.

230 Minutes of the Freiburg University Senate.

231 Letter No.132 in: Schwabe and Reichardt 1984, p.408 f.

232 Letter No.133, *ibid.*, p.409.

233 From the personal papers of Gröber in the Archiepiscopal Archive in Freiburg (Gröber/54).

234 These rulings are recorded in the personal file on Heidegger in the Central State Archive in Stuttgart, under 'Kultusministerium Baden-Württemberg EA III/1'.

235 This episode is recorded in the personal papers of Clemens Bauer.

236 The manuscript of this lecture is preserved among the personal papers of Clemens Bauer.

237 Detailed source references for this and the account that follows will be found in Ott 1988d.

238 For further details see Ott 1988d.

239 The full story is told in the 1949 exchange of letters between Guardini and Gadamer (preserved in the personal papers of Guardini in the Bavarian State Library).

240 *Lexis. Studien zur Sprachphilosophie, Sprachgeschichte und Begriffsforschung*, Vol. II 1.2., Lahr-Baden 1949–51.

241 I am indebted to Gerd Tellenbach for the opportunity to reconstruct these events from his private papers and files – and specifically from his correspondence with Heidegger of June 1947. The following account is largely based on this source, supplemented by additional material from the personal papers of Clemens Bauer (already cited above).

242 Supplement No.101 of 16 July 1949 (p.280) to the *Badisches Gesetz- und Verordnungsblatt*, No.27, 16 July 1949.

243 Record of proceedings of the Faculty of Philosophy, 2 May 1949.

244 For further details see Ott 1985.

245 From the personal papers of Romano Guardini (Bavarian State Library).

246 Cf. Petzet 1983, pp.59 f. and 71 f.

247 Now in the Complete Works, Volume 8.

248 Cf. *Martin Heidegger. Erhart Kästner. Briefwechsel 1953–1974*, ed. Heinrich W. Petzet, Frankfurt a.M. 1986.

249 Letter from Heidegger to Gerhart Baumann, quoted in: Gerhart Baumann, *Erinnerungen an Paul Celan*, Frankfurt a.M. 1986, p.59 f.

250 Cf. Otto Pöggeler, *Spur des Worts. Zur Lyrik Paul Celans*, Freiburg/Munich 1986, p.259.

251 As quoted in Clemens Podewil, 'Die nachbarlichen Stämme', in: Neske 1977, p.211.

252 '. . . as though Heraclitus were looking over our shoulder.' See Neske 1977, pp.223–228.

253 Quoted from the *Introduction to Metaphysics*, Tübingen 1953, p.113.

254 First published in: *Christ in der Gegenwart*, 1976. Reprinted in: Neske 1977, pp.253–256.

255 Westdeutscher Verlag, Opladen.

256 Jacques Le Rider, 'Le dossier Heidegger des archives du Ministère des Affaires Étrangères', in: *Allemagnes d'Aujourd'hui*, No. 107, 1989, pp. 97–109.

257 Extracts from the address 'Science under threat' (revised by Hartmut Tietjen) appear in *Zur philosophischen Aktualität Heideggers* (being the proceedings of a symposium held by the Alexander von Humboldt Foundation on 24–28 April 1989 in Bonn–Bad Godesberg), Vol. 1: *Philosophie und Politik*, ed. Dietrich Papenfuss and Otto Pöggeler, Frankfurt a.M. 1991, pp. 5–27.

258 *Heideggers Kritik am Nationalsozialismus und an der Technik*, Tübingen 1989.

259 *Briefwechsel* 1920-1963, ed. Walter Biemel and Hans Saner, Frankfurt a.M.–Munich–Zurich 1990.

260 *Briefwechsel 1918-1962*, ed. Joachim W. Storck, Marbach a.N. 1989 (Vol. 33 in the series *Marbacher Schriften*).

261 'Die Verjudung des deutschen Geistes. Ein unbekannter Brief Heideggers', an article in *Die Zeit*, No. 52, 22 December 1989.

262 Due to be published in 1993 by the University of California Press, Berkeley.

263 *Heidegger and the Political*, edited by Marcus Brainard with David Jacobs and Rick Lee (*Graduate Faculty Philosophy Journal*, New School for Social Research, Vol. 14, No. 2–Vol. 15, No. 1, 1991).

BIBLIOGRAPHY

Arendt, Hanna *and* Jaspers, Karl (1985), *Briefwechsel 1926–1969*, ed. Lotte Köhler, Hans Sauer, Munich

Biemel, Walter (1973), *Martin Heidegger in Selbstzeugnissen und Bilddokumenten*, Reinbek

Casper, Bernhard (1980), 'Martin Heidegger und die Theologische Fakultät Freiburg 1909–1923', *Freiburger Diözesan-Archiv*, No. 100, pp. 534–541

Dahms, Hans-Joachim (1987), 'Aufstieg und Ende der Lebensphilosophie: Das Philosophische Seminar der Universität Göttingen zwischen 1917 und 1950', in: *Die Universität Göttingen unter dem Nationalsozialismus. Das verdrängte Kapitel ihrer 250jährigen Geschichte*, ed. Heinrich Becker *et al*, Munich

Farias, Victor (1987), *Heidegger et le nazisme. Morale et politique*, trans. from Spanish and German, Lagrasse

Franzen, W. (1976), *Martin Heidegger*, Stuttgart

—— (1988), 'Die Sehnsucht nach Härte und Schwere', in: Gethmann-Siefert and Pöggeler 1988, pp. 78 ff.

Gadamer, Hans-Georg (1977), *Philosophische Lehrjahre. Eine Rückschau*, Frankfurt a.M.

Gethmann-Siefert, Annemarie *and* Otto Pöggeler (eds.) (1988), *Heidegger und die praktische Philosophie*, Frankfurt a.M.

Gethmann-Siefert, Annemarie *and* Kurt Rainer Meist (eds.) (1988), *Philosophie und Poesie. Spekulation und Erfahrung* (Otto Pöggeler zum 60. Geburtstag gewidmet), Stuttgart

Grassi, Ernesto (1970) *Macht des Bildes: Ohnmacht der rationalen Sprache*, Munich

Haecker, Theodor (1933), *Was ist der Mensch*, Leipzig

Haeffner, Gerd (1981), 'Martin Heidegger (1889–1979)', in: Otfried Höffe (ed.), *Klassiker der Philosophie*, Vol. II, Munich, pp. 361–384

Heidegger, Martin (1983), *Die Selbstbehauptung der deutschen Universität.*

Das Rektorat 1933/34 – Tatsachen und Gedanken, ed. Hermann Heidegger, Frankfurt a.M.

Hollerbach, Alexander (1986), 'Im Schatten des Jahres 1933. Erik Wolf und Martin Heidegger', *Freiburger Universitätsblätter*, No. 92, pp. 33 ff.

Jaspers, Karl (1977), *Philosophische Autobiographie*, expanded edn., Munich

—— (1978), *Notizen zu Martin Heidegger*, ed. Hans Saner, Munich (reissued 1988)

Kisiel, Theodore (1988a), 'The Missing Link in the Early Heidegger', in: Kockelmans, Joseph (ed.), *Hermeneutic Phenomenology: Lectures and Essays*, Washington D. C., pp. 1–40

—— (1988b), 'War der frühe Heidegger tatsächlich ein "christlicher Theologe"?' in: Gethmann-Siefert, Meist 1988

Löwith, Karl (1986), *Mein Leben in Deutschland vor und nach 1933*, Stuttgart

Marten, Rainer (1988), 'Heideggers Geist', *Allmende*, No. 20, pp. 82–95ff.

Martin, Bernd *and* Schramm, Gottfried (1986), 'Ein Gespräch mit Max Müller', *Freiburger Universitätsblätter*, No. 92, pp. 13–31

Moehling, Karl August (1972), *Martin Heidegger and the Nazi Party: An Examination*, Diss. phil. Northern Illinois University

Mörchen, Hermann (1984), 'Zur Offenhaltung der Kommunikation zwischen dem Theologen Rudolf Bultmann und dem Denken Martin Heideggers', in: Bernd Jaspert (ed.) *Rudolf Bultmanns Werk und Wirkung*, Darmstadt 1988, pp. 234–252ff.

Mussgnug, Dorothee (1985), 'Die Universität zu Beginn der nationalsozialistischen Herrschaft', in: *Semper Apertus* 1985, Vol. III, pp. 464–503

Neske, Günter (ed.) (1977), *Erinnerung an Martin Heidegger*, Pfullingen

Ochwadt, Curd *and* Tecklenborg, Erwin (1981), *Das Mass des Verborgenen. Heinrich Ochsner zum Gedächtnis*, Hanover

Ott, Hugo (1983), 'Martin Heidegger als Rektor der Universität Freiburg i. Br. 1933/34, I. Die Übernahme des Rektorats der Universität Freiburg i. Br. durch Martin Heidegger im April 1933', *Zeitschrift des Breisgau-Geschichtsvereins* ('Schau-ins-Land'), No. 102, pp. 121–136

—— (1984a), 'Martin Heidegger als Rektor der Universität Freiburg i. Br. 1933/34, II. Die Zeit des Rektorats von Martin Heidegger (23. April 1933 bis 23. April 1934)', *Zeitschrift des Breisgau-Geschichtsvereins* ('Schau-ins-Land'), No. 103, pp. 107–130

—— (1984b), 'Martin Heidegger als Rektor der Universität Freiburg i. Br. 1933/34', *Zeitschrift für die Geschichte des Oberrheins*, No. 132, pp. 343–358

—— (1984c), 'Der junge Martin Heidegger. Gymnasial-Konviktszeit und Studium', *Freiburger Diözesan-Archiv*, No. 104, pp. 315–325

—— (1984d), 'Der Philosoph im politischen Zwielicht. Martin

Heidegger und der Nationalsozialismus', *Neue Zürcher Zeitung*, 3/4 November 1984

—— (1985), 'Martin Heidegger und die Universität Freiburg nach 1945. Ein Beispiel für die Auseinandersetzung mit der politischen Vergangenheit', *Historisches Jahrbuch*, No. 105, pp. 95–128

—— (1986), 'Der Habilitand Martin Heidegger und das von Schaezler'sche Stipendium. Ein Beitrag zur Wissenschaftsförderung der katholischen Kirche', *Freiburger Diözesan-Archiv*, No. 106, pp. 141–160

—— (1987a), 'Edith Stein (1891–1942) und Freiburg. Ein Beitrag anlässlich der Seligsprechung am 1. Mai 1987', *Freiburger Diözesan-Archiv*, No. 107, pp. 253–274

—— (1988a), 'Die Weltanschauungsprofessuren (Philosophie und Geschichte) an der Universität Freiburg – besonders im Dritten Reich', *Historisches Jahrbuch* No. 108, pp. 157–173

—— (1988b), 'Martin Heidegger und der Nationalsozialismus', in: Gethmann-Siefert *and* Pöggeler 1988, pp. 64 ff.

—— (1988c), 'Edmund Husserl und die Universität Freiburg', in: *Edmund Husserl und die phänomenologische Bewegung. Zeugnisse in Text und Bild*, on behalf of the Husserl Archive, in Freiburg i. Br., ed. Hans Reiner Sepp, Freiburg and Munich 1988, pp. 95–102

—— (1988d), 'Um die Nachfolge Martin Heideggers nach 1945', in: Gethmann-Siefert, Meist 1988

Petzet, H. W. (1983), *Auf einen Stern zugehen. Begegnungen und Gespräche mit Martin Heidegger 1929 bis 1976*, Frankfurt a.M.

Pöggeler, Otto (1983), *Der Denkweg Martin Heideggers*, Pfullingen (1st edn. 1963)

—— (1985), 'Den Führer führen; Heidegger und kein Ende', collected reviews in *Philosophische Rundschau*, Year 32, pp. 26 ff.

—— (1988), 'Heideggers politisches Selbstverständnis', in: Gethmann-Siefert *and* Pöggeler 1988, pp. 17–63

Poliakov, L. and J. Wulf (1983), *Das Dritte Reich und seine Denker*, Frankfurt, Vienna and Berlin (1st edn. 1959)

Richardson, William F. (1963), *Heidegger. Through Phenomenology to Thought*, The Hague

Schneeberger, Guido (1962), *Nachlese zu Heidegger*, Bern

Schuhmann, Karl (1977), *Husserl-Chronik*, The Hague

—— (1978), 'Zu Heideggers Spiegel-Gespräch über Husserl', *Zeitschrift für Philosophische Forschung* 32

Schwabe, K. *and* R. Reichardt (ed.) (1984), *Gerhard Ritter. Ein politischer Historiker in seinen Briefen*, Schriften des Bundesarchivs, Vol. 33, Boppard

Schwan, Alexander (1965), *Politische Philosophie im Denken Heideggers*, Opladen

Semper Apertus. Sechshundert Jahre Ruprechts-Karl-Universität Heidelberg 1386–1986, Vol. III: 1918–1986, Heidelberg 1985
Sheehan, Thomas (1981), *Heidegger. The Man and the Thinker*, Precedent
—— (1988), 'Heidegger and the Nazis', *The New York Review of Books*, 15 June 1988, pp. 38–47
Stadelmann, Rudolf (1942), *Vom Erbe der Neuzeit*, Vol. 1, Leipzig
Sternberger, Dolf (1984), 'Die grossen Worte des Rektors Heidegger. Eine philosophische Untersuchung', *Frankfurter Allgemeine Zeitung*, 2 March 1984
Willms, Bernard (1977), 'Politik als Geniestreich? Bemerkung zu Heideggers Politikverständnis', in: *Martin Heidegger. Fragen an sein Werk*, Stuttgart, pp. 16–20

Recent publications by the present author (since 1989)

Martin Heidegger: Sentieri Biografici (Italian edn., trans. by Flavio Cassinari, foreword by Carlo Sini), SugarCo Edizioni, Milan 1990.
Martin Heidegger: Eléments pour une biographie (French edn., trans. by Jean-Michel Belail, afterword by Jean-Michel Palmier), Editions Payot, Paris 1990.
Fallet Heidegger (Swedish edn., trans. by Yngre Andersson), Daidalos, Göteborg 1991.
'Der junge Martin Heidegger als Lyriker. Zu zwei (veröffentlichten) unbekannten Gedichten (1911 und 1915)', in: *Geteilte Sprache. Festschrift für Rainer Marten*, Amsterdam 1988, pp. 85–90.
'A Distant Command. The Political Dimension of Heidegger's Philosophy', in: *German Comments*, 17 Jan 1990, pp. 82–87.
'Zu den katholischen Wurzeln im Denken Martin Heideggers. Der theologische Philosoph.' Lecture given at the Heidegger Symposium organized by the Goethe Institute in Rome, 29–31.5.1989, Proceedings to be published.
'Martin Heidegger – Mentalität der Zerrissenheit', *Freiburger Diözesan-Archiv* 110, 1990, pp. 427–448.
'Biographische Gründe für Heideggers "Mentalität der Zerrissenheit"', in: *Faszination und Erschrecken*, ed. Peter Kemper, Campus 1990.
'Martin Heidegger und die Politik.' Lecture given at the Heidegger Symposium: 'Heidegger in der philosophischen Kultur Europas', Turin, 6–8.11.1990, Proceedings to be published in: *Rivista di filosofia.*
'Phänomenologie und Ontologie. Edith Stein zwischen Edmund Husserl und Martin Heidegger.' Husserl-Schütz-Symposium, 15–17.11.1990, Vienna. Proceedings to be published.

'Martin Heidegger. Ein schwieriges Verhältnis zur Politik', in: *Heidegger – Technik, Politik*, Würzburg 1991, pp. 215–228.

'Preface: Heidegger and the Denazification Proceedings', in: *Heidegger and the Political* (Graduate Faculty Philosophy Journal 14, 1991), pp. 523–527.

'Preface: Martin Heidegger's Contributions to "Der Akademiker"', in: *Heidegger and the Political* (Graduate Faculty Philosophy Journal 14, 1991), pp. 481–485.

'"Herkunft aber bleibt stets Zukunft". Zum katholischen Kontinuum im Leben Martin Heideggers', in: *Annales Universitatis Saraviensis*, Heidelberg (forthcoming).

'Zu den katholischen Wurzeln im Denken Martin Heideggers' (Lecture given at the Heidegger colloquium, 13–15.10.1989, Yale University, New Haven, Conn.), in: *Kunst-Politik-Technik* (2745), ed. Jammes *et al.* 1992, pp. 225–239 (forthcoming).

INDEX